Photovoltaics – Fundamentals, Technology, and Practice

Photovoltaics – Fundamentals, Technology, and Practice

Konrad Mertens
Muenster University of Applied Sciences
Steinfurt, Germany

Second Edition

This edition first published 2019

Edition History
Photovoltaik © 2015 Carl Hanser Verlag, Munich, FRG

Registered Office(s)
John Wiley & Sons, Inc., 111 River Street, Hoboken, NJ 07030, USA
John Wiley & Sons Ltd, The Atrium, Southern Gate, Chichester, West Sussex, PO19 8SQ, UK

Editorial Office
The Atrium, Southern Gate, Chichester, West Sussex, PO19 8SQ, UK

For details of our global editorial offices, customer services, and more information about Wiley products visit us at www.wiley.com.

Wiley also publishes its books in a variety of electronic formats and by print-on-demand. Some content that appears in standard print versions of this book may not be available in other formats.

Library of Congress Cataloging-in-Publication Data applied for
Hardback ISBN: 9781119401049

Cover design: Wiley
Cover Images: (Background) © RapidEye/Gettyimages;
(Front cover image) Courtesy of Solar-Fabrik AG

Set in 10/12pt WarnockPro by SPi Global, Chennai, India

Contents

Preface to the First International Edition

A steadily growing number of requests for an English version led to the decision to publish this international version of the Photovoltaics textbook. For this – besides the translation of the German text – several figures, tables, and solar radiation maps were extended with worldwide data and information. Moreover, as the photovoltaic market and technology have developed very quickly, numerous updates have found room in this first international version.

I would like to express my thanks to John Wiley & Sons, Ltd. for making this book possible. My special thanks to Gunther Roth for the translation, Richard Davies for managing the whole book project, and Laura Bell for taking care of all the small problems that arose.

A new English website has been created at *www.textbook-pv.org*.
Here the reader can find most of the figures of the book as a free download, supporting software and solutions to the exercises (and possibly corrections to the book).

I am very glad that the book is now accessible for a worldwide readership. We all know that the transformation of the current worldwide energy system into an environmentally friendly and sustainable one is a giant challenge. Photovoltaics can become an important part in the future energy supply. I hope that this book will help to deepen the understanding of the technology and the possibilities for photovoltaics, and can therefore support this development.

Steinfurt, January 2014 *Konrad Mertens*

Preface to the Second International Edition

The large demand makes it possible that the second edition of this textbook can be published already. Explicitly, I thank the readers for the positive comments on the first edition. Many thanks to John Wiley & Sons, Ltd. who made it possible that this edition is now published in full color.

As the development of photovoltaics advances quickly, this new edition shows a considerable increase in volume. Besides the usual actualizations, it contains a completely new chapter about "Storage of Solar Power." Moreover, the chapter dealing with photovoltaic metrology was amended by the description of up-to-date on-site measurement technologies such as outdoor-electroluminescence and detection of potential induced degradation (PID) of solar modules.

The chapter "Future Development" now also looks on the cooperation of the different renewable energies in the present and in the future.

Particularly, I want to recommend the accompanying website *www.textbook-pv.org*.

Here you can find the figures of the book as a free download, supporting software, the solutions to the exercises, and corrections to the book.

I wish much joy and success to all the readers through learning the fascinating technology of photovoltaics.

Steinfurt, May 2018 *Konrad Mertens*

Abbreviations

AC	alternating current
ALB	Albedo
AM	air mass
a-Si	amorphous silicon
CAES	compressed air energy storage
CCCV	constant current, constant voltage
CdTe	cadmium telluride
CET	Central European Time
CID	current interrupt device
CIGS	copper indium gallium sulfide
CIGSe	copper indium gallium selenide
CIS	copper indium sulfide
CISe	copper indium selenide
c-Si	crystalline silicon
DC	direct current
DOD	depth of discharge
DSM	demand side management
EEG	Renewable Energy Law (Erneuerbare-Energien-Gesetz)
EMC	electromagnetic compatibility
EVA	ethyl-vinyl-acetate
FF	fill factor
GaAs	gallium arsenide
GaN	gallium nitride
GCB	generator connection box
HIT	heterojunction with intrinsic thin layer
IBC	interdigitated back contact
ITO	indium tin oxide
LST	local solar time
MPP	maximum power point
NAS	sodium sulfur
NOCT	nominal operating cell temperature
PECVD	plasma enhanced chemical vapor deposition
PERC	passivated emitter and rear cell
PERL	passivated emitter rear locally diffused
PID	potential induced degradation

PR	performance ratio
PWM	pulse width modulation
RCD	residual current device
SiC	silicon carbide
SOC	state of charge
SR	sizing ratio
STC	standard test conditions
TC	temperature coefficient
tce	tons of coal equivalent
TCO	transparent conducting oxide
toe	tons of oil equivalent
UTC	coordinated universal time
VRF	vanadium redox flow

1
Introduction

The supply of our industrial community with electrical energy is indispensable on the one hand, but, on the other hand, it is accompanied by various environmental and safety problems. In this chapter, therefore, we will look at the present energy supply and will familiarize ourselves with renewable energies as feasible future alternatives. At the same time, photovoltaics will be presented in brief and its short but successful history will be considered.

1.1 Introduction

In the introduction, we will explain why we are occupying ourselves with photovoltaics and who should read this book.

1.1.1 Why Photovoltaics?

In past years, it has become increasingly clear that the present method of generating energy has no future. Thus, finiteness of resources is noticeably reflected in the rising prices of oil and gas. At the same time, we are noticing the first effects of burning fossil fuels. The melting of the glaciers, the rise of the ocean levels, and the increase in weather extremes, as well as the nuclear catastrophe in Fukushima, all show that nuclear energy is not the path to follow in the future. Besides the unsolved final storage question, fewer and fewer people are willing to take the risk of large parts of their country being radioactive.

Fortunately, there is a solution with which a sustainable energy supply can be assured: Renewable energy sources. These use infinite sources as a basis for energy supplies and can ensure a full supply with a suitable combination of different technologies such as biomasses, photovoltaics, wind power, and so on. A particular role in the number of renewable energies is played by photovoltaics. It permits an emission-free conversion of sunlight into electrical energy and, because of its great potential, will be an important pillar in future energy systems.

However, the changeover of our energy supply will be a huge task that can only be mastered with the imagination and knowledge of engineers and technicians. The object of this book is to increase this technical knowledge in the field of photovoltaics. For this purpose, it will deal with the fundamentals, technologies, practical uses, and commercial framework conditions of photovoltaics.

Photovoltaics – Fundamentals, Technology, and Practice, Second Edition. Konrad Mertens.

1.1.2 Who Should Read This Book?

This book is meant mostly for students of the engineering sciences who wish to deepen their knowledge of photovoltaics. Nevertheless, it is written in such a way that it is also suitable for technicians, electricians, and the technically interested layman. Furthermore, it can be of use to engineers in the profession to help them to gain knowledge of the current technical and commercial position of photovoltaics.

1.1.3 Structure of the Book

In the introduction, we will first deal with the subject of energy: What is energy and into what categories can it be divided? From this base, we will then consider the present energy supply and the problems associated with it. A solution to these problems is renewable energies and will be presented next in a brief overview. As we are primarily interested in photovoltaics in this book, we will finish with the relatively young but stormy history of photovoltaics.

Chapter 2 deals with the availability of solar radiation. We become familiar with the features of sunlight and investigate how solar radiation can be used as efficiently as possible. Then in the Sahara Miracle, we will consider what areas would be necessary to cover the whole of the world's energy requirements with photovoltaics.

Chapter 3 deals with the basics of semiconductor physics. Here we will concentrate on the structure of semiconductors and an understanding of the p–n junction. Besides this, the phenomenon of light absorption will be explained, without which no solar cell can function. Those familiar with semiconductors can safely skip this chapter.

Chapter 4 gets to the details: We learn of the structure, method of operation, and characteristics of silicon solar cells. Besides this, we will view in detail the parameters and degree of efficiency on which a solar cell depends. Based on world records of cells, we will then see how this knowledge can be successfully put to use.

Chapter 5 deals with cell technologies: What is the path from sand, via silicon solar cell, to the solar module? What other materials are there and what does the cell structure look like in this case? Besides these questions, we will also look at the ecological effects of the production of solar cells.

Chapter 6 deals with the structures and properties of solar generators. Here we will deal with the optimum interconnection of solar modules in order to minimize the effects of shading. Besides this, we will present various types of plants such as pitched roof and ground-mounted plants.

Chapter 7 deals with system technology of grid-connected plants. At the start, there is the question of how to convert direct current efficiently into alternating current. Then we will become familiar with the various types of inverters and their advantages and disadvantages.

Chapter 8 deals with the storage of solar power, the very hot topic of the chapter. We learn to know different battery types together with their operating modes. Moreover, it is about systems who can enhance the self-consumption of solar power in domestic households or commercial enterprises. In an own subchapter off-grid systems are considered.

Chapter 8 concentrates on photovoltaic metrology. Besides the acquisition of solar radiation, we deal especially with the determination of the real power of solar modules.

Furthermore, we become familiar with modern methods of quality analysis such as thermography and electroluminescence metrology.

Chapter 9 presents design and operation of grid-coupled plants. Besides the optimum planning and dimensioning of plants, methods of profitability calculation are also discussed. In addition, methods for monitoring plants are shown and the operating results of particular plants are presented.

Chapter 10 provides a view of the future of photovoltaics. First, we will estimate power generation potential in Germany. This is followed by a consideration of price development and the coaction of the different energies in the current electrical power system. Finally, we will reflect on how the future energy system will look like and what role photovoltaics will thereby play.

Each chapter has exercises associated with it, which will assist in repeating the material and deepening the knowledge of it. Besides, they provide a control of the students' own knowledge. The solutions to the exercises can be found in the Internet under *www.textbook-pv.org*.

1.2 What Is Energy?

We take the use of energy in our daily lives as a matter of course, whether we are operating the coffee machine in the morning, using the car during the day, or returning to a warm home in the evening. In addition, the functionality of our whole modern industrial community is based on the availability of energy: Production and transport of goods, computer-aided management, and worldwide communication are inconceivable without a sufficient supply of energy.

At the same time, the recognition is growing that the present type of energy supply is partly uncertain, environmentally damaging, and available only to a limited extent.

1.2.1 Definition of Energy

What exactly do we understand about the term *energy*? Maybe a definition of energy from a famous mouth will help us. Max Planck (founder of quantum physics: 1858–1947) answered the question as follows:

Energy is the ability of a system to bring outside effects (e.g. heat, light) to bear.

For instance, in the field of mechanics, we know the potential energy (or stored energy) of a mass m that is situated at a height h (Figure 1.1a):

$$W_{\text{Pot}} = m \cdot g \cdot h \tag{1.1}$$

with g: Earth's gravity, $g = 9.81\ \text{m s}^{-2}$.

If a bowling partner drops his 3 kg bowling ball, then the "1-m-high ball" system can have a distinct effect on his foot.

If, on the other hand, he propels the ball as planned forward, then he performs work on the ball. With this work, energy is imparted to the ball system. Thus, we can say in general:

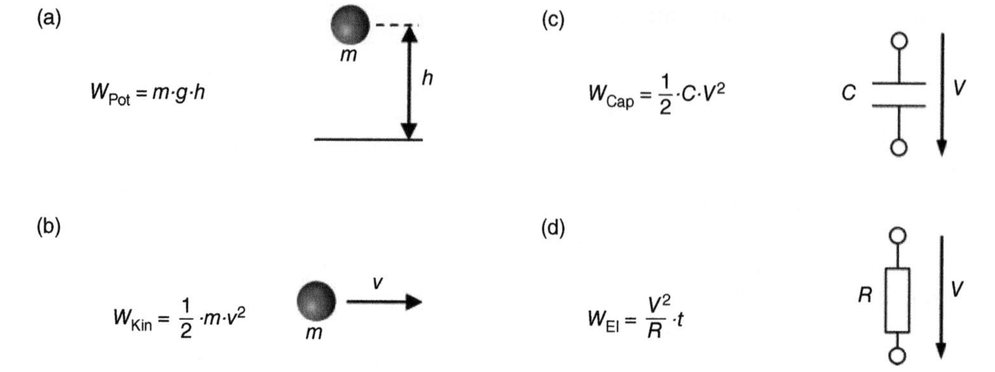

Figure 1.1 Depiction of different forms of energy. (a) Potential energy. (b) Kinetic energy. (c) Capacitor energy. (d) Energy at resistor.

The energy of a system can be changed with the addition or transfer of work. To put it another way, energy is stored work.

In the case of the bowling partner, the ball obtains kinetic energy W_{Kin} (or movement energy, see Figure 1.1b) in being propelled forward:

$$W_{\text{Kin}} = \frac{1}{2} \cdot m \cdot v^2 \tag{1.2}$$

with v: velocity of the ball.

A similar equation describes the electrotechnics of the energy stored in a capacitor W_{Cap} (see Figure 1.1c)

$$W_{\text{Cap}} = \frac{1}{2} \cdot C \cdot V^2 \tag{1.3}$$

with C: capacity of the capacitor; V: voltage of the capacitor.

If, again, there is a voltage V at an ohmic resistor R, then in the time t it will be converted into electrical work W_{El} (Figure 1.1d):

$$W_{\text{El}} = P \cdot t = \frac{V^2}{R} \cdot t. \tag{1.4}$$

The power P shows what work is performed in the time t:

$$P = \frac{\text{Work}}{\text{Time}} = \frac{W}{t}. \tag{1.5}$$

1.2.2 Units of Energy

Unfortunately, many different units are in use to describe energy. The most important relationship is

$$1\ \text{J(Joule)} = 1\ \text{W s} = 1\ \text{N m} = 1\ \text{kg m s}^{-2}. \tag{1.6}$$

Table 1.1 Prefixes and prefix symbols.

Prefix	Prefix symbol	Factor	Number
Kilo	k	10^3	Thousand
Mega	M	10^6	Million
Giga	G	10^9	Billion
Tera	T	10^{12}	Trillion
Peta	P	10^{15}	Quadrillion
Exa	E	10^{18}	Quintillion

Example 1.1 ***Lifting a sack of potatoes***

If a sack of 50 kg of potatoes is lifted by 1 m, then this provides it with stored energy of

$$W_{Pot} = m \cdot g \cdot h = 50\ \text{kg} \cdot 9.81\ \text{m s}^{-2} \cdot 1\ \text{m} = 490.5\ \text{N m} = 490.5\ \text{W s}.$$ ■

In electrical engineering, the unit of the kilowatt hour (kW h) is very useful and results in

$$1\ \text{kW h} = 1000\ \text{W h} = 1000\ \text{W} \cdot 3600\ \text{s} = 3.6 \times 10^6\ \text{W s} = 3.6\ \text{MW s} = 3.6\ \text{MJ}. \tag{1.7}$$

Due to the fact that in the energy industry, very large quantities are often dealt with, a listing of units that prefixes into factors of 10 is useful; see Table 1.1.

1.2.3 Primary, Secondary, and End Energy

Energy is typically stored in the form of energy carriers (coal, gas, wood, etc.). This form of energy is typically called primary energy. In order to use it for practical purposes, it needs to be converted. If one wishes to generate electricity, then for instance, coal is burned in a coal-fired power station in order to generate hot steam. The pressure of the steam is again used to drive a generator that makes electrical energy available at the exit of the power station (Figure 1.2). This energy is called secondary energy. This process chain is associated with relatively high conversion losses. If the energy is transported on to a household, then further losses are incurred from the cables and transformer stations. These are added together under distribution losses. The end energy finally arrives at the end customer.

With a petrol-driven car, the oil is the primary energy carrier. It is converted to petrol by means of refining (secondary energy) and then brought to the petrol station. As soon as the petrol is in the tank, it becomes end energy. This must again be differentiated from useful energy, and in the case of the car, it is the mechanical movement of the vehicle. As a car engine has an efficiency of less than 30%, only a small fraction of the applied primary energy arrives on the road. In the case of electrical energy, the useful energy would be light (lamp) or heat (stove plates).

In order that end energy is available at the socket, the conversion and distribution chain shown in Figure 1.2 must be passed through. As the efficiency of a conventional

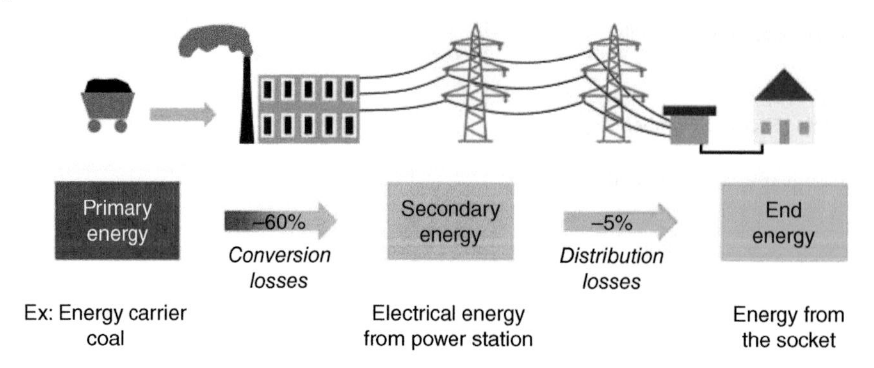

Figure 1.2 Depiction of the types of energy as an example of coal-fired power: Only about one-third of the applied primary energy arrives at the socket by the end customer.

power station with approximately 40% is relatively small, the overall degree of efficiency η_{Over} up to the socket of the end user is

$$\eta_{\mathrm{Over}} = \eta_{\mathrm{Powerstation}} \cdot \eta_{\mathrm{Distr}} \approx 0.4 \cdot 0.95 \approx 0.38. \tag{1.8}$$

Thus, we can state that in the case of conventional electrical energy, only about one third of the applied primary energy arrives at the socket.

And yet electrical energy is used in many fields as it is easy to transport and permits the use of applications that could hardly be realized with other forms of energy (e.g. computers, motors). At the same time, however, there are uses for which the valuable electricity should not be used. Thus, for the case of electric space heating, only a third of the applied primary energy is used, whereas with modern gas energy, it is more than 90%.

1.2.4 Energy Content of Various Substances

The conversion factors in Table 1.2 are presented in order to estimate the energy content of various energy carriers.

In the energy industry, the unit toe is often used. This means tons of oil equivalent and refers to the conversion factor of 1 kg crude oil in Table 1.2. Thus, 1 toe is

Table 1.2 Conversion factors of various energy carriers [1, Wikipedia].

Energy carrier	Energy content (kW h)	Remarks
1 kg coal	8.14	—
1 kg crude oil	11.63	Petrol: 8.7 kW h l^{-1}, diesel: 9.8 kW h l^{-1}
1 m^3 natural gas	8.82	—
1 kg wood	4.3	(at 15% moisture)

1000 kg ·11.63 (kW h) kg^{-1} = 11.630 kW h. Correspondingly, there is the conversion of tons of coal equivalent (tce) with the factor for coal in Table 1.2.

We can remember the very approximate rule:

$$1\ m^3\ \text{natural gas} \approx 1\ l\ \text{oil} \approx 1\ l\ \text{petrol} \approx 1\ \text{kg coal} \approx 1\ \text{kg wood} \approx 10\ \text{kWh}.$$

1.3 Problems with Today's Energy Supply

The present worldwide energy supply is associated with a series of problems; the most important aspects will be presented in the following sections.

1.3.1 Growing Energy Requirements

Figure 1.3 shows the development of worldwide primary energy usage in the last 40 years. In the period considered this more than doubled, the average annual growth being 2.2%. While at first mainly Western industrial countries made up the greatest part, emerging countries, especially China, caught up rapidly.

One reason for the growth in energy requirements is the growth of the world population. This has almost doubled in the past 40 years from 3.7 billion to the present 7 billion people. By the year 2030, a further rise to more than 8 billion people is expected [3].

The second cause for this development is the rising standards of living. Thus, the requirement of primary energy in Germany is approximately 45 000 kW h per head; in a weak industrialized country such as Bangladesh, on the other hand, it is only 1500 kW h per head. With the growing standards of living in the developing countries, the per-head consumption will increase substantially. In China, as a very dynamic emerging nation, it is above 26 000 kW h per head. The International Energy Agency (IEA) assumes that China will increase its energy requirement in the next 25 years by 75% and India by even 100%.

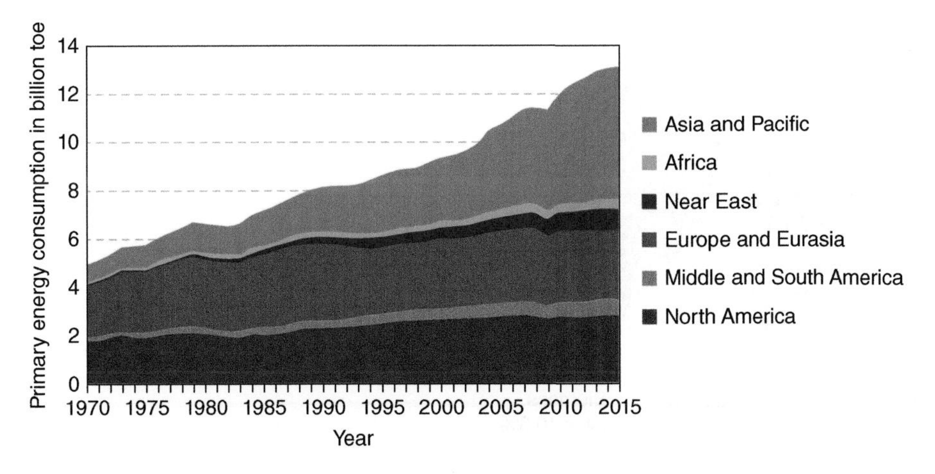

Figure 1.3 Development of worldwide primary energy requirements since 1971 [2].

The growing energy requirement would not be so grave if this did not cause a series of problems:

1. Tightening of resources
2. Climate change
3. Hazards/disposals

These will now be looked at in more detail.

1.3.2 Tightening of Resources

The worldwide requirement for energy is covered today mainly by the fossil fuels: oil, natural gas, and coal. From Figure 1.4, it can be seen that they make up a portion of more than 80%, while biomass, hydro, and renewable energies (wind, photovoltaics, solar heat, etc.) up to now have only reached 14%.

Meanwhile, the strong usage of fossil sources has led to scarcity. Table 1.3 shows the individual extraction quantities in 2001 and 2008. Already, in 2001, the estimated reserves of oil were estimated to last 43 years and natural gas 64 years. Only coal reserves were estimated to last for a relatively long period of 215 years. By 2008, more oil reserves were found but by then the annual consumption had increased substantially. Thus, the reserves have reduced from 140 to 41 years.

If one assumes that the world energy consumption continues to grow as previously, then reserves will be reduced drastically in 30–65 years (see also Exercise 1.3). The scarcity of fuels will lead to strongly rising prices and distribution wars.

In the past, a start has also been made with the extraction of oil from oil sands and oil shales. This has been carried out particularly in Canada and the United States. However, much engineering effort is required for the generation of synthetic oil. Extraction in open-cast mining leads to the destruction of previously intact ecosystems. Therefore, the use of these additional fossil sources is no real future option.

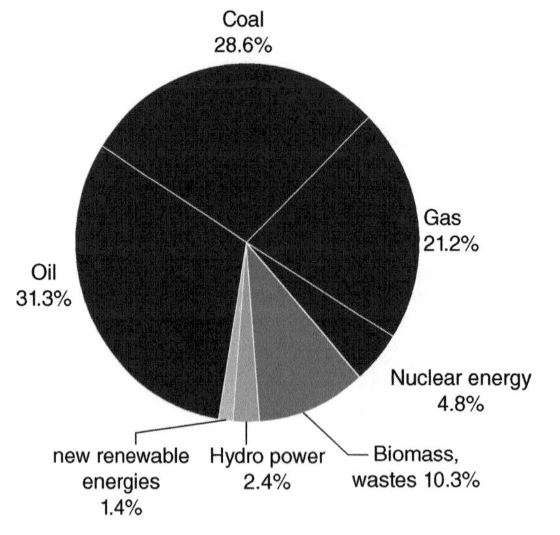

Figure 1.4 Distribution of worldwide primary energy consumption in 2014 according to energy carriers [4].

Table 1.3 Extraction and reserves of fossil fuels [5].

	Oil		Natural gas		Coal	
	2001	2008	2001	2008	2001	2008
Extraction (EJ a^{-1})	147	163	80	121	91	151
Reserves (EJ)	6351	6682	5105	7136	19 620	21 127
Lasting for	43 a	41 a	64 a	59 a	215 a	140 a
Lasting with an annual growth of 2.2%		30 a		38 a		65 a

1.3.3 Climate Change

The decomposition of biomasses (wood, plants, etc.) causes carbon dioxide (CO_2) to be released into the atmosphere. At the same time, plants grow due to photosynthesis and take up CO_2 from the air. In the course of the history of the Earth, this has equalized itself and has led to a relatively constant CO_2 concentration in the atmosphere.

CO_2 is also created when wood, coal, natural gas, or oil are burned and is released into the atmosphere. In the case of wood, this is not tragic as long as felled trees are replanted. The newly growing wood binds CO_2 from the air and uses it for building up the existing biomasses.

In the case of energy carriers, however, this looks different. These were formed millions of years ago from biomass and have been burned up in one to two centuries. Figure 1.5 shows the course of CO_2 concentrations in the atmosphere in the last 20 000 years.

Apparently, in earlier times, there were already fluctuations in these concentrations but really disturbing is the steep rise since the start of industrialization. In 2017, the concentration was slightly above 400 ppm (parts per million) – a value that has not been reached for millions of years.

Why is the CO_2 concentration in the atmosphere so important for us? The reason is that CO_2, besides the other trace gases (e.g. methane, CH_4), affects the temperature of

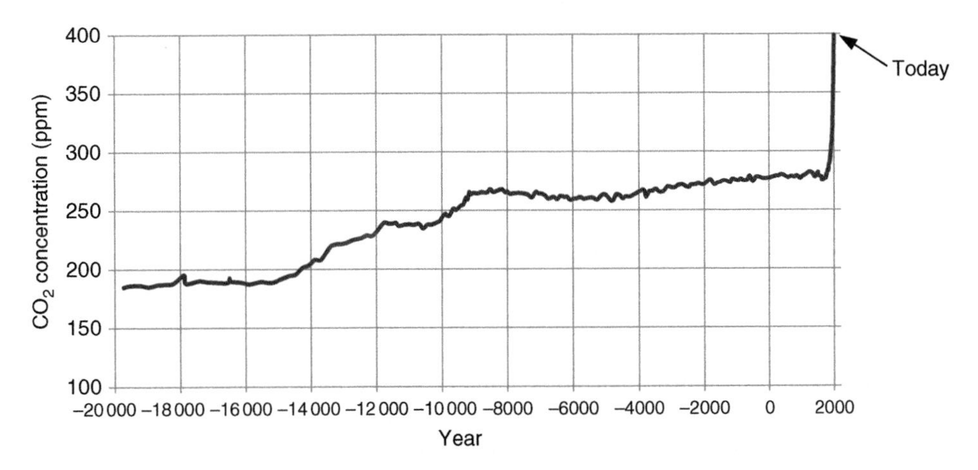

Figure 1.5 Development of the CO_2 content in the atmosphere in the last 22 000 years: Noticeable is the steep rise since the start of industrialization [6–8].

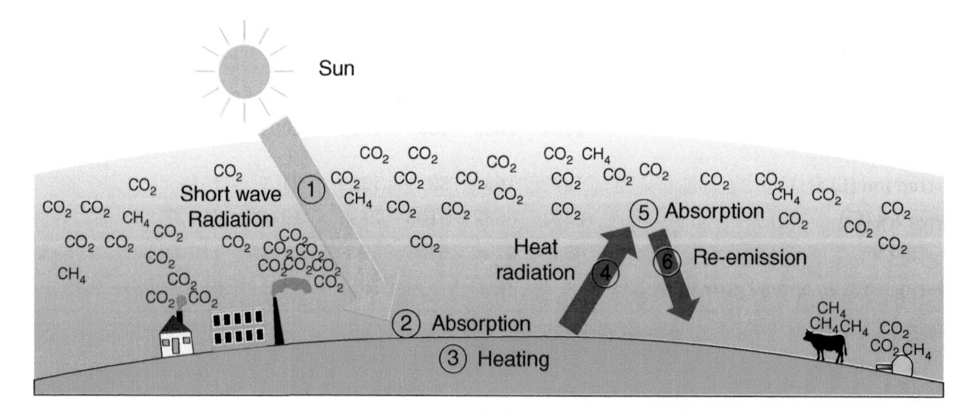

Figure 1.6 Depiction in principle of the greenhouse effect: The heat radiation reflected from the ground is held back by the greenhouse gases.

the Earth through the greenhouse effect. For clarity, consider Figure 1.6. The light from the Sun (visible and infrared radiation ①) arrives almost unhindered through the atmosphere and is absorbed by the ground ②. This causes the surface ③ to be warmed and emits heat radiation as the so-called blackbody source (see Chapter 2) ④. This radiation is again absorbed by the trace gases ⑤ and released to the environment as heat ⑥. The heat energy thus remains in the atmosphere to a large extent, and only a small amount is returned back into space.

The comparison to a greenhouse is thus fitting: The atmosphere with trace gases acts as the glass of a greenhouse that allows the Sun's rays to pass into the greenhouse but holds back the internally resulting heat radiation. The result is a heating up of the greenhouse.

Now we should be happy that this greenhouse effect even exists. Without it, the temperature on the Earth would be −18 °C, but because of the natural greenhouse effect, the actual average temperature is approximately 15 °C. However, the additional emissions of CO_2, methane, and so on, caused by people as an anthropogenic greenhouse effect, lead to additional heating. Since the start of industrialization, this temperature rise has been approximately 0.8 °C. The Intergovernmental Panel on Climate Change (IPCC) expects that this will increase up to 2 °C until the end of the twenty-first century if the emissions of greenhouse gases will not be reduced (*www.ipcc.ch*).

The results of the temperature increase can already be seen in the reduction of glaciers and melting of the ice in the North Polar Sea. Besides this, extreme weather phenomena (hurricanes, drought periods in some regions) are connected to the rise in temperatures. In the long term, further rises in temperatures are expected with a significant rise in water levels and displacement of weather zones.

In order to slow down the climate change, the Kyoto agreement was promulgated in 1997 at the World Climate Conference in Kyoto, Japan. There the industrial countries obligated themselves to lower their greenhouse emissions by 5.5% below the 1990 level by 2012. The declared aim was the limitation of the rise of temperature caused by people by 2 °C. Of its own will, Germany obligated itself to reduce emissions by 21%. After Germany had achieved its aim, in 2010, the Federal Government decided on a reduction of 40% by 2020 (compared to 1990). Important elements to achieving this goal, besides the increase in energy efficiency, are the extensions of renewable energies. After the catastrophe in Fukushima, Germany also decided to completely change over to renewable energies by 2050.

In 2015, the United Nations Framework Convention on Climate Change (short COP 21) was held in Paris. It had the aim to limit the global warming to below 2 °C, preferably to 1.5 °C. To achieve this, the net greenhouse gas emissions have to be reduced to zero from 2045 to 2060. The future will show if this aim will be reached.

1.3.4 Hazards and Disposal

An almost CO_2-free generation of electricity is presented by nuclear energy. However, it is associated with a number of other problems. The reactor catastrophe in Fukushima in 2011 showed that the risk of a super catastrophe (largest expected accident) can never be fully excluded. Even when no tsunami is expected in Germany, there is still a great danger as the nuclear power stations here are only insufficiently protected against terrorist attack.

Added to this is the unsolved problem of final storage of radioactive waste. At present, there is no final storage for highly radioactive waste in the world. This must be safely stored for thousands of years. There is also the question as to whether it is ethically correct to saddle future generations with such a burdensome legacy.

Here, too the availability should be taken into account. The known reserves of uranium including estimated stores are approximately 4.6 million tons. Included in these are the ones with relatively poor concentrations of uranium that are difficult to extract. If we accept the current annual requirements of 68 000 $t\,a^{-1}$ as a basis, then the stocks will last for 67 years [9]. Assuming the increase of energy from Table 1.3, then the reserves will last for approximately 40 years. If the whole of today's energy requirement were changed over to nuclear energy, then the stocks of uranium would last for just 4 years.

1.4 Renewable Energies

1.4.1 The Family of Renewable Energies

Before we turn to photovoltaics in more detail, we should allocate it into the family of renewable energies. The term renewable (or regenerative) means that the supply of energy is not used up. The wind blows every year again and again. The Sun rises every day, and plants grow again after the harvest. In the case of geothermic energy, the Earth is cooling off, but this will only be noticeable thousands of years in the future.

As Figure 1.7 shows, the actual primary energies of the renewable energies are the movements of the planets, the heat of the Earth, and solar radiation. Whereas movements of the planets are used only in the somewhat exotic tidal power plants, the heating of the Earth as well as the heating of buildings can be achieved with the aid of a heat pump as well as for generating electrical energy in a geothermal power station.

Solar radiation is the basis for a surprising range of energies. Thus, the use of hydropower is only possible by the condensation of water and subsequent precipitation onto land. Atmospheric movement originates mostly due to solar radiation, which is also the basis for the use of wind power. In the case of biomass products, it is again sunlight that causes photosynthesis, and thus the growth of biomass is conditioned by it.

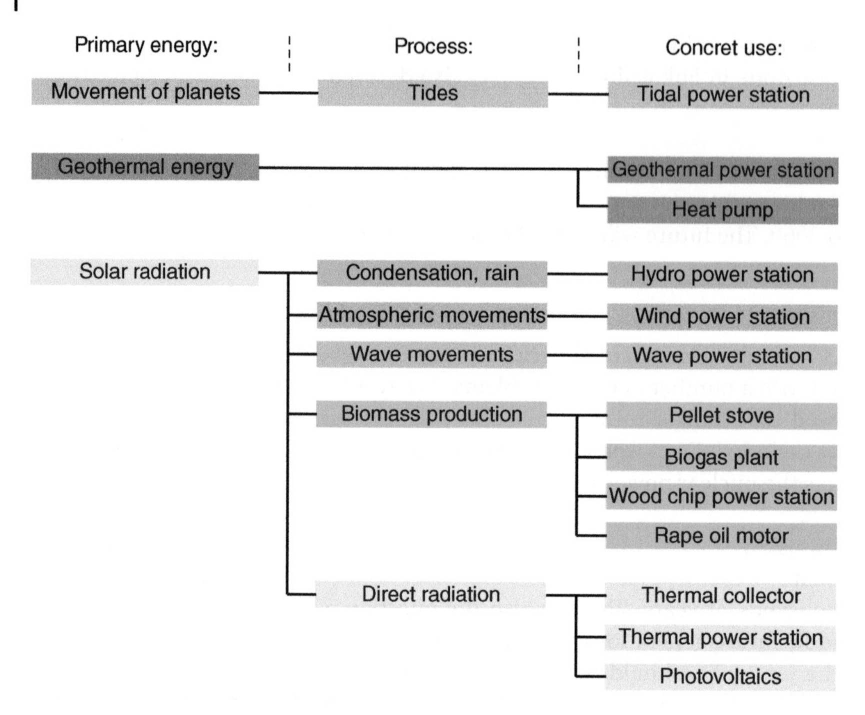

Figure 1.7 Various possibilities for the use of renewable energies.

Solar radiation can also be used directly for the generation of heat, for instance in a thermal collector for domestic water or domestic space heating. Thermal solar power stations generate and process heat from concentrated sunlight in order to drive generators for production of electricity. Last but not least, with photovoltaics, solar radiation is directly converted into electrical energy.

Thus, we can consider photovoltaics as the young daughter of the large family of renewable energies. However, they have a very special attraction: They are the only one able to convert sunlight directly into electrical energy without complicated intermediate processes and without using mechanical converters that can wear out.

1.4.2 Advantages and Disadvantages of Renewable Energies

Renewable energies have certain common characteristics. Their greatest advantage is that in contrast to all other energy carriers, they are practically inexhaustible. Added to this, they are almost free of any emissions and with only few environment effects and hazards.

A further important advantage is in the fact that there are practically no fuel costs. The Sun shines for free, the wind blows irrespectively, and the heat of the Earth is an almost inexhaustible reservoir. On the other hand, the energy densities in which the renewable energies are available are small. Large areas are needed (solar module area for photovoltaics, rotor area for wind turbines, etc.) in order to "collect" sufficient energy. This means that typically large investment costs are incurred as the large areas require the use of a lot of material.

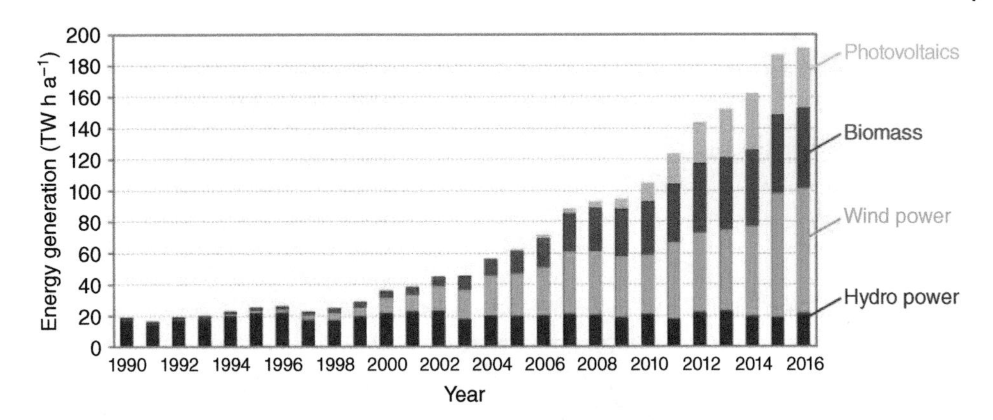

Figure 1.8 Development of renewable energies in Germany since 1990: In 2016, the contribution of energy generation was already more than 191 TW h a^{-1}, which corresponded to approximately 33% of the overall electricity demand [10].

A further large disadvantage is the varying energy supply. Photovoltaics and wind power are especially affected by this. As a result, further power stations (backup power stations) must be kept on reserve in order to ensure a constant supply. Geothermal power is not affected by this; it can provide energy practically independently of time of day or year. A special case is biomass which is the only renewable energy that is easy to store (branches in the woods, biogas in the tank, etc.).

In many developing countries, there is no power grid. There, a further advantage of renewable energies can be used: Their decentralized availability and utility. Thus autarchic village power supplies can be installed far from large towns without an overland grid being necessary.

1.4.3 Previous Development of Renewable Energies

Figure 1.8 shows the contribution of renewable energies to the power supply in Germany over the past 20 years. In addition to the classical hydropower, wind power and biomass have also made great strides. In recent years, photovoltaics, with growth rates of more than 30%, has also made much progress. In 2016, the renewable energies generated 191 TW h of electrical power, which corresponds to about 32% of German energy demand. Photovoltaics provided a share of 6.5%, and wind power was around 13.5%.

1.5 Photovoltaics – The Most Important in Brief

In the following chapters, we will work through some fundamentals in which some may perhaps question their necessity. For this reason, in order to increase motivation and for the sake of clarity, we will briefly consider the most important aspects of photovoltaics.

1.5.1 What Does "Photovoltaics" Mean?

The term photovoltaics is a combination of the Greek word *phós*, *phōtós* (light, of the light), and the name of the Italian physicist Alessandro Volta (1745–1825).

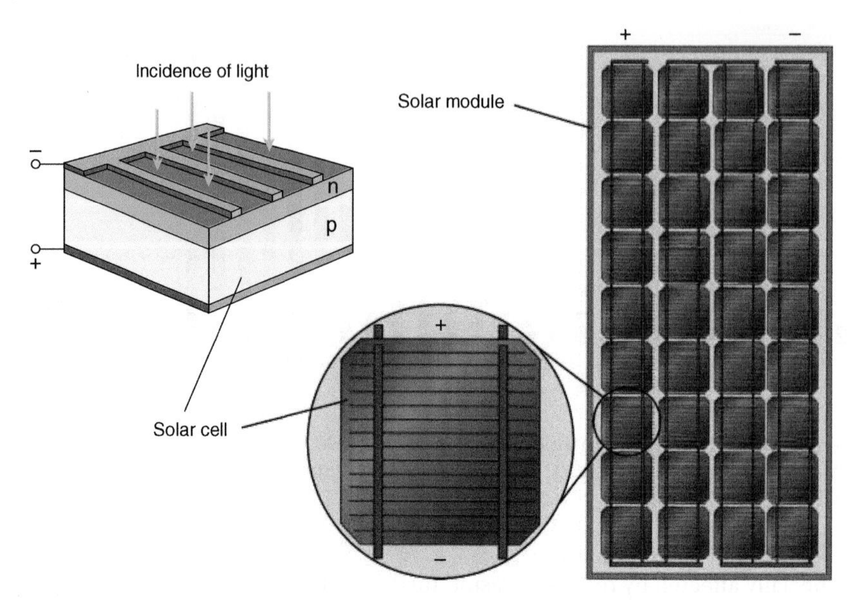

Figure 1.9 The solar cell and solar module as basic components of photovoltaics.

He discovered the first functional electrochemical battery, and the unit of electricity, Volt, is named after him. Thus, a translation of the word ***photovoltaic*** could be light battery or also light source. More generally, we understand the word photovoltaics as the direct conversion of sunlight into electric energy.

1.5.2 What Are Solar Cells and Solar Modules?

The basic component of every photovoltaic plant is the solar cell (see Figure 1.9). This consists in most cases of silicon, a semiconductor that is also used for diodes, transistors, and computer chips. With the introduction of foreign atoms (doping) a p–n junction is generated in the cell that "installs" an electrical field in the crystal. If light falls on the solar cell, then charge carriers are dissolved out of the crystal bindings and moved by the electrical field to the outer contacts. The result at the contacts of the solar cells is the creation of a voltage of approximately 0.5 V. The released current varies depending on radiation and cell area, and lies between 0 and 10 A.

In order to achieve a usable voltage in the region of 20–50 V, many cells are switched together in series in a solar module (Figure 1.9). Besides this, the solar cells in the modules are mechanically protected and sealed against environmental influences (e.g. entrance of moisture).

1.5.3 How Is a Typical Photovoltaic Plant Structured?

Figure 1.10 shows the structure of a classical grid-coupled plant typical for Germany. Several solar modules are switched in series into a string and connected to an inverter. The inverter converts the direct current delivered by the modules into alternating current and feeds it into the public grid. A feed meter measures the generated electricity in

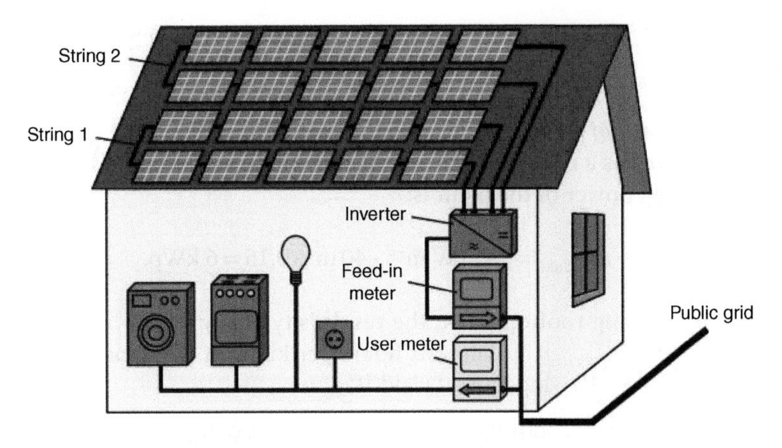

Figure 1.10 Structure of a classical grid-coupled photovoltaic plant. An inverter converts the direct current supplied by the solar modules into alternating current and feeds it into the public grid.

order to collect payment for the energy generated. The user meter counts the current consumption of the household separately.

The plant is financed on the basis of the Renewable Energy Law (EEG). This guarantees that the fed-in electric energy is paid for by the energy supply company for 20 years at a guaranteed price. To a certain extent, the owner of the plant becomes a power station operator.

Examples of plants where the produced solar energy is partly used in the household ("self-consumption") are presented in Chapter 8.

1.5.4 What Does a Photovoltaic Plant "Bring?"

For the owner of a solar power plant, the power of his plant is of interest and so is the quantity of energy fed into the grid during the course of a year.

The power of a solar module is measured according to the Standard Test Conditions (STC) and defined by three limiting conditions:

1. Full Sun radiation (irradiance $E = \text{ESTC} = 1000\ \text{W m}^{-2}$)
2. Temperature of the solar module: $\vartheta_{\text{Module}} = 25\,°\text{C}$
3. Standard light spectrum AM 1.5 (for more details see Chapter 2).

The capacity of the solar module under these conditions is the rated power (or nominal power) of the module. It is given in Watt–Peak (Wp) as it actually describes the peak power of the module under optimal conditions.

The degree of efficiency η_{Module} of a solar module is the relationship of delivered electric rated power PSTC referenced to incidental optical power P_{Opt}:

$$\eta_{\text{Module}} = \frac{P_{\text{STC}}}{P_{\text{Opt}}} = \frac{P_{\text{STC}}}{E_{\text{STC}} \cdot A} \tag{1.9}$$

with A: module surface.

The efficiency of silicon solar modules is in the range of 15–22%. Besides silicon there are also other materials such as cadmium telluride or copper–indium-selenide, which go under the name of thin-film technologies (see Chapter 5). These reach module efficiencies of 7–15%.

Example 1.2 ***Power and yield of a roof plant***
Assume that the house owner has a roof area of $40\,\mathrm{m}^2$ available. He buys modules with an efficiency of 15%. The rated power of the plant is:

$$P_{\mathrm{STC}} = P_{\mathrm{Opt}} \cdot \eta_{\mathrm{Module}} = E_{\mathrm{STC}} \cdot A \cdot \eta_{\mathrm{Module}} = 1000\,\mathrm{W\,m^{-2}} \cdot 40\,\mathrm{m}^2 \cdot 0.15 = 6\,\mathrm{kWp}.$$

In Germany with a South-facing rooftop plant, the result is typically a specific yield W_{Year} of approximately $900\,\mathrm{kW\,h\,(kWp)^{-1}}$ (kilowatt hours per kilowatt peak) per year. This brings our houseowner the following annual yield W_{year}:

$$W_{\mathrm{Year}} = P_{\mathrm{STC}} \cdot w_{\mathrm{Year}} = 6\ \mathrm{kWp} \cdot 900\,\mathrm{kW\,h}\ (\mathrm{kWp\,a})^{-1} = 5400\,\mathrm{kW\,h\,a^{-1}}.$$

In comparison to the typical electric power consumption of a household of 3000–4000 kW h per year, the energy quantity is not bad. ■

After this quick course on the subject of photovoltaics, we will now consider the quite recent history of solar power generation.

1.6 History of Photovoltaics

1.6.1 How It all Began

In the year 1839, the French scientist Alexandre Edmond Becquerel (the father of Antoine Henri Becquerel, after whom the unit of the activity of radioactive material is named) discovered the photoelectric effect while carrying out electrochemical experiments. He placed two coated platinum electrodes in a container with an electrolyte and determined the current flowing between them (see Figure 1.11a). Becquerel found that the strength of the current changed when exposed to light [13]. In this case, it was a matter of the outside photoeffect in which electrons move out of a fixed body when exposed to light.

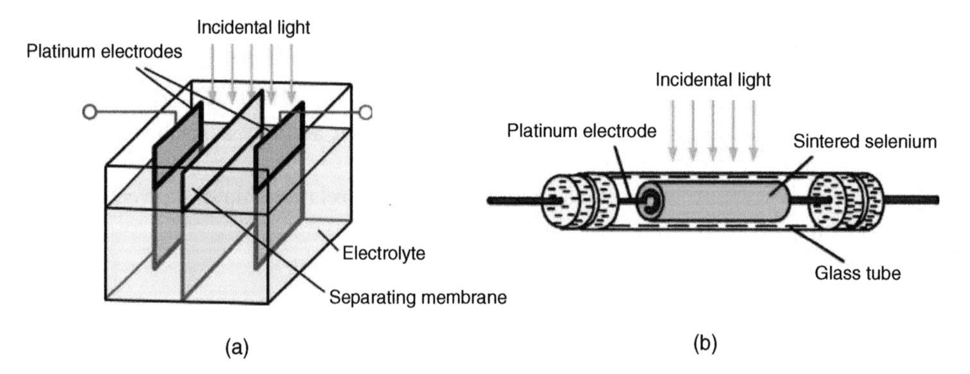

Figure 1.11 The beginnings of photovoltaics: (a) Electrochemical experiment of Becquerel and (b) the first solar cell by Adams and Day [11, 12].

In 1873, the British engineer Willoughby Smith and his assistant Joseph May discovered that the semiconductor selenium changed its resistance when exposed to light. They thus saw for the first time the internal photoeffect relevant for photovoltaics in which electrons in the semiconductor are torn from their bindings by light and are thus available as free charge carriers in the solid-state body.

Three years later, the Englishmen William Adams and Richard Day found out that a selenium rod provided with platinum electrodes can produce electrical energy when it is exposed to light (see Figure 1.11b). With this, it was proved for the first time that a solid body can directly convert light energy into electrical energy.

In 1883, the New York inventor Charles Fritts built a small "Module" of selenium cells with a surface area of approximately 30 cm^2 that had an efficiency of almost 1%. For this purpose, he coated the selenium cells with a very thin electrode of gold. He sent a module to Werner von Siemens (German inventor and entrepreneur, 1816–1892) for assessment. Siemens recognized the importance of the discovery and declared to the Royal Academy of Prussia that with this "The direct conversion of light into electricity has been shown for the first time" [14]. As a result, Siemens developed a lighting-measuring instrument based on selenium.

In the following years, the physical background of the photoeffect became better explained. In part, this was particularly due to Albert Einstein (1879–1955) who presented his light quantum theory in 1905, for which he was awarded the Nobel Prize 16 years later. At the same time, there were technological advances: In 1916, the Polish chemist Jan Czochralski at the AEG Company discovered the crystal growth process named after him. With the Czochralski process, it became possible to produce semiconductor crystals as single crystals of high quality.

1.6.2 The First Real Solar Cells

In 1950, the coinventor of the transistor, the American Nobel laureate William B. Shockley (1910–1989) presented an explanation of the method of functioning of the p–n junction and thus laid the theoretical foundations of the solar cells used today. On this basis, Daryl Chapin, Calvin Fuller, and Gerald Pearson in the Bell Labs developed the first silicon solar cell with an area of 2 cm^2 and an efficiency of up to 6% and presented it to the public on 25 April 1954 (Figure 1.12) [15]. The *New York Times* published this on its front page the next day and promised its readers "The fulfillment of one of the greatest desires of mankind – the use of the almost limitless energy of the sun."

The Bell cell combined for the first time the concept of the p–n junction with the internal photoeffect. In this, the p–n junction serves as conveyor that removes the released electrons. Thus, this effect can be more accurately described as the depletion layer photoeffect or also as the photovoltaic effect.

In the following years, the efficiency was raised to 10%. Because of the high prices of the solar modules (the price per Watt was around 1000 times more than today's price), they were only used for special applications. On 17 March 1958, the first satellite with solar cells on board was launched: The American satellite Vanguard 1 with two transmitters on board (Figure 1.13). Transmitter 1 was operated by mercury batteries and ceased operation after 20 days. Transmitter No. 2 drew its energy from six solar cells attached to the outer skin of the satellite and operated until 1964.

Figure 1.12 The inventors of the first "real" solar cell: Chapin, Fuller, and Pearson. The right-hand figure shows the first "solar module" in the world, a mini module of eight solar cells. *Source:* Courtesy of AT&T Archives and History Center.

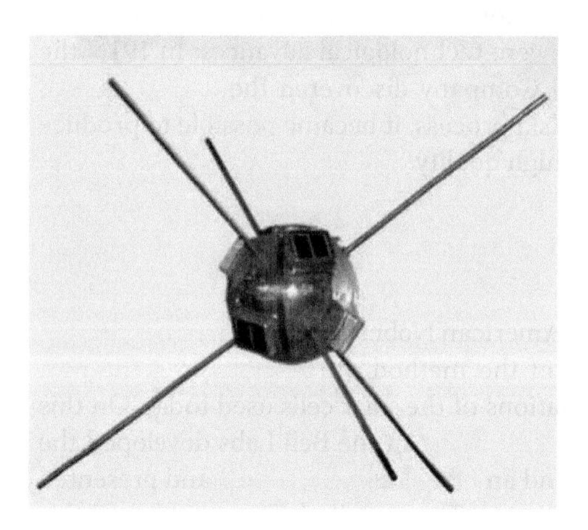

Figure 1.13 View of the Vanguard 1 satellite: Because of the diameter of 16 cm, it was also called "Grapefruit." *Source:* NASA.

The success of this project led to photovoltaics being used as the energy source for satellites. The developments in the 1960s were thus advanced by space flight. Besides the silicon cells, the first solar cells of gallium arsenide (GaAs) and other alternative materials were presented.

As a comparison to Vanguard I, Figure 1.14 shows one of the two solar arrays of the 2007 spacecraft Dawn. The probe has the task to examine the dwarf planet Ceres, which is about 400 Mio km away from the Sun. In 2015, the spacecraft managed to reach the planet and broadcast photos of the heavenly body since then. The name "Dawn" is well chosen:

Due to the large distance to the Sun, the intensity of the sunlight is only about 1/10 of that on Earth. Therefore, large efforts were made for the solar energy power supply: The array of 5 kW power is built up of high-capacity three-layer stacked cells of InGaP/InGaAs/Ge (see Chapter 5) that achieve an efficiency of 27.5% in space [16].

Figure 1.14 Modern solar array of the space probe *Dawn* with a power of 5 kW. *Source:* NASA.

1.6.3 From Space to Earth

The terrestrial use of photovoltaics limited itself first to applications in which the nearest electrical grid was very far away: Transmitter stations, signal systems, remote mountain huts, and so on (Figure 1.15). However, a change in thinking was brought about with the oil crisis in 1973. Suddenly, alternative sources of energy became the center of interest.

Figure 1.15 Examples of photovoltaic island plants: Telephone booster of 1955 with the legendary Bell Solar Battery and modern solar-driven lighting tower in Australia. *Source:* Courtesy of AT&T Archives and History Center, Erika Johnson.

In 1977, in the Sandia Laboratories in New Mexico, a solar module was developed with the aim of producing a standard product for economical mass production.

The accident in the nuclear power plant in Harrisburg (1979) and especially the reactor catastrophe in Chernobyl (1986) increased the pressure on governments to find new solutions in energy supply.

1.6.4 From Toy to Energy Source

From the end of the 1980s, research in the field of photovoltaics intensified especially in the United States, Japan, and Germany. In addition, research programs were started in the construction of grid-coupled photovoltaic plants that could be installed on single-family homes. In Germany, this was first the 1000-roof program of 1990–1995 that provided valuable knowledge on the reliability of modules and inverters as well as on the questions of grid-feeding [17].

The Energy Feed-in Law of 1991 obligated energy suppliers to accept power from small renewable power stations (wind, photovoltaics, etc.). Whereas the wind industry developed in a storm of activity, the subsidy of 17 Pfennigs per kilowatt hour was not nearly enough for an economical operation of photovoltaic plants. For this reason, environmentalists demanded a higher subsidy for solar power. A key role in this was played by the German Aachen Association for the Promotion of Solar Power (SFV). This association achieved that in 1995 the cost-covering subsidy at a rate of 2 DM per kW h for power from photovoltaics was introduced, which throughout the Federal Republic became known as the Aachen Model.

The 2000 promulgated EEG was introduced based on this model. This successor law of the Energy Feed-in Law defined cost-covering subsidies for various renewable energy sources and led to an unexpected boom in photovoltaics (see Figure 1.16). Thus, the cumulated installed capacity in Germany rose from 100 MWp in 2000 to about 41 GWp in 2016 (dark red bars in Figure 1.16). This represents an annual average growth of 45%. The annual installations reached 7.5 GWp a^{-1} in 2010 to 2012 and dropped to 1.5 GWp a^{-1} due to deteriorated funding conditions (blue bars in Figure 1.16).

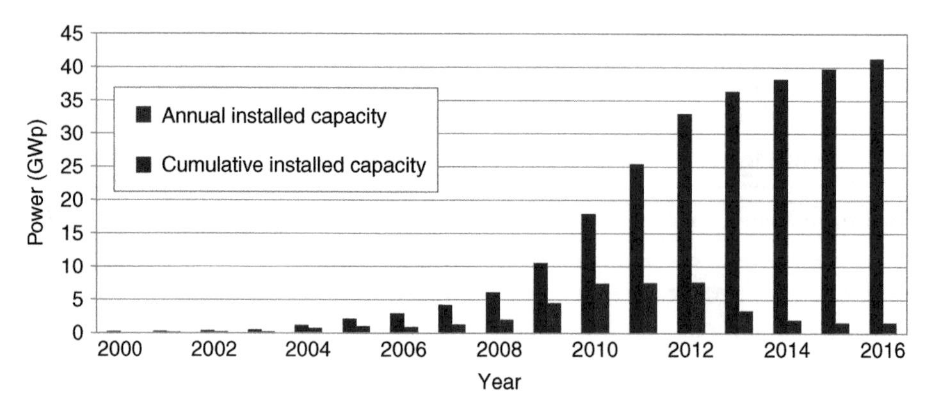

Figure 1.16 Development of the photovoltaics market in Germany: The total installed capacity, meanwhile, reaches 41 GWp.

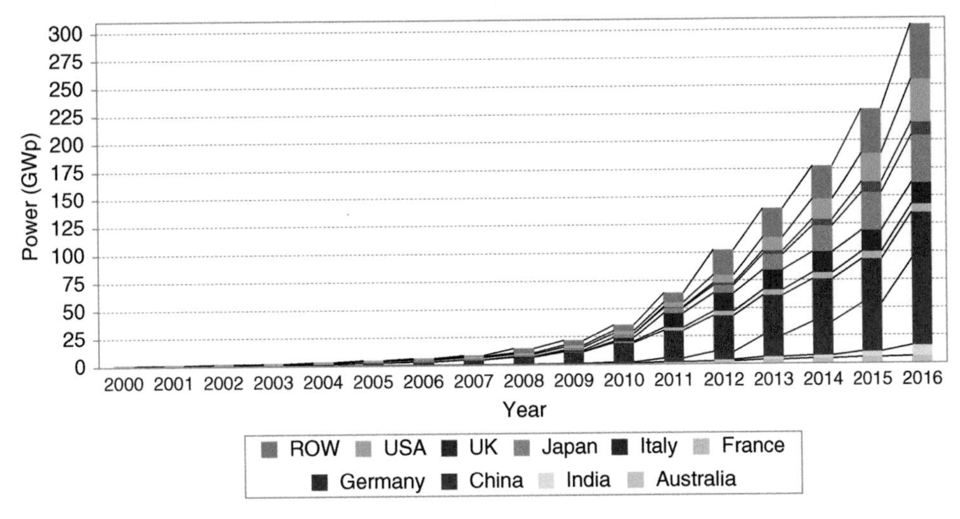

Figure 1.17 Development of the worldwide cumulative installed photovoltaics capacity: The most dynamic markets show China and the United States.

Figure 1.17 shows the worldwide installed photovoltaic capacities. After Germany became the driving factor for some time, now other countries are catching up more and more.

The strongest growth is shown from China with more than 30 GWp installed only in 2015. The total installed capacity of China, meanwhile, attains 80 GWp. The United States already reached about 40 GWp of cumulated capacity and will surpass Germany in the next years. After the reactor catastrophe of Fukushima, Japan strongly intensified the expansion of photovoltaics. This is obvious in a significant growth of installations, which led to a cumulated value of 43 GWp at the end of 2016. Spain and Italy had a real boom some years ago; now, the market there only develops in a restrained manner.

A strong newcomer is India with total installations of 10 GWp reached in only five years. The Indian government has announced its plan to reach an installed capacity of 22 GWp up to 2022.

All the other countries not shown individually in Figure 1.16 (Rest of World – ROW) have a total PV capacity of 50 GWp.

Independently of the view of the individual countries, Figure 1.17 emphasizes the fast growth of photovoltaics in the past years. The cumulated installed PV capacity has grown from 700 MWp in 2000 to 305 GWp in 2016 (the 435-fold of the starting value!). It can be assumed that this development will go on. As the prices of solar modules still go down from year to year (see Chapter 11), photovoltaics more and more not only becomes an ecological but also an economical alternative to fossil energies.

The possible future development in photovoltaics and its contribution to electric energy consumption is dealt with in Chapter 11.

2

Solar Radiation

The basis of all life on the Earth is the radiation from the Sun. The utilization of photovoltaics also depends on the availability of sunlight. For this reason, we will devote this chapter to the characteristics and possibilities of solar radiation.

2.1 Properties of Solar Radiation

2.1.1 Solar Constant

The Sun represents a gigantic fusion reactor inside which each of four hydrogen atoms is melted into one helium atom. In this atomic fusion, temperatures of around 15 million °C are created. The energy freed is released into space in the form of radiation.

Figure 2.1 shows the Sun–Earth system more or less to scale. The distance between the two space bodies is approximately 150 million km and other dimensions can be taken from Table 2.1.

The Sun continuously radiates an amount of $P_{Sun} = 3.845 \times 10^{26}$ W in all directions, of which the Earth receives only a small fraction.

In order to calculate this value, we assume there is a sphere around the Sun that has a radius of $r = r_{SE}$. At this distance, the amount of radiation from the Sun has already spread over the whole area of the sphere. Thus, at the position of the Earth, we get the following power density or irradiance.

$$Es = \frac{\text{Radiation power}}{\text{Area of sphere}} = \frac{P_{Sun}}{4 \cdot \pi \cdot r_{SE}^2} = \frac{3.845 \times 10^{26}\ \text{W}}{4 \cdot \pi \cdot (1.496 \times 10^{11}\ \text{m})^2} = 1367\ \text{W m}^{-2}. \tag{2.1}$$

The result of 1367 W m^{-2} is called the solar constant.

The solar constant is 1367 W m^{-2}. It denotes the irradiance outside the Earth's atmosphere.

2.1.2 Spectrum of the Sun

Every hot body gives off radiation to its surroundings. According to Planck's law of radiation, the surface temperature determines the spectrum of the radiation. In the case

Photovoltaics – Fundamentals, Technology, and Practice, Second Edition. Konrad Mertens.

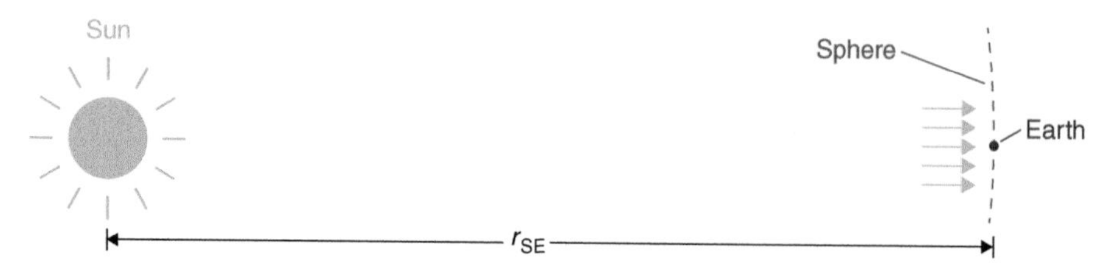

Figure 2.1 Determination of the solar constant.

Table 2.1 Characteristics of the Sun and the Earth.

Properties	Sun	Earth
Diameter	$d_{Sun} = 1\,392\,520$ km	$d_{Earth} = 12\,756$ km
Surface temperature	$T_{Sun} = 5778$ K	$T_{Earth} = 288$ K
Temperature at center	15 000 000 K	6700 K
Radiated power	$P_{Sun} = 3.845 \times 10^{26}$ W	–
Distance Sun–Earth	$r_{SE} = 149.6$ Mio km	–

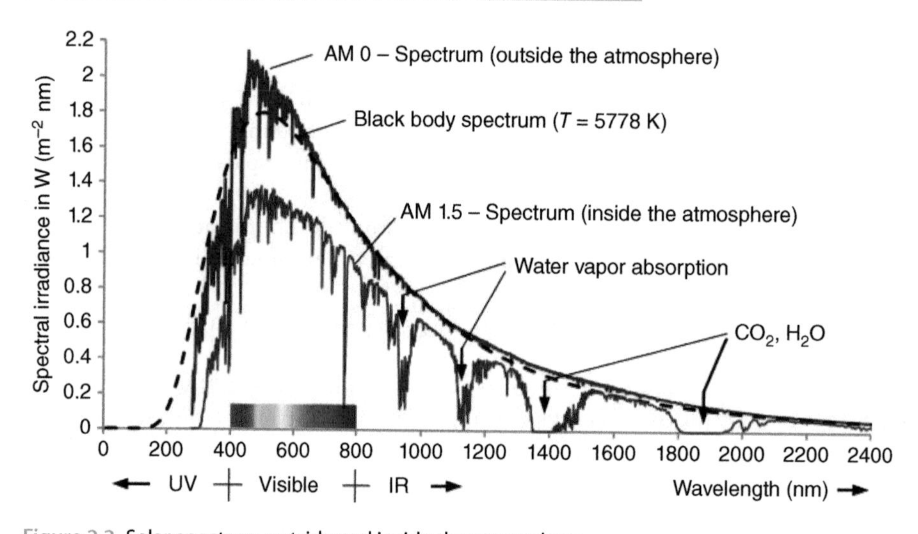

Figure 2.2 Solar spectrum outside and inside the atmosphere.

of the Sun, the surface temperature is 5778 K, which leads to the idealized Black Body Spectrum shown in Figure 2.2 (dashed line). The actual spectrum measured outside the Earth's atmosphere (AM 0) approximately follows this idealized line. The term *AM* 0 stands for Air Mass 0 and means that this light has not passed through the atmosphere. If the individual amounts of this spectrum are added together in Figure 2.2, then the result is an irradiance of 1367 W m^{-2}, which is the previously mentioned solar constant.

However, the spectrum changes when sunlight passes through the atmosphere. There are various reasons for this:

1. *Reflection of light*: Sunlight is reflected in the atmosphere, and this reduces the radiation reaching the Earth.

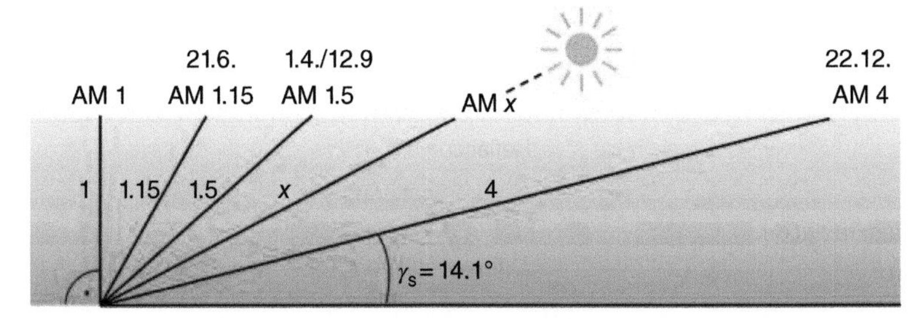

Figure 2.3 Explanation of the term *Air Mass*: The number *x* represents the extension of the path compared to the vertical distance through the atmosphere, here for the location Berlin [18].

2. *Absorption of light*: Molecules (O_2, O_3, H_2O, CO_2, …) are excited at certain wavelengths and absorb a part of the radiation causing "gaps" in the spectrum especially in the infrared region (see, for instance, Figure 2.2 at $\lambda = 1400$ nm).
3. *Rayleigh scattering*: If light falls on particles that are smaller than the wavelength, then Rayleigh scattering occurs. This is strongly dependent on wavelength ($\sim 1/\lambda^4$) so shorter wavelengths are scattered particularly strongly.
4. *Scattering of aerosols and dust particles*: This concerns particles that are large compared to the wavelength of light. In this case, one speaks of Mie scattering. The strength of Mie scattering depends principally on the location; it is greatest in industrial and densely populated areas.

2.1.3 Air Mass

As we have seen, the spectrum changes when passing through the atmosphere. The effect is greater, the longer the path of the light. For this reason, one designates the different spectra according to the path of the rays through the atmosphere. Figure 2.3 shows the principle: The term *AM* 1.5 means, for example, that the light has traveled 1.5 times the distance in comparison to the vertical path through the atmosphere.

At a known Sun height angle γ_S of the Sun, the AM value x gives:

$$x = \frac{1}{\sin \gamma_S}. \tag{2.2}$$

The Sun has different heights, depending on the time of day and year. Figure 2.3 shows on which days in Berlin the respective AM values are reached (always a noon Sun position).

The standard spectrum for measuring solar modules has established itself at the AM 1.5 spectrum as it arrives in spring and autumn and can be viewed as an average year's spectrum.

2.2 Global Radiation

2.2.1 Origin of Global Radiation

Various effects, such as scattering and absorption, cause a weakening of the AM 0 spectrum from space. In the summation of the AM 1.5 spectrum in Figure 2.2,

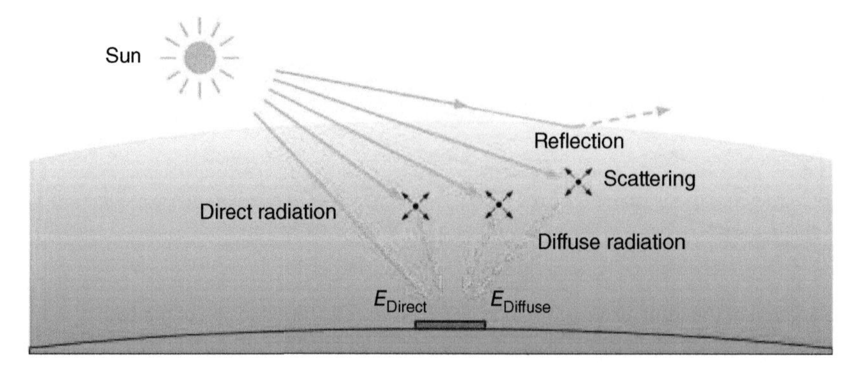

Figure 2.4 Origin of global radiation: It is the sum of the direct and diffuse radiation.

we receive only 835 W m^{-2}. Thus, of the originally available 1367 W m^{-2}, the Earth receives just 61% as so-called direct radiation. However, due to the scattering of light in the atmosphere, there exists a further portion: Diffuse radiation (see Figure 2.4).

Weak radiation portions arrive from all directions of the sky and are added to diffuse radiation. The sum of both types of radiation is called global radiation.

$$E_G = E_{Direct} + E_{Diffuse}. \tag{2.3}$$

On a nice, clear summer's day it is possible to measure on a surface vertical to the radiation of the Sun a global radiation value of $E_G = E_{STC} = 1000$ W m^{-2}. This is the reason why, in the definition of the standard test conditions for solar modules (see Section 1.5), one uses an enhanced AM 1.5 spectrum by the factor 1000/835 = 1.198. This, then, has a total irradiance of exactly $E_{STC} = 1000$ W m^{-2} and is thus suitable for determining the peak power of a solar module.

In reality, why do not irradiances higher than 1000 W m^{-2} occur?

In individual cases, higher global radiation can occur. This is mostly the case in mountainous regions such as the Alps. Besides the reduced atmospheric density, reflection from snow and ice can occur. However, one also sometimes measures radiation values up to 1300 W m^{-2} in flat country. This occurs in sunny weather, and bright light clouds around the Sun increase the diffuse portion. This effect is called cloud enhancement.

2.2.2 Contributions of Diffuse and Direct Radiation

The contribution of diffuse radiation to global radiation is often underrated. In Germany, when viewed over the whole year, diffuse radiation makes a greater contribution than direct radiation. For evidence of this, see Table 2.2. This shows the average monthly radiation H (irradiated energy) on a flat surface in various locations.

Table 2.2 Total radiation over the year on a horizontal level for various places in kW h (m² d)⁻¹ [19].

Place		Jan	Feb	Mar	Apr	May	Jun	Jul	Aug	Sep	Oct	Nov	Dec	Σ
Glasgow	H_{Direct}	0.06	0.23	0.49	1.18	1.66	1.60	1.38	1.03	0.60	0.25	0.10	0.03	0.71
55.7 °N	$H_{Diffuse}$	0.39	0.81	1.45	2.23	2.83	3.11	2.97	2.46	1.73	1.00	0.51	0.29	1.65
4.5 °W	H	**0.45**	**1.04**	**1.94**	**3.41**	**4.49**	**4.71**	**4.35**	**3.49**	**2.33**	**1.25**	**0.61**	**0.32**	**2.36**
Hamburg	H_{Direct}	0.13	0.37	0.74	1.49	2.18	2.32	2.01	1.82	1.10	0.52	0.18	0.10	1.08
53.6 °N	$H_{Diffuse}$	0.40	0.78	1.35	2.04	2.55	2.79	2.67	2.26	1.63	0.99	0.51	0.31	1.52
10.0 °E	H	**0.53**	**1.15**	**2.09**	**3.53**	**4.73**	**5.11**	**4.68**	**4.08**	**2.73**	**1.51**	**0.69**	**0.41**	**2.60**
London	H_{Direct}	0.17	0.36	0.82	1.36	1.88	2.08	1.91	1.72	1.24	0.61	0.26	0.12	1.04
51.6 °N	$H_{Diffuse}$	0.48	0.84	1.43	2.06	2.56	2.79	2.68	2.28	1.70	1.08	0.61	0.38	1.57
0.0 °W	H	**0.65**	**1.20**	**2.25**	**3.42**	**4.44**	**4.87**	**4.59**	**4.00**	**2.94**	**1.69**	**0.87**	**0.50**	**2.61**
Munich	H_{Direct}	0.36	0.75	1.28	1.83	2.43	2.62	2.69	2.26	1.71	0.89	0.38	0.24	1.45
48.4 °N	$H_{Diffuse}$	0.67	1.05	1.60	2.18	2.61	2.81	2.71	2.35	1.82	1.24	0.75	0.55	1.70
11.7 °E	H	**1.03**	**1.80**	**2.88**	**4.01**	**5.04**	**5.43**	**5.40**	**4.61**	**3.53**	**2.13**	**1.13**	**0.79**	**3.15**
Marseille	H_{Direct}	1.01	1.34	2.40	3.24	4.03	4.78	5.03	4.24	3.05	1.76	1.05	0.79	2.72
43.3 °N	$H_{Diffuse}$	0.79	1.11	1.49	1.90	2.16	2.18	2.02	1.85	1.58	1.24	0.87	0.70	1.49
5.4 °E	H	**1.80**	**2.45**	**3.89**	**5.14**	**6.19**	**6.96**	**7.05**	**6.09**	**4.63**	**3.00**	**1.92**	**1.49**	**4.21**
New York	H_{Direct}	0.84	1.35	1.88	2.43	3.04	3.25	3.17	3.03	2.34	1.68	0.74	0.57	2.03
40.8 °N	$H_{Diffuse}$	1.03	1.37	1.85	2.31	2.62	2.74	2.68	2.37	1.98	1.51	1.12	0.91	1.87
74.0 °W	H	**1.87**	**2.72**	**3.73**	**4.74**	**5.66**	**5.99**	**5.85**	**5.40**	**4.32**	**3.19**	**1.86**	**1.48**	**3.90**
Cairo	H_{Direct}	2.16	2.94	3.80	4.60	5.41	5.95	5.82	5.34	4.50	3.56	2.48	1.92	4.04
30.1 °N	$H_{Diffuse}$	1.26	1.47	1.76	1.99	2.05	2.01	1.99	1.89	1.73	1.50	1.30	1.18	1.68
31.2 °E	H	**3.42**	**4.41**	**5.56**	**6.59**	**7.46**	**7.96**	**7.81**	**7.23**	**6.23**	**5.06**	**3.78**	**3.10**	**5.72**
Miami	H_{Direct}	2.03	2.50	3.16	3.82	3.54	3.07	3.36	3.23	2.72	2.50	2.10	1.90	2.82
25.9 °N	$H_{Diffuse}$	1.46	1.71	1.96	2.17	2.42	2.54	2.47	2.36	2.17	1.83	1.53	1.37	2.00
80.1 °W	H	**3.49**	**4.21**	**5.12**	**5.99**	**5.96**	**5.61**	**5.83**	**5.59**	**4.89**	**4.33**	**3.63**	**3.27**	**4.82**
Quito	H_{Direct}	1.97	2.07	2.20	2.10	2.06	2.05	2.27	2.33	1.98	1.96	2.12	1.87	2.08
0.2 °S	$H_{Diffuse}$	2.16	2.27	2.35	2.23	2.06	1.96	1.99	2.13	2.28	2.28	2.18	2.11	2.17
78.5 °W	H	**4.13**	**4.34**	**4.55**	**4.33**	**4.12**	**4.01**	**4.26**	**4.46**	**4.26**	**4.24**	**4.30**	**3.98**	**4.25**
Sidney	H_{Direct}	3.38	3.01	2.74	2.25	1.81	1.72	1.86	2.54	3.17	3.57	3.48	3.67	2.76
33.9 °S	$H_{Diffuse}$	2.56	2.27	1.81	1.34	0.98	0.80	0.86	1.07	1.48	1.97	2.43	2.61	1.68
151.2 °E	H	**5.94**	**5.28**	**4.55**	**3.59**	**2.79**	**2.52**	**2.72**	**3.61**	**4.65**	**5.54**	**5.91**	**6.28**	**4.44**
Buenos Aires	H_{Direct}	4.38	3.57	2.99	2.20	1.68	1.26	1.41	1.88	2.62	3.11	3.93	4.32	2.77
34.6 °S	$H_{Diffuse}$	2.39	2.17	1.74	1.31	0.97	0.84	0.90	1.17	1.56	2.03	2.35	2.50	1.66
53.4 °W	H	**6.77**	**5.74**	**4.73**	**3.51**	**2.65**	**2.10**	**2.31**	**3.05**	**4.18**	**5.14**	**6.28**	**6.82**	**4.43**

In Hamburg, the average diffuse radiation $H_{Diffuse}$ is 1.52 kW h (m² d)⁻¹ against a H_{Direct} of 1.08 kW h (m² d)⁻¹. Thus, diffuse radiation contributes almost 60% to the annual global radiation. In Munich, the position is somewhat different. Diffuse radiation here only contributes 54%.

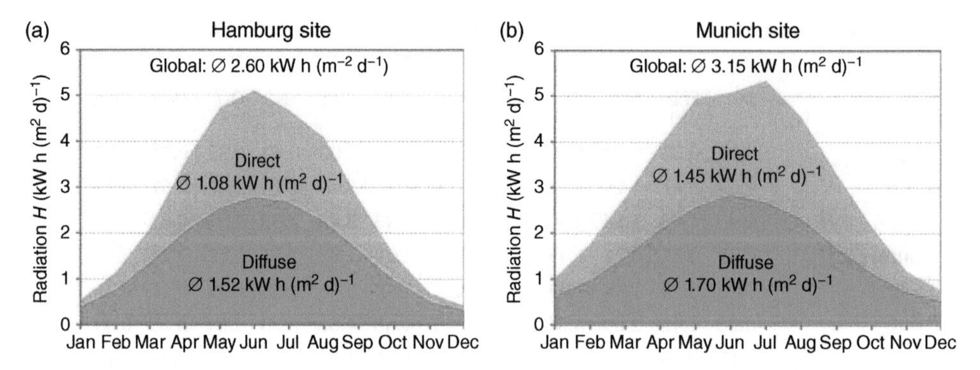

Figure 2.5 Radiation for a year on a horizontal level for the sites of Hamburg and Munich.

Thus, to summarize:

In Germany, diffuse radiation makes a slightly higher contribution to global radiation than direct radiation.

In locations to the south of Germany, the position is different: In Marseille and Cairo, the direct radiation of 65% and 71%, respectively, contributes to the main portion of global radiation.

In Figure 2.5, the data for Hamburg and Munich are shown again graphically. What can be derived from this? First, it is obvious that it is more worthwhile to operate a photovoltaic plant in Munich than in Hamburg. The average global radiation of $H = 3.15\,\text{kW h (m}^2\,\text{d)}^{-1}$ provides over a whole year (365 days) an annual total of

$$H = 3.15\ \text{kW h (m}^2\ \text{d)}^{-1} \cdot 365\ \text{d a}^{-1} = 1150\ \text{kW h (m}^2\ \text{a)}^{-1}.$$

The corresponding year's total in Hamburg is only $949\,\text{kW h m}^{-2}$ per annum.

Furthermore, it can be seen that the diffuse radiation in Munich is only slightly higher than in Hamburg. The higher global radiation in Munich is achieved mainly due to greater direct radiation. The reason for this is easy to guess: The higher position of the Sun in Munich. But obviously the height of the Sun has hardly any influence on the diffuse radiation.

The extent of the difference in daily courses of direct and diffuse radiation is shown in Figure 2.6. Here the radiation for a sunny day and a cloud-covered summer's day are shown.

On a sunny day, direct radiation clearly dominates, whereas on a cloud-covered day, it plays practically no role compared to diffuse radiation. Yet the cloud-covered day with $3.7\,\text{kW h m}^{-2}$ still brings more than half the radiation of the sunny day. This shows how profitable even cloud-covered days can be for photovoltaic usage.

2.2.3 Global Radiation Maps

In order to be able to estimate at planning stage the yield of a photovoltaic plant, it is necessary to obtain the data of the global radiation at the planned site. The most important characteristic for this is the year's total H of the global radiation

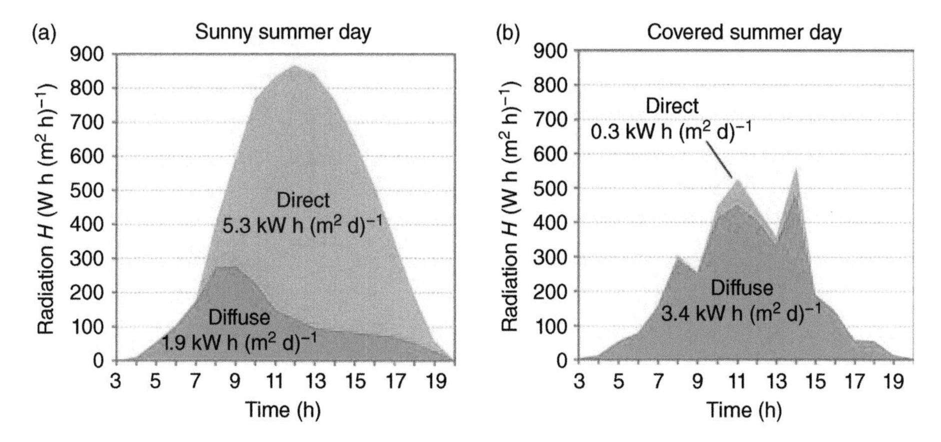

Figure 2.6 Radiation on two summer days in Braunschweig: The cloud-covered day provides a good half of the radiation energy of the sunny day [20].

on a horizontal level. Nowadays, global radiation maps are available that show this characteristic in high resolution. The basis of these is many years of measurements by a dense network of measuring stations, satellite pictures, and simulation tools. Figure 2.7 shows an example of this type of map from the German Weather Services.

It can be clearly seen that the annual radiation energy increases from the north to the south. The values range from 900 to 1150 kW h (m^2 a)$^{-1}$. On average, one can assume 1000 kW h (m^2 a)$^{-1}$ for Germany. This unusual unit can be bypassed by means of a very clear model, the Model of Sun full load hours:

We assume that the Sun can only have two conditions:

1. It shines with "Full load": $E = E_{STC} = 1000\ \mathrm{W\ m^{-2}}$
2. It is fully "Switched off": $E = 0$.

For how long must the Sun shine with full load for it to provide a radiation of $H = 1000\ \mathrm{kW\,h\ (m^2\ a)^{-1}}$ on the Earth's surface, for instance?

$$\frac{H}{E_{STC}} = \frac{1000\ \mathrm{kW\ h\ (m^2\ a)^{-1}}}{1000\ \mathrm{W\ m^{-2}}} = 1000\ \mathrm{h\ a^{-1}}. \tag{2.4}$$

The Sun would need 1000 full load hours to provide the same optical energy as it actually provides in a year (8760 h).

In Germany, the Sun provides approximately 1000 full load hours.

In other countries, the situation is in part much better. Figure 2.8 shows the radiation situation in Europe. Radiation is mostly in the region of 1000–1500 kW h (m^2 a)$^{-1}$. Extreme values are found in Scotland with only 700 and in southern Spain with 1800 kW h (m^2 a)$^{-1}$.

An overall picture is provided in Figure 2.9, which shows a world map of radiation. The highest radiation is naturally found in the region of the equator with values of up to 2500 full load hours.

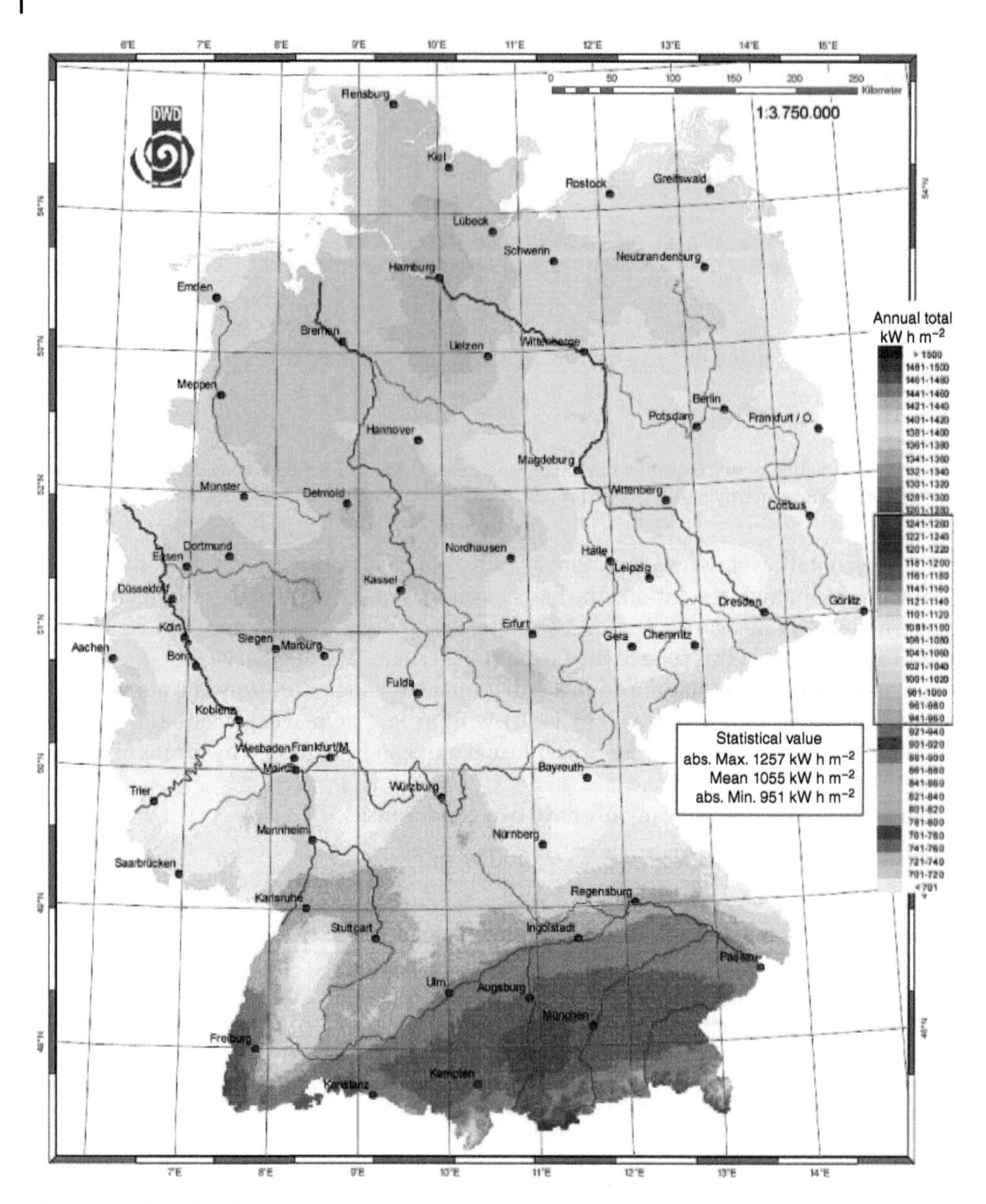

Figure 2.7 Global horizontal radiation in Germany [21].

2.3 Calculation of the Position of the Sun

2.3.1 Declination of the Sun

Within a year, the Earth travels around the Sun in an almost perfect circle. Because the axis of the Earth is tilted, the height of the Sun changes over the course of a year. Figure 2.10 shows this connection for the summer and winter changes.

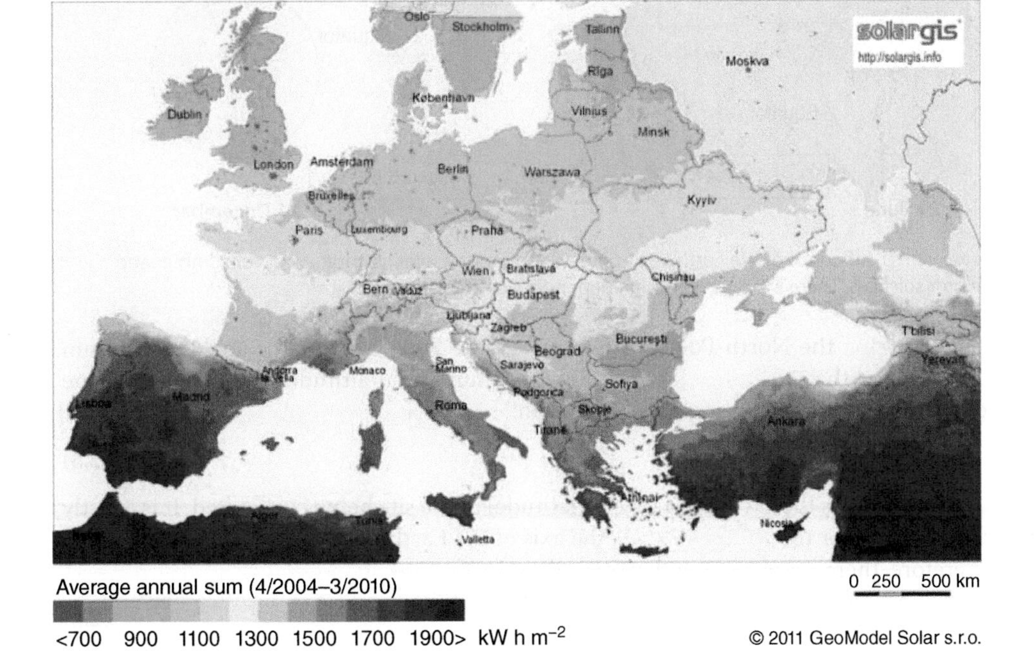

Figure 2.8 Global radiation map of Europe. Radiation is between 700 kW h m^{-2} in Scotland and 1800 kW h m^{-2} per year in southern Spain. Source: Reproduced with permission of GHI Solar Map © 2017 Solargis.

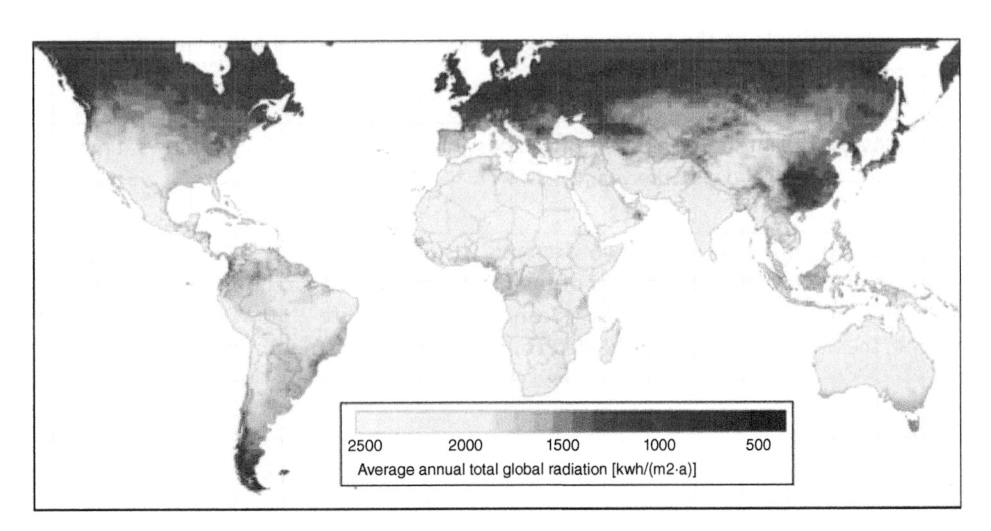

Figure 2.9 World map of global radiation. Source: Created by Meteonorm, www.meteonorm.com.

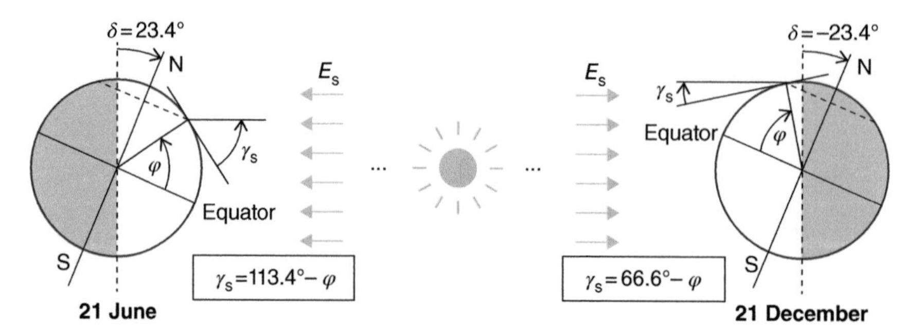

Figure 2.10 Influence of the Earth's axis tilt on the noonday solar altitude γ_{S_Max} for summer and winter solstices. There is a different value depending on the latitude ϕ.

In summer, the North Pole is tilted toward the Sun so that large angles of the sun (often called the solar altitude) exist. The maximum solar altitude γ_{S_Max} (noon) can be determined with simple angle consideration:

$$\gamma_{S\_Max} = 113.4^\circ - \phi. \tag{2.5}$$

The angle ϕ is the geographic width (latitude) of the site being considered. It is exactly the opposite for the winter solstice, the axis of the Earth slants away from the Sun, and therefore, there is only a noonday solar altitude of

$$\gamma_{S\_Min} = 66.6^\circ - \phi. \tag{2.6}$$

Example 2.1 ***Solar altitude at Münster***
A PV plant is to be erected at Münster (latitude $\phi = 52^\circ$). In the middle of June, the solar altitude is $113.4^\circ - 52^\circ = 61.4^\circ$. In the middle of December, the value is $66.6^\circ - 52^\circ = 14.6^\circ$. ∎

The solar altitude of the winter solstice is especially important in planning photovoltaic plants.

A photovoltaic plant should be structured if possible so that there are no shadows at noon even on the shortest day of the year (21.12).

Figure 2.10 shows only the two extreme cases of the Sun declination δ. This is understood to be the respective tilt of the Earth's axis in the direction of the Sun. This changes continuously over the year as can be seen in Figure 2.11. With the aid of this picture, the Sun declination can be determined for each day of the year.

2.3.2 Calculating the Path of the Sun

In the case of possible shadows, it is helpful in planning the details of the plant to know the path of the Sun on certain days. In order to make the calculation simple, we use the so-called local solar time (LST) or solar time. This is the time in which at noon the Sun is exactly in the south and has thus reached its highest point of the day. In principle, it is therefore the time that a sundial would show at the site.

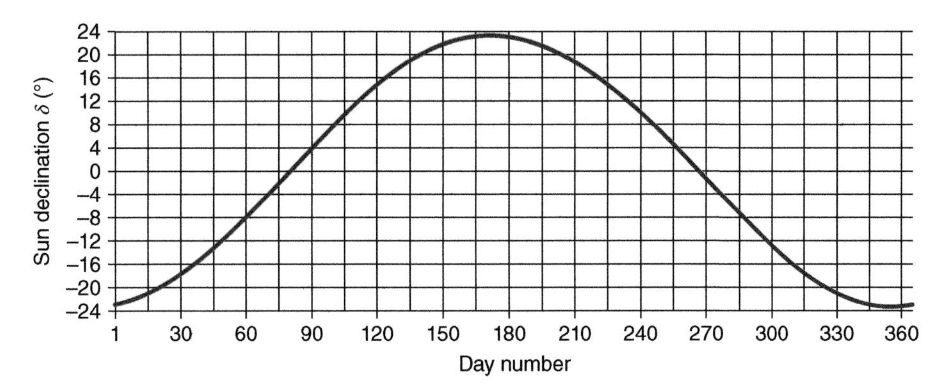

Figure 2.11 Sun declination for the year (365 days).

Coordinated Universal Time (*UTC*) would, however, be referenced to the zero meridian at Greenwich. Thus, it would only be correct with the LST at those places that are situated along 0° longitude.

The difference between both times can be roughly determined from the respective geographic location Λ as the Earth rotates once in 24 h (360°) and thus 1 h corresponds to just 15°.

$$\text{LST} = \text{UTC} + 1\,\text{h} \cdot \frac{\Lambda}{15^\circ}. \tag{2.7}$$

What is ignored here is that LST does not only depend on the rotation of the Earth but also on the Sun's declination and the elliptical path of the Earth around the Sun.

Example 2.2 ***LST in Belfast***

You are in Belfast and would like to find out the LST. Your wristwatch shows noon. Belfast has a longitude of $\Lambda = -6.0^\circ$.

$$\text{LST} = 12\,\text{h} + 1\,\text{h} \cdot \frac{-6^\circ}{15^\circ} = 12\,\text{h} - 60\,\text{min} \cdot 0.4 = 11\,:\,36\,\text{h}.$$

Thus LST is 11:36 h: You will have to wait almost another half hour until the Sun reaches its highest point.

In the case of other time zones, the respective clock time must first be converted to world time in order to use Equation (2.7). Thus, for instance, noon Middle European Time (MET) would correspond to a World Time of 11:00. ■

We will now calculate the path of the Sun for certain days. For this purpose, Figure 2.12 shows the angle definitions for describing the position of the Sun. This shows the Sun's azimuth α_S as well as the solar altitude γ_S and depicts the displacement of the Sun from the south. Positive values show western deviations and negative values the eastern displacements.

Remarks: This definition of the Sun's azimuth is the most common one in the field of photovoltaics. However, some simulations make use of the definition according to DIN. There, north is defined with 0° and then added in a clockwise direction (east – 90°, south – 180°, etc.). We will also insert the hour angle ω as an abbreviation: This calculates

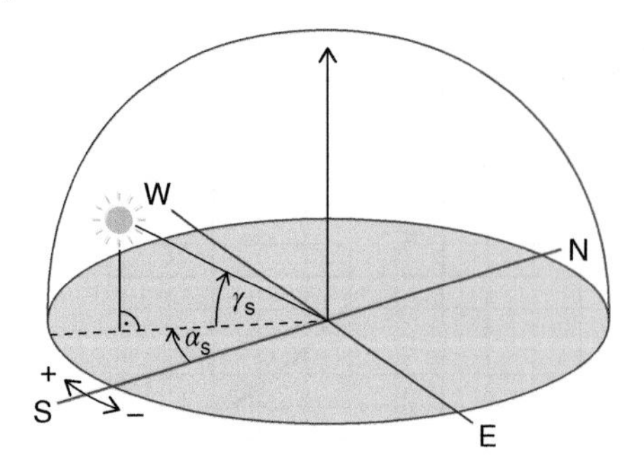

Figure 2.12 Calculation of the Sun's position with the dimension of solar altitude γ_S and Sun's azimuth α_S.

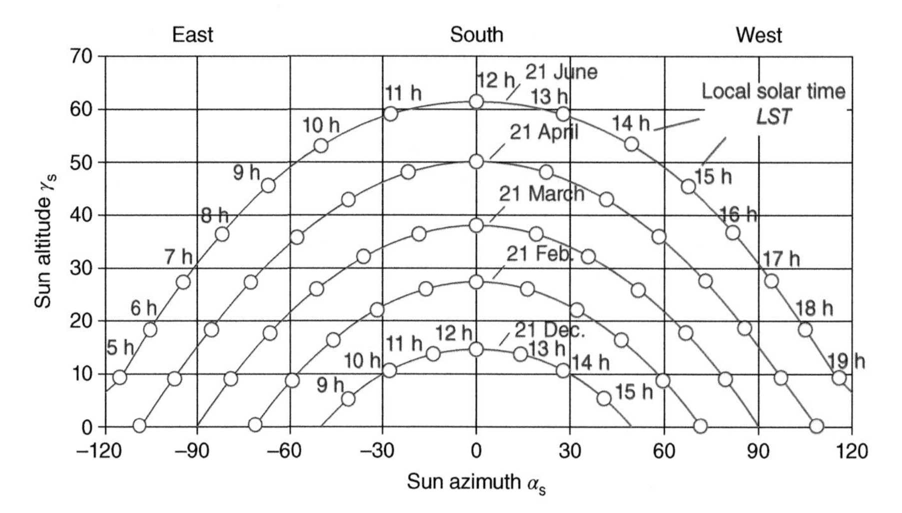

Figure 2.13 Sun path diagram for the town of Münster (geographic latitude $\phi = 52°$): The time used is local solar time (LST).

the LST into the respective rotation position of the Earth:

$$\omega = (\text{LST} - 12) \cdot 15°. \tag{2.8}$$

Now the position of the Sun can be determined from the latitude ϕ and the Sun declination δ:

$$\sin \gamma_S = \sin \phi \cdot \sin \delta + \cos \phi \cdot \cos \delta \cdot \cos \omega, \tag{2.9}$$

$$\sin \alpha_S = \frac{\cos \delta \cdot \sin \omega}{\cos \gamma_S}. \tag{2.10}$$

Figure 2.13 shows a Sun path diagram calculated according to this equation. It shows the path of the Sun for the town of Münster on different days of the year. On 21 June and 21 December, the determined maximum and minimum solar altitudes already shown in Example 2.1 are reached. The time entered is the LST so the Sun reaches its highest position at noon.

These types of diagrams help to show up possible shadows caused by trees, houses, or similar and to estimate their effects on the plant's yield (see Chapter 10).

The website www.textbook-pv.org contains an Excel file that can be used to determine and print out the Sun path diagram for any desired latitude.

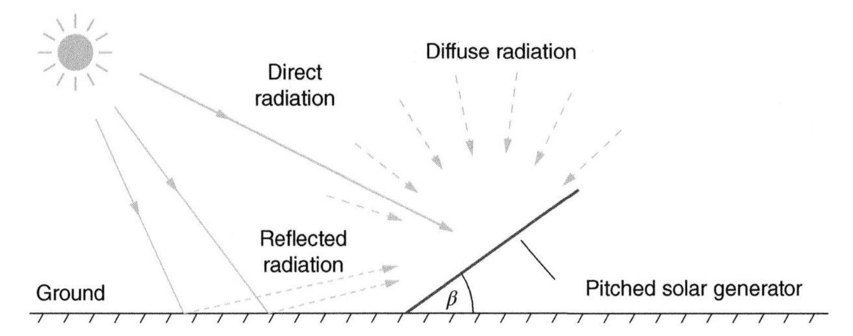

Figure 2.14 Radiation situation with tilted surfaces: The radiation is made up of the direct, diffused, and reflected radiation.

2.4 Radiation on Tilted Surfaces

Photovoltaic plants are mostly installed on pitched roofs so that the module is at an angle of attack β to the horizontal. Also in the case of flat roofs or ground-mounted plants, it is usual to tilt the modules in order to achieve a higher annual yield.

2.4.1 Radiation Calculation with the Three-component Model

Figure 2.14 shows the radiation relationships in the case of a suitable solar module surface (or more generally, a solar generator). Besides the direct and diffuse radiation, there is still a further radiation component: The radiation reflected from the ground. These add themselves to an overall radiation E_{Gen} on the tilted generator.

$$E_{\text{Gen}} = E_{\text{Direct\_Gen}} + E_{\text{Diffuse\_Gen}} + E_{\text{Refl\_Gen}}. \tag{2.11}$$

We will now look at the calculation of the individual components in more detail.

2.4.1.1 Direct Radiation

Let us consider the case that direct sunlight shines on a tilted solar module. For this case, the left sketch of Figure 2.15 shows how solar radiation impinges on a horizontal surface A_{H}. The optical power P_{Opt} of the impinging radiation is

$$P_{\text{Opt}} = E_{\text{Direct-H}} \cdot A_{\text{H}}. \tag{2.12}$$

If a solar generator were arranged exactly vertically to the solar radiation, then it would be possible to take up the same power on a smaller surface A_{Vertical}:

$$P_{\text{Opt}} = E_{\text{Direct\_H}} \cdot A_{\text{H}} = E_{\text{Direct\_Vertical}} \cdot A_{\text{Vertical}}. \tag{2.13}$$

The strength of the radiation $E_{\text{Direct\_Vertical}}$ is, therefore, increased by the factor $A_{\text{H}}/A_{\text{Vertical}}$ compared to the horizontal strength of radiation. This increase can be seen in Figure 2.15 in that the light rays are closer together on the vertical surface in comparison to the horizontal surface.

Figure 2.15b shows the general case: a solar generator tilted by the angle β. For determining the strength of the radiation in the generator level, a pair of trigonometric

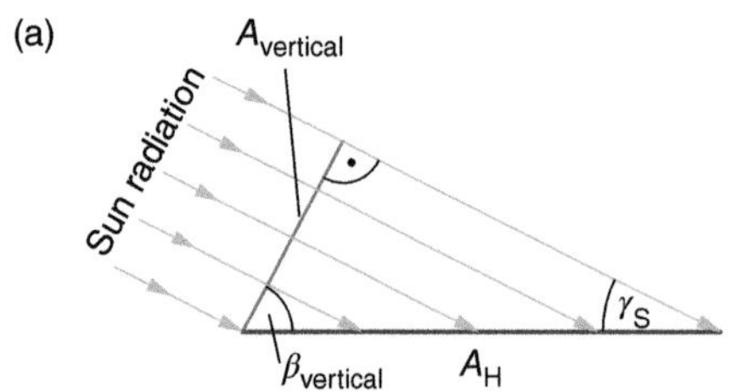

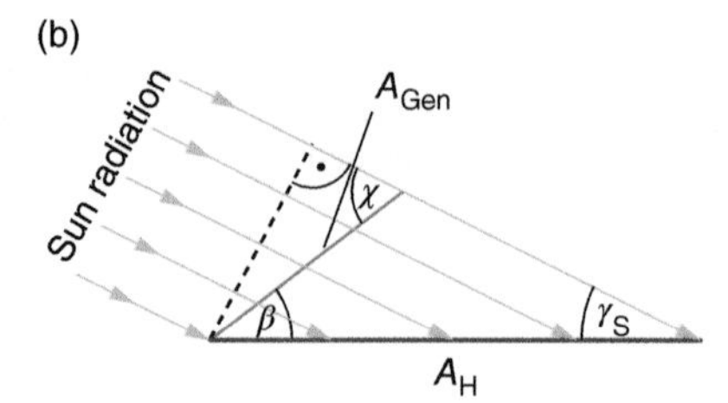

A_H: Horizontal surface
$A_{vertical}$: Surface vertical to incidental direction
A_{Gen}: Surface in generator direction
γ_S: Solar altitude angle
β: Elevation angle of solar generator
χ: Auxiliary angle

Figure 2.15 Influence of the solar generator tilt on direct radiation.

equations are used:

$$A_{\text{Vertical}} = A_{\text{H}} \cdot \sin \gamma_{\text{S}}, \tag{2.14}$$

$$A_{\text{Vertical}} = A_{\text{Gen}} \cdot \sin \chi. \tag{2.15}$$

The complementary angle χ can be calculated by the sum of the angles in the triangle and as subsidiary angle:

$$\chi = \gamma_{\text{S}} + \beta. \tag{2.16}$$

Using Equations (2.13)–(2.16), we then derive

$$E_{\text{Direct\_Gen}} = E_{\text{Direct\_H}} \cdot \frac{\sin(\gamma_{\text{S}} + \beta)}{\sin \gamma_{\text{S}}}. \tag{2.17}$$

It must be noted that this equation applies only for direct radiation.

Example 2.3 ***Radiation on a tilted solar module***
Assume that on a clear summer day the Sun is at a solar altitude of 40° in the sky. On flat ground, the irradiance is measured at $E_{\text{Direct\_H}} = 700\ \text{W m}^{-2}$. What angle of attack is ideal for a solar module, and what yield can be expected when the diffuse radiation and reflection of the ground are ignored? ■

For this case, Figure 2.16 shows Equation (2.17) as a diagram. With an increasing angle of the solar module, the irradiance increases continuously up to an angle of $\beta = 50°$ (thus an angle $90° - \gamma_{\text{S}}$). Here the irradiance reaches a maximum value of $E_{\text{Direct\_Gen}} = 1089\ \text{W m}^{-2}$. The solar module power can thus be increased by more than 50%.

2.4.1.2 Diffuse Radiation

The calculation of the diffuse radiation of tilted surface can be much simplified. For this purpose, we make a simple assumption that the diffuse radiation from the whole sky is approximately of the same strength (the isotropic assumption: Figure 2.17a). Thus the strength of radiation of a solar generator at an angle β is calculated as follows:

$$E_{\text{Diffus\_Gen}} = E_{\text{Diffus\_H}} \cdot \frac{1}{2} \cdot (1 + \cos \beta). \tag{2.18}$$

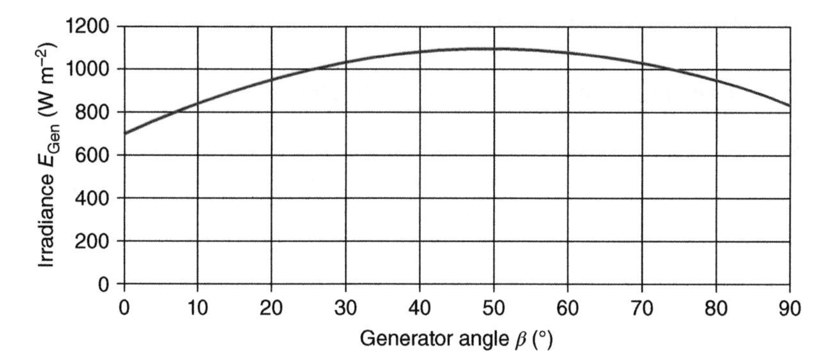

Figure 2.16 Example of dependency of the irradiance (direct radiation) on a tilted solar module for a solar altitude of $\gamma_S = 40°$.

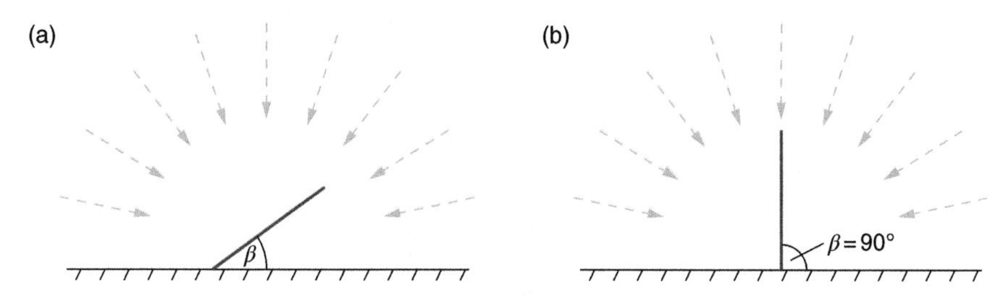

Figure 2.17 Isotropic assumption for diffuse radiation on a tilted surface. Only half the radiation can be used in the case of a vertically standing solar generator.

Starting with a horizontal generator ($\beta = 0°$), the radiation is reduced until at ($\beta = 90°$) it is:

$$E_{Diffus\_Gen} = \frac{E_{Diffus\_H}}{2}. \tag{2.19}$$

In this case, the solar generator is vertical so that only the left-hand side of the sky can be used (Figure 2.17b).

The isotropic assumption is only to be understood as a rough approximation. Thus, the sky around the Sun is mostly brighter than in the region of the horizon. More refined models are used in modern simulation programs in order to achieve greater degrees of accuracy.

2.4.1.3 Reflected Radiation

As shown in Figure 2.14, a part of the global radiation is reflected from the ground and can act as an additional radiation contribution to the solar generator.

In the calculation of this portion, the main problem is that every ground material reflects (or more exactly: scatters) differently. The so-called albedo (ALB) value describes the resulting reflection factor. Table 2.3 lists the ALB value of some types of ground.

The large range of the given values shows that simulation of the reflected radiation is accompanied by large uncertainties. If the ground is not known, then the standard value of ALB = 0.20 is often entered into the simulation program.

Table 2.3 Albedo value of different types of ground [22].

Material	Albedo (ALB)	Material	Albedo (ALB)
Grass (July, August)	0.25	Asphalt	0.15
Lawn	0.18, …, 0.23	Concrete, clean	0.30
Unmown fields	0.26	Concrete, weathered	0.20
Woods	0.05, …, 0.18	Snow cover, new	0.80, …, 0.90
Heath surfaces	0.10, …, 0.25	Snow cover, old	0.45, …, 0.70

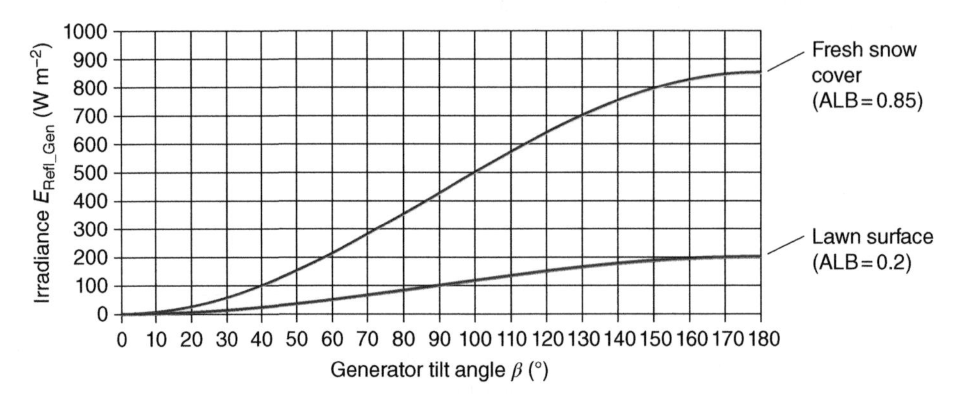

Figure 2.18 Reflected radiation as an example of lawn and fresh snow for various degrees of tilt of the solar module ($E_G = 1000\,\text{W m}^{-2}$).

An isotropic assumption is again made for the calculation of the reflected radiation on the tilted generator.

$$E_{\text{Relf\_Gen}} = E_G \cdot \frac{1}{2} \cdot (1 - \cos\beta) \cdot \text{ALB}. \tag{2.20}$$

Figure 2.18 shows the results for the case of ground covered with lawn and fresh snow. In the case of flat solar modules ($\beta = 0$), the portion of the radiation reflected by the ground is zero and then rises continuously. At $\beta = 90°$, half the available reflective radiation reaches the solar generator. This is the case of facade plants where solar modules are fixed to the vertical walls of a house. If one goes beyond 90°, then the part of the reflection radiation continues to rise, but the top face of the solar module now faces down, which obviously is not the optimum for overall radiation.

2.4.2 Radiation Estimates with Diagrams and Tables

The equations and characteristic values described are meant as an aid to understanding the features and limits of the use of solar radiation. However, simulation programs are always used today for detailed planning of photovoltaic plants, and they work with

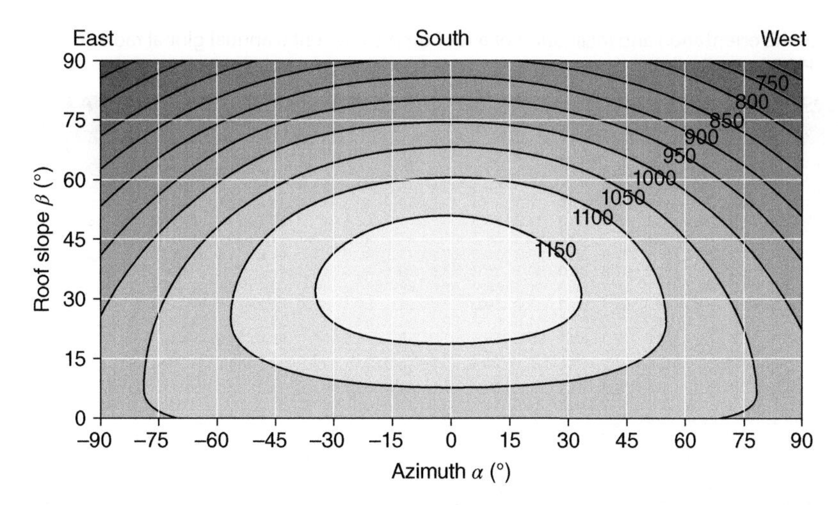

Figure 2.19 Radiation diagram for estimating the suitability of a roof for the city of Berlin given as the radiation sum H in kW h (m^2 a)$^{-1}$ [23].

refined models and detailed weather information in order to create very exact yield forecasts (see Chapter 10).

Radiation diagrams and tables are of help for a rough estimate of the radiation on a pitched roof. Figure 2.19 shows such a diagram for the city of Berlin.

A relatively broad maximum of the radiation total for a south-facing roof (azimuth $\alpha = 0°$) and a roof slope of approximately 35° can be clearly seen. The direction does not make much difference in the case of a lesser slope (e.g. = 15°). If, however, the roof is relatively steep (e.g. = 60°), then a south–west direction leads to a reduction in radiation to less than 1050 kW h (m^2 a)$^{-1}$.

A somewhat more accurate estimate of the deviation of the roof slope and orientation from the optimum can be found in Table 2.4 for the example of Berlin. The radiation amount here was standardized on the maximum values $\alpha = 0°$ and $\beta = 35°$ so that the deviations can be read directly in percentages.

For instance, the yield of a facade plant ($\beta = 90°$) that is aligned directly south is only 72% of the optimum. A horizontal surface, however, still yields 86% of the optimum.

Even a northern roof can be used for solar energy due to the relatively high diffuse portion of the radiation. In the case of a roof orientated exactly to the north with a roof inclination of 35°, for example, it still receives about 57% of the optimally orientated south roof.

Appendix A lists more diagrams like those in Table 2.4 for different locations worldwide.

Figure 2.19 and Table 2.4 only show the different incidental radiation dependent on the alignment and the roof pitch. In reality, though, soiling also plays a certain role. Thus, soiling of the solar module is reduced with increasing inclination angle. Added to this is the fact that a larger angle of attack ensures that snow in winter can slide off more easily and thus more solar energy can be harvested. In the case of facade plants, shading by trees and other houses also plays a large role.

Table 2.4 Effect of the orientation and inclination of a roof on the incidental annual global radiation (Simulation with PV-SOL for Berlin).

	Azimuth α	Angle of slope β 0°	5°	10°	15°	20°	25°	30°	35°	40°	45°	50°	55°	60°	65°	70°	75°	80°	85°	90°
	−180°	88.4	82.6	78.3	73.8	69.5	65.2	61.0	56.8	52.8	49.0	45.4	42.0	39.3	37.3	35.7	34.3	33.0	31.6	30.4
North	−175°	88.4	82.6	78.3	73.8	69.5	65.2	61.0	56.8	52.8	48.9	45.3	42.0	39.3	37.4	35.8	33.4	33.1	31.7	30.5
	−170°	88.4	82.6	78.4	73.9	69.5	65.3	61.1	57.0	52.9	49.2	45.6	42.4	39.7	37.8	36.2	34.8	33.4	32.1	30.8
	−165°	88.4	82.6	78.4	74.1	69.8	65.5	61.4	57.3	53.4	49.6	46.1	42.9	40.4	38.6	37.0	35.5	34.1	32.7	31.5
	−160°	88.4	82.7	78.6	74.3	70.1	65.9	61.8	57.8	53.9	50.2	46.8	43.8	41.5	39.7	38.0	36.5	35.1	33.6	32.2
	−155°	86.4	82.8	78.9	74.7	70.6	66.4	62.4	58.5	54.7	51.1	47.8	45.1	43.0	41.0	39.3	37.8	36.2	34.6	33.2
	−150°	86.4	83.0	79.1	75.2	71.1	67.0	63.1	59.3	55.6	52.2	49.2	46.6	44.5	42.6	40.9	39.2	37.5	35.9	34.4
	−145°	86.4	83.1	79.4	75.7	71.7	67.8	64.0	60.3	56.8	53.6	50.8	48.4	46.3	44.4	42.4	40.7	39.0	37.2	35.6
	−140°	86.4	83.3	79.9	76.2	72.5	68.7	65.0	61.5	58.2	55.3	52.7	50.3	48.2	46.1	44.2	42.4	40.5	38.8	37.0
North east	−135°	86.4	83.5	80.3	76.8	73.3	69.7	66.3	63.0	59.9	57.1	54.6	52.3	50.2	48.2	46.1	44.2	42.2	40.4	38.4
	−130°	86.4	83.7	80.7	77.5	74.2	70.9	67.6	64.6	61.7	59.1	56.6	54.4	52.2	50.1	48.1	46.1	44.0	42.0	40.0
	−125°	86.4	84.0	81.2	78.3	75.2	72.1	69.1	66.3	63.6	61.2	58.7	56.5	54.4	52.2	50.2	48.0	45.9	43.8	41.7
	−120°	86.4	84.2	81.7	79.0	76.3	73.5	70.7	68.1	65.6	63.3	61.0	58.7	56.5	54.4	52.3	50.1	47.9	45.7	43.5
	−115°	86.4	84.5	82.3	79.9	77.3	74.8	72.4	70.0	67.6	65.4	63.3	61.0	58.9	56.6	54.4	52.3	49.9	47.7	45.3
	−110°	86.4	84.8	82.9	80.7	78.5	76.3	74.1	71.9	69.7	67.6	65.4	63.3	61.2	59.0	56.7	54.4	52.0	49.6	47.2
	−105°	86.4	85.1	83.5	81.6	79.7	77.7	75.8	73.7	71.8	69.8	67.7	65.6	63.5	61.3	59.0	56.6	54.2	51.6	49.2
	−100°	86.4	85.4	84.1	82.6	80.9	79.2	77.4	75.7	73.8	72.0	70.0	68.0	65.8	63.6	61.2	58.8	56.4	53.7	50.9
	−95°	86.4	85.7	84.7	83.5	82.1	80.7	79.1	77.6	75.8	74.1	72.2	70.2	68.1	65.9	63.5	61.0	58.5	55.8	52.9
East	−90°	86.4	86.0	85.3	84.4	83.3	82.1	80.8	79.4	77.9	76.3	74.4	72.5	70.4	68.0	65.8	63.2	60.4	57.7	54.9
	−85°	86.4	86.3	86.0	85.3	84.5	83.6	82.5	81.3	79.9	78.4	76.6	74.7	72.6	70.2	67.9	65.3	62.4	59.6	56.6
	−80°	86.4	86.7	86.6	86.2	85.7	85.0	84.1	83.1	81.8	80.4	78.7	76.8	74.7	72.4	69.9	67.4	64.5	61.4	58.3
	−75°	86.4	87.0	87.2	87.2	86.8	86.4	85.7	84.8	83.7	82.3	80.7	78.9	76.8	74.5	71.9	69.2	66.4	63.3	60.0
	−70°	86.4	87.2	87.8	88.0	88.0	87.8	87.2	86.5	85.5	84.2	82.7	80.9	78.8	76.5	73.9	71.0	68.0	64.9	61.6
	−65°	86.4	87.6	88.3	88.8	89.1	89.1	88.8	88.2	87.2	86.1	84.6	82.8	80.6	78.4	75.8	72.9	69.7	66.5	63.1
	−60°	86.4	87.8	88.9	89.7	90.1	90.4	90.2	89.7	88.9	87.8	86.3	84.6	82.6	80.0	77.5	74.6	71.4	68.0	64.4
	−55°	86.4	88.1	89.5	90.5	91.1	91.5	91.5	91.2	90.5	89.4	88.0	86.3	84.3	81.8	79.0	76.2	72.9	69.4	65.7
	−50°	86.4	88.4	90.0	91.2	92.1	92.6	92.8	92.5	91.9	91.0	89.6	87.8	85.9	83.5	80.6	77.5	74.2	70.7	66.9
South east	−45°	86.4	88.6	90.4	91.9	93.0	93.6	94.0	93.9	93.3	92.4	91.2	89.3	87.2	84.9	82.1	78.9	75.5	71.8	67.9
	−40°	86.4	88.8	90.9	92.5	93.8	94.6	95.1	95.1	94.5	93.8	92.5	90.9	88.6	86.2	83.4	80.2	76.6	72.8	68.9
	−35°	86.4	89.0	91.3	93.1	94.5	95.5	96.0	96.1	95.7	94.9	93.7	92.1	89.9	87.3	84.5	81.2	77.7	73.7	69.6
	−30°	86.4	89.3	91.6	93.6	95.2	96.2	96.9	97.1	96.8	96.0	94.7	93.1	91.0	88.5	85.5	82.1	78.5	74.5	70.2
	−25°	86.4	89.3	91.9	94.0	95.7	96.9	97.7	97.9	97.7	97.0	95.7	94.0	91.9	89.3	86.4	83.0	79.2	75.2	70.8
	−20°	86.4	89.5	92.2	94.5	96.2	97.5	98.2	98.6	98.4	97.7	96.6	94.9	92.6	90.0	87.1	83.6	79.8	75.7	71.2
	−15°	86.4	89.6	92.4	94.7	96.6	98.0	98.8	99.2	99.0	98.3	97.1	95.6	93.4	90.7	87.6	84.1	80.3	76.0	71.5
	−10°	86.4	89.7	92.6	95.0	96.9	98.3	99.2	99.6	99.4	98.7	97.6	95.9	93.8	91.2	88.1	84.5	80.6	76.3	71.8
	−5°	86.4	89.8	92.7	95.1	97.1	98.5	99.5	99.9	99.7	99.1	97.9	96.2	94.0	91.4	88.4	84.9	80.9	76.5	72.0
South	0°	86.4	89.8	92.7	95.2	97.1	98.7	99.6	100.0	99.9	99.3	98.2	96.5	94.3	91.6	88.5	85.0	81.0	76.8	72.1
	5°	86.4	89.8	92.7	95.1	97.1	98.5	99.5	99.9	99.7	99.1	97.9	96.2	94.0	91.4	88.4	84.9	80.9	76.5	72.0
	10°	86.4	89.7	92.6	95.0	96.9	98.3	99.2	99.6	99.4	98.7	97.6	95.9	93.8	91.2	88.1	84.5	80.6	76.3	71.8
	15°	86.4	89.6	92.4	94.7	96.6	98.0	98.8	99.2	99.0	98.3	97.1	95.6	93.4	90.7	87.6	84.1	80.3	76.0	71.5
	20°	86.4	89.5	92.2	94.5	96.2	97.5	98.2	98.6	98.4	97.7	96.6	94.9	92.6	90.0	87.1	83.6	79.8	75.7	71.2
	25°	86.4	89.3	91.9	94.0	95.7	96.9	97.7	97.9	97.7	97.0	95.7	94.0	91.9	89.3	86.4	83.0	79.2	75.2	70.8
	30°	86.4	89.3	91.6	93.6	95.2	96.2	96.9	97.1	96.8	96.0	94.7	93.1	91.0	88.5	85.5	82.1	78.5	74.5	70.2
	35°	86.4	89.0	91.3	93.1	94.5	95.5	96.0	96.1	95.7	94.9	93.7	92.1	89.9	87.3	84.5	81.2	77.7	73.7	69.6
	40°	86.4	88.8	90.9	92.5	93.8	94.6	95.1	95.1	94.5	93.8	92.5	90.9	88.6	86.2	83.4	80.2	76.6	72.8	68.9
South west	45°	86.4	88.6	90.4	91.9	93.0	93.6	94.0	93.9	93.3	92.4	91.2	89.3	87.2	84.9	82.1	78.9	75.5	71.8	67.9
	50°	86.4	88.4	90.0	91.2	92.1	92.6	92.8	92.5	91.9	91.0	89.6	87.8	85.9	83.5	80.6	77.5	74.2	70.7	66.9
	55°	86.4	88.1	89.5	90.5	91.1	91.5	91.5	91.2	90.5	89.4	88.0	86.3	84.3	81.8	79.0	76.2	72.9	69.4	65.7
	60°	86.4	87.8	88.9	89.7	90.1	90.4	90.2	89.7	88.9	87.8	86.3	84.6	82.6	80.0	77.5	74.6	71.4	68.0	64.4
	65°	86.4	87.6	88.3	88.8	89.1	89.1	88.8	88.2	87.2	86.1	84.6	82.8	80.6	78.4	75.8	72.9	69.7	66.5	63.1
	70°	86.4	87.2	87.8	88.0	88.0	87.8	87.2	86.5	85.5	84.2	82.7	80.9	78.8	76.5	73.9	71.0	68.0	64.9	61.6
	75°	86.4	87.0	87.2	87.2	86.8	86.4	85.7	84.8	83.7	82.3	80.7	78.9	76.8	74.5	71.9	69.2	66.4	63.3	60.0
	80°	86.4	86.7	86.6	86.2	85.7	85.0	84.1	83.1	81.8	80.4	78.7	76.8	74.7	72.4	69.9	67.4	64.5	61.4	58.3
	85°	86.4	86.3	86.0	85.3	84.5	83.6	82.5	81.3	79.9	78.4	76.6	74.7	72.6	70.2	67.9	65.3	62.4	59.6	56.6
West	90°	86.4	86.0	85.3	84.4	83.3	82.1	80.8	79.4	77.9	76.3	74.4	72.5	70.4	68.0	65.8	63.2	60.4	57.7	54.9
	95°	86.4	85.7	84.7	83.5	82.1	80.7	79.1	77.6	75.8	74.1	72.2	70.2	68.1	65.9	63.5	61.0	58.5	55.8	52.9
	100°	86.4	85.4	84.1	82.6	80.9	79.2	77.4	75.7	73.8	72.0	70.0	68.0	65.8	63.6	61.2	58.8	56.4	53.7	50.9
	105°	86.4	85.1	83.5	81.6	79.7	77.7	75.8	73.7	71.8	69.8	67.7	65.6	63.5	61.3	59.0	56.6	54.2	51.6	49.2
	110°	86.4	84.8	82.9	80.7	78.5	76.3	74.1	71.9	69.7	67.6	65.4	63.3	61.2	59.0	56.7	54.4	52.0	49.6	47.2
	115°	86.4	84.5	82.3	79.9	77.3	74.8	72.4	70.0	67.6	65.4	63.3	61.0	58.9	56.6	54.4	52.3	49.9	47.7	45.3
	120°	86.4	84.2	81.7	79.0	76.3	73.5	70.7	68.1	65.6	63.3	61.0	58.7	56.5	54.4	52.3	50.1	47.9	45.7	43.5
	125°	86.4	84.0	81.2	78.3	75.2	72.1	69.1	66.3	63.6	61.2	58.7	56.5	54.4	52.2	50.2	48.0	45.9	43.8	41.7
	130°	86.4	83.7	80.7	77.5	74.2	70.9	67.6	64.6	61.7	59.1	56.6	54.4	52.2	50.1	48.1	46.1	44.0	42.0	40.0
	135°	86.4	83.5	80.3	76.8	73.3	69.7	66.3	63.0	59.9	57.1	54.6	54.3	50.2	48.2	46.1	44.2	42.2	40.4	38.4
	140°	86.4	83.3	79.9	76.2	72.5	68.7	65.0	61.5	58.2	55.3	52.7	50.3	48.2	46.1	44.2	42.4	40.5	38.8	37.0
North west	145°	86.4	83.1	79.4	75.7	71.7	67.8	64.0	60.3	56.8	53.6	50.8	48.4	46.3	44.4	42.4	40.7	39.0	37.2	35.6
	150°	86.4	83.0	79.1	75.2	71.1	67.0	63.1	59.3	55.6	52.2	49.2	46.6	44.5	42.6	40.9	39.2	37.5	35.9	34.4
	155°	86.4	82.8	78.9	74.7	70.6	66.4	62.4	58.5	54.7	51.1	47.8	45.1	43.0	41.0	39.3	37.8	36.2	34.6	33.2
	160°	86.4	82.7	78.6	74.3	70.1	65.9	61.8	57.8	53.9	50.2	46.8	43.8	41.5	39.7	38.0	36.5	35.1	33.6	32.2
	165°	86.4	82.6	78.4	74.1	69.8	65.5	61.4	57.3	53.4	49.6	46.1	42.9	40.4	38.6	37.0	35.5	34.1	32.7	31.5
	170°	86.4	82.6	78.4	73.9	69.5	65.3	61.1	57.0	52.9	49.2	45.6	42.4	39.7	37.8	36.2	34.8	33.4	32.1	30.8
North	175°	86.4	82.6	78.3	73.8	69.5	65.2	61.0	56.8	52.8	48.9	45.3	42.0	39.3	37.4	35.8	34.4	33.1	31.7	30.5
	180°	86.4	82.6	78.3	73.8	69.5	65.2	61.0	56.8	52.8	49.0	45.4	42.0	39.3	37.3	35.7	34.3	33.0	31.6	30.4

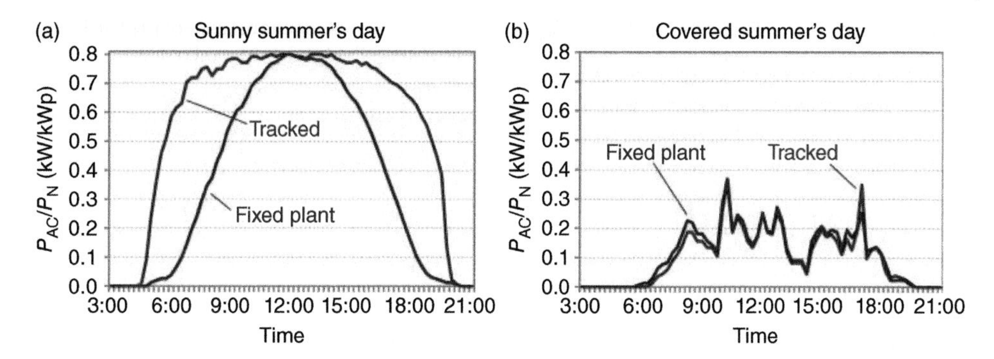

Figure 2.20 Yields of fixed and tracked photovoltaic plants on two different days. Source: IBC Solar AG.

2.4.3 Yield Gain through Tracking

Basically, it is possible to increase the yield of a photovoltaic plant in that the solar generator actively tracks the Sun (see Chapter 6). The tracking, however, only increases the direct radiation portion, whereas the diffuse radiation remains almost the same. Figure 2.20 shows this in an example of the daily yield of a tracking and a fixed photovoltaic plant on two different days. On the sunny day, the tracking brings an energy yield of almost 60%. In the case of the overcast day, the yield of the tracking plant provides approximately 10% less than the fixed plant. The reason is that the tracking modules are relatively steep in the morning and afternoons and, therefore, receive less diffuse radiation.

As we have seen, the portion of the diffuse radiation of the yearly global radiation in Germany is more than half. For this reason, the yield of tracking in our lines of latitude of approximately 30% is limited. For this reason, the utilization of tracking should be well thought through because of the substantial mechanical and electrical effort required. A compromise may be a single-axis tracking plant that can easily achieve a yield of 20% (see Chapter 6). In southern countries with a high degree of direct radiation, the situation is much better: Here, two-axis tracking plants can achieve more than a 50% increase in yield.

2.5 Radiation Availability and World Energy Consumption

In concluding this chapter, we will also consider the potential of solar radiation.

2.5.1 The Solar Radiation Energy Cube

As we have seen in Section 2.1, the Sun shines continuously on the Earth with a power density of 1367 W m^{-2}. Approximately 1000 W m^{-2} of this arrives inside the atmosphere. We can simply roughly calculate the energy arriving on Earth W_{Earth}. For this, we calculate the cross-sectional area A_{Earth} of the Earth's sphere as shown in Figure 2.21.

The total optical power P_{Earth} radiated by the Sun on the Earth is then

$$P_{\text{Earth}} = E_{\text{STC}} \cdot A_{\text{Earth}} = E_{\text{STC}} \cdot \frac{\pi \cdot d_{\text{Earth}}^2}{4} = 1.278 \times 10^{17}\ \text{W}. \tag{2.21}$$

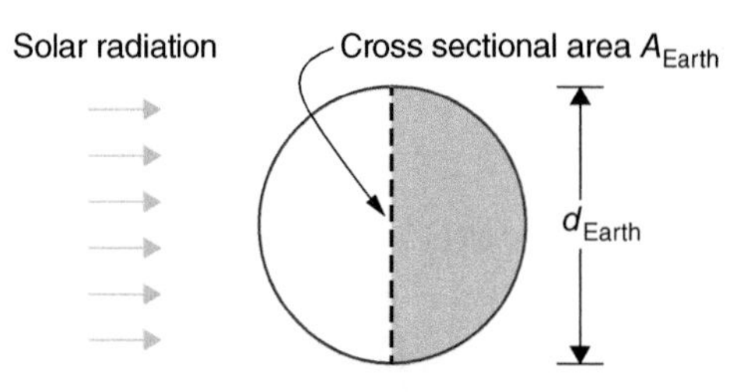

Figure 2.21 Cross-sectional area of the Earth for determining the total incidental radiation energy.

Over the whole year, the Earth receives radiation energy of

$$W_{Earth} = P_{Earth} \cdot t = 1.278 \times 10^{17}\ \mathrm{W} \cdot 8760\ \mathrm{h} = 1.119 \times 10^{18}\ \mathrm{kW\ h}. \tag{2.22}$$

This number only tells us something if, for instance, we use it in relationship to the current yearly world energy consumption. From Figure 1.3, we can see that this is approximately 13 billion tons of oil equivalent. After conversion into kilowatt hour (Table 1.2) we get

$$\frac{W_{Earth}}{W_{World}} = \frac{1.119 \times 10^{18}\ \mathrm{kW\ h}}{1.512 \times 10^{14}\ \mathrm{kW\ h}} = 7401. \tag{2.23}$$

The Sun sends us more than 7000 times the energy that we use in a year.

This relationship is clearly seen in the energy cube depicted in Figure 2.22. The yearly solar incidental radiation is represented by the large cube; the small blocks of world energy usage at the bottom-right look tiny compared with the large solar cube. Also interesting in the same figure is the comparison of the annual solar radiation with the reserves of fossil and nuclear energy carriers. It must be mentioned here that with these cubes at the bottom left, the reserves still available are included, whereas the large solar radiation cube is available again every new year.

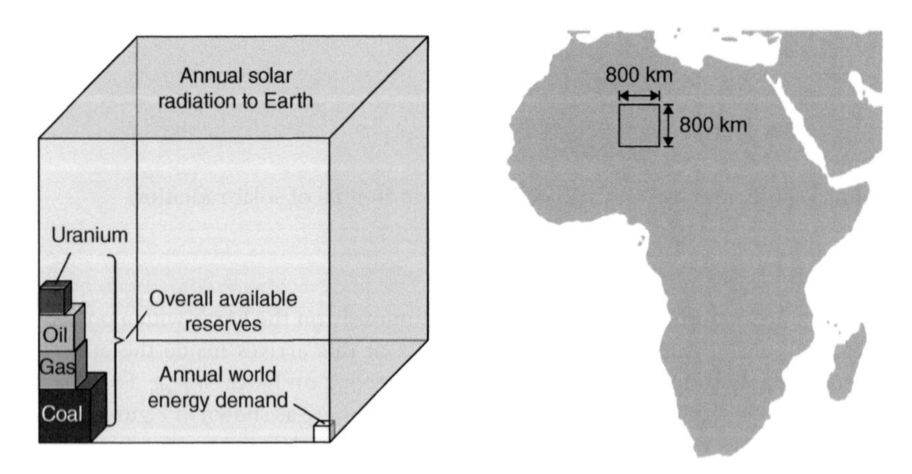

Figure 2.22 Estimate of the potential of solar energy: The yearly solar radiation outnumbers worldwide energy demand more than 7000-fold [18].

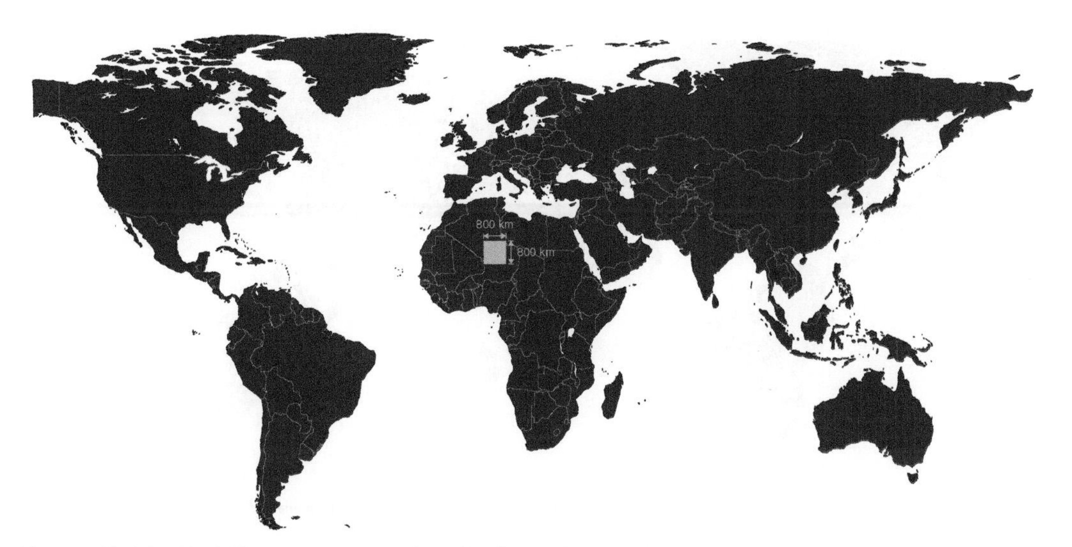

Figure 2.23 The Sahara Miracle: The area necessary to meet the total world energy demand is only 800 km × 800 km.

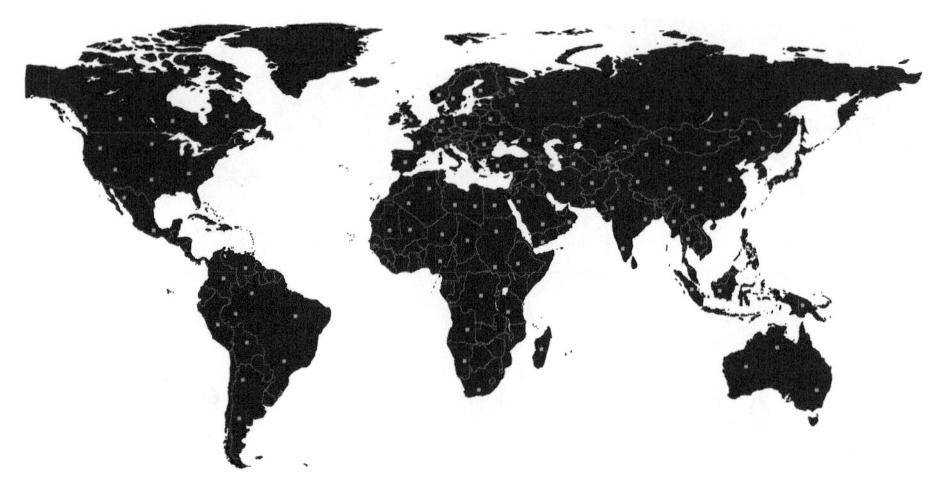

Figure 2.24 Distribution of the necessary area over the whole world: For example, 200 squares with side lengths of 85 km are needed.

2.5.2 The Sahara Miracle

Now we will not be able to make use of all the energy radiating onto the Earth. Therefore, we will look at the case from another point of view and ask ourselves:

What area would be necessary in order to supply the whole primary consumption of the world with photovoltaics?

To find a solution, let us assume that the solar modules would be erected in the Sahara. The best solar modules available on the market have an efficiency of 22%. As a precautionary measure, we will take a total system efficiency of $\eta_{\text{Total}} = 10\%$. In this way, the losses from cables, inverters, and transmission lines as well as the distance between module rows are more than covered.

According to Figure 2.9, the Sahara annually supplies approximately 2500 kW h m^{-2} radiation energy. With an efficiency of 10%, it is thus possible to obtain electrical energy of approximately 250 kW h m^{-2}.

For covering the worldwide primary energy consumption W_{World}, we would thus need an area of

$$A = \frac{W_{\text{World}}}{250\ \text{kW h m}^{-2}} = \frac{1.512 \times 10^{14}\ \text{kW h}}{250\ \text{kW h m}^{-2}} = 6.048 \times 10^{11}\ \text{m}^2 = 6.048 \times 10^5\ \text{km}^2. \tag{2.24}$$

This would result, for instance, in a square of approximately 778 km per side. In order to estimate the size, Figure 2.23 shows a square of 800 km × 800 km, which is about 7% of the size of the Sahara's surface. Thus, this area is sufficient to cover the Earth's primary energy consumption with photovoltaics. In this way, one could really speak of a "Sahara Miracle."

In practice, of course, it would make no sense to concentrate the photovoltaic power plants at one site in the world. In that case, humankind would only have energy during daytime in the Sahara. To get a continuous energy flow, the installation of PV plants along the equator could be a good idea. A rough calculation shows that this strip only needs a width of about 15 km.

More practical is the scenario in Figure 2.24. Here the PV plants are distributed over many countries to achieve short cable lengths and a decentralized structure. As the mean yearly radiation sum a value of 1000 kW h (m^2 a)$^{-1}$ can be assumed. This results in, for example, 200 squares with side lengths of 85 km.

In Chapter 11, we will consider the future role of photovoltaics in more detail.

3

Fundamentals of Semiconductor Physics

Typically, solar cells consist of semiconductors. In order to understand how solar cells work, we will first deal with the structure and properties of semiconductors. This is associated with the consideration of the p–n junction and the optical features of semiconductors.

3.1 Structure of a Semiconductor

3.1.1 Bohr's Atomic Model

To start off, we will consider an individual atom. According to Bohr's atomic model, an atom consists of a nucleus and a shell. The nucleus contains protons and neutrons, whereas the shell contains electrons, which orbit the nucleus. The protons are positively and electrically charged with an elementary charge of $+q$, and the electrons are negatively charged with a charge of $-q$. The size of the elementary charge is 1.6×10^{-19} A s. As the number of protons in the nucleus is equal to the number of electrons (the so-called atomic number), an atom is electrically neutral on its outside.

The simplest atom we know of is the hydrogen atom (Figure 3.1). It has the atomic number 1, and thus has only one proton in the nucleus and one electron in the shell.

Niels Bohr recognized that electrons can only circulate in very particular paths (so-called shells) around the nucleus and defined this in his first postulate.

Bohr's first postulate:
There are only certain discrete shells permitted for an electron.

Each of these shells stands for a particular path radius that represents the respective energy state of the electron. The shells are designated with the letters of K, L, M, and so on. In Figure 3.1, the possible energy states for the hydrogen atom are shown. In the basic state, the electron is situated on the K-shell. If the electron is moved to the L-shell, then energy of 10.2 eV (electron volts) is necessary. In order to separate the electron completely from the atom (thus to transport it into "infinity"), the so-called ionizing energy W_∞ of 13.6 eV must be used.

Photovoltaics – Fundamentals, Technology, and Practice, Second Edition. Konrad Mertens.

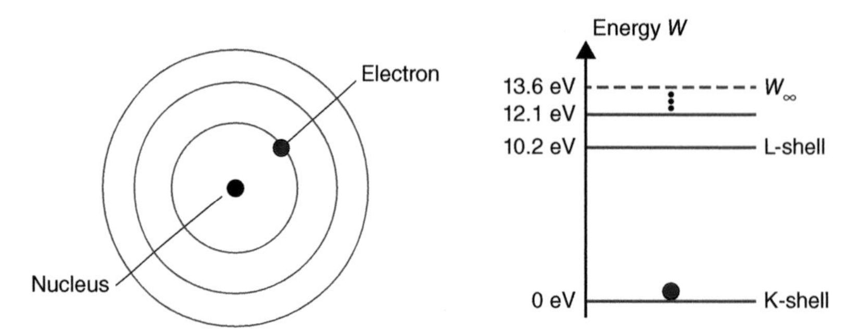

Figure 3.1 Structure and energy model of the hydrogen atom.

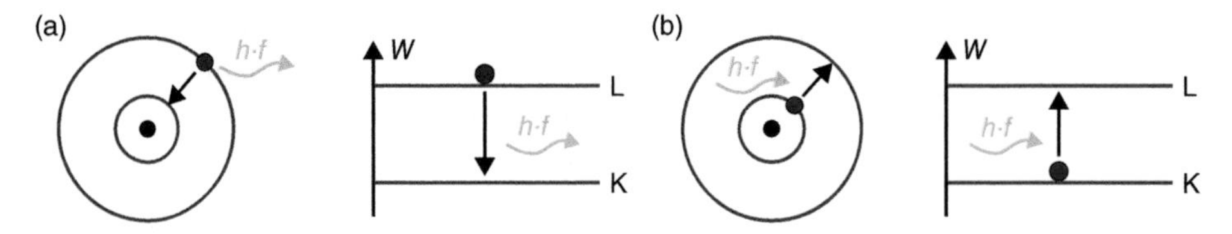

Figure 3.2 Schematic depiction of the emission (a) and absorption (b) of light.

What happens in the transfer from one shell to another? This is described in the following postulate:

Bohr's second postulate:
The transfer of an electron from one shell to another occurs under the emission or absorption of electromagnetic radiation.

The frequency f of this radiation is thus determined by the following equation:

$$\Delta W = |W_2 - W_1| = h \cdot f \tag{3.1}$$

with W_1, energy before the transfer; W_2, energy after the transfer; h, Planck's constant; $h = 6.6 \times 10^{-34}$ W s^2

In order to determine the wavelength λ from the frequency f, the following equation is used:

$$\lambda = \frac{c_0}{f} \tag{3.2}$$

with c_0: speed of light in a vacuum, $c_0 = 299.792$ km s$^{-1} \approx 3 \times 10^8$ m s^{-1}.

In order to better understand Bohr's second postulate, consider Figure 3.2: Panel (a) shows how the electron from the L-shell falls onto the K-shell. The energy released by this is radiated in the form of light as a photon. This process is called light emission. A light packet of a particular wavelength ("light particle") is known as a photon.

The opposite case is shown in Figure 3.2b: A light particle hits an electron and is "swallowed." The energy released in this lifts the electron from the K-shell to the L-shell. This process is called the absorption of light.

Example 3.1 ***Light emission***
An electron of a hydrogen atom falls from the M-shell to the L-shell. What will be the wavelength of the radiated light?

Calculation:

$$\Delta W = W_1 - W_2 = 12.1 - 10.2\ \text{eV} = 1.9\ \text{eV} = h \cdot f$$

$$\Rightarrow f = \frac{1.9\,\text{eV}}{h} = \frac{1.9\ \text{V} \cdot 1.6 \times 10^{-19}\ \text{A s}}{6.6 \times 10^{-34}\ \text{W s}^2} = 0.461 \times 10^{15}\ \text{s}^{-1} = 461 \times 10^{12}\ \text{Hz}.$$

The wavelength is calculated again by

$$\lambda = \frac{c_o}{f} = \frac{3 \cdot 10^8\ \text{m s}^{-1}}{461 \times 10^{12}\ \text{Hz}} = 6.508 \times 10^{-7}\ \text{m} = 650.8 \times 10^{-9}\ \text{m} \approx 651\ \text{nm}.$$

The light radiates at 651 nm and thus in the red region. ■

3.1.2 Periodic Table of Elements

Table 3.1 shows a section of the Periodic Table of elements. The rows of the table provide the highest shells that are occupied by electrons. The value is also obtained from the respective column of an element. This is understood to be the number of electrons in the outer shell and is often called the valence. For instance, we recognize that the noble gas helium (He) has two electrons and thus fully occupies the K-shell. In the following lithium (Li), the K-shell is also occupied; the third electron is situated on the L-shell. The electrons of the outermost shell are called valence electrons as they are decisive in the bonding of atoms.

Example 3.2 ***Valence electrons of silicon***

The element silicon (Si), which is extremely important for photovoltaics, has the atomic number 14 and is situated in main Group IV. The K- and L-shells are full, and there are four electrons in the topmost shell. The Si atom thus possesses four valence electrons. We can also say that silicon is four-valent or "tetravalent." ■

3.1.3 Structure of the Silicon Crystal

When the electrons of neighboring atoms make fixed connections, then a regular lattice structure can be formed. Such a structure is called a crystal. With silicon, each valence electron makes a connection with an electron of the neighboring atom. The lattice formed in this way is shown in Figure 3.3 as a sphere and as a two-dimensional depiction.

In this, the atomic nucleus together with all internal shells is drawn as a circle. The correct designation is Si^{4+}, as the 14 protons in the nucleus together with the 10 electrons on the inner shells result in 4× positive charge. The combination of the Si^{4+} ion with the four surrounding electrons forms the whole silicon atom. At the same time, one recognizes that each Si atomic nucleus is surrounded by eight valence electrons. This is called the noble gas configuration as they are comparable with the noble gas argon that also has eight electrons in the outer shell (see Table 3.1).

3.1.4 Compound Semiconductors

The lattice considered up to now is made up exclusively of silicon, an element in Group IV. It is possible, however, to combine elements of different main groups. A

Table 3.1 Extract from the Periodic Table of elements.

Main Group/valence								Shell
I	II	III	IV	V	VI	VII	VIII	
H Hydrogen 1							He Helium 2	K
Li Lithium 3	Be Beryllium 4	B Boron 5	C Carbon 6	N Nitrogen 7	O Oxygen 8	F Fluorine 9	Ne Neon 10	L
Na Sodium 11	Mg Magnesium 12	Al Aluminum 13	Si Silicon 14	P Phosphorous 15	S Sulfur 16	Cl Chlorine 17	Ar Argon 18	M
K Potassium 19	Ca Calcium 20	Ga Gallium 31	Ge Germanium 32	As Arsenic 33	Se Selenium 34	Br Bromine 35	Kr Krypton 36	N
Rb Rubidium 37	Sr Strontium 38	In Indium 49	Sn Tin 50	Sb Antimony 51	Te Telluride 52	J Iodine 53	Xe Xenon 54	O

The number under the element name is the atomic number.

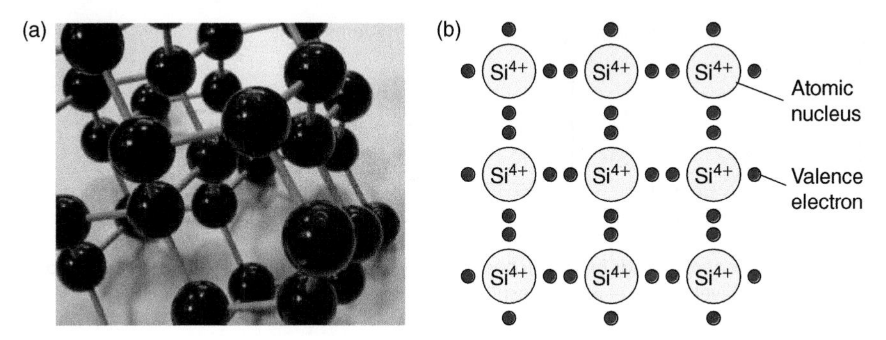

Figure 3.3 Structure of a silicon crystal: (a) Spherical model and (b) two-dimensional depiction.

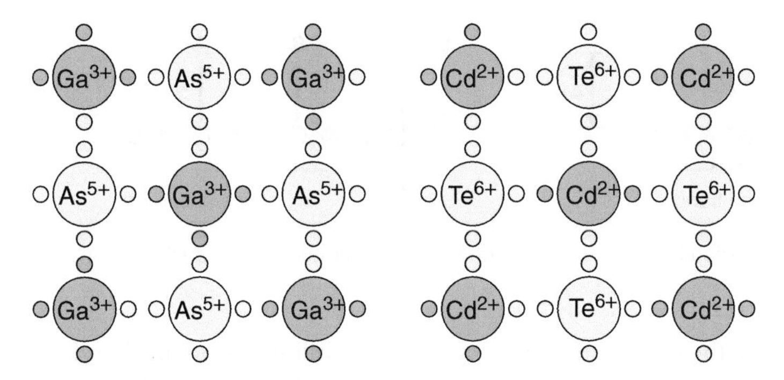

Figure 3.4 Lattice of compound semiconductors of the example GaAs and CdTe.

well-known representative of the material connection is gallium-arsenide (GaAs), which can ensure high efficiencies in solar cells. It consists of trivalent gallium and pentavalent arsenic atoms and is thus called a III/V-semiconductor. Figure 3.4 shows the structure: The crystal contains at the same in equal parts gallium and arsenic atoms that always include their valence electrons in the connection so that the result is again the, especially, stable noble gas configuration. Besides III/V-semiconductors, II/VI-semiconductors are also of interest. Figure 3.4 shows this in the example of cadmium-telluride (CdTe).

3.2 Band Model of a Semiconductor

3.2.1 Origin of Energy Bands

Meanwhile, we know that there are defined, discrete levels of energy for the electrons of an individual atom. What happens if we make a thought experiment and bring two atoms close together? There occurs a mutual coupling of the atoms. The result is that the energy conditions change and each state divides into two individual states (see Figure 3.5). An analogy to this phenomenon can be seen from classical mechanics. If one couples two harmonic oscillators (e.g. two guitar strings), then the result is two new resonance frequencies. In the case of three coupled atoms, the result is always three new levels. If one views a semiconductor crystal, then a practically infinite number of

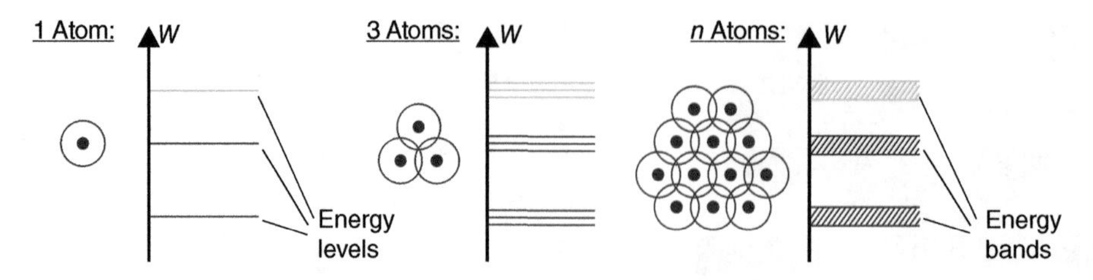

Figure 3.5 Origin of the energy bands in a semiconductor crystal: The coupling of the atoms leads to a spreading of the energy levels. For $n \to \infty$, this results in continuous energy bands.

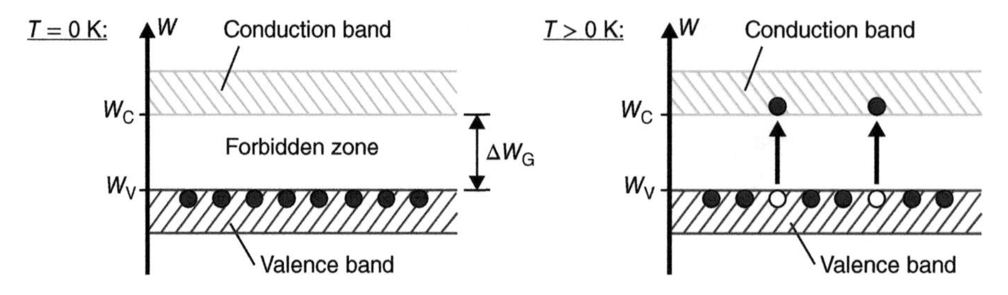

Figure 3.6 Valence and conduction bands for silicon: With rising temperatures, individual electrons rise into the conduction band.

atoms can be coupled together. Individual levels will hardly still be recognized and in this case one speaks of energy bands. The energy bands show all the energy states that are permitted for an electron.

The highest band still occupied by electrons is decisive for the electrical relationship of a solid-state body. As this is occupied by the valence electrons, it is called the valence band (see Figure 3.6). The first unoccupied band is called the conduction band. In order to enter into the conduction band, an electron must first overcome the forbidden zone. The width of the forbidden zone decides the amount of energy needed to move out of the valence band into the conduction band. This is also called the bandgap ΔW_G. It is the result of the difference of the lowest allowed level of the conduction band W_C and the highest allowed level of the valence band W_V. In the case of silicon, the bandgap is at $\Delta W_G = 1.12$ eV. The index "G" stands for the term *bandgap*.

What does this mean for the electrical behavior of the crystals? In the case of the zero absolute temperature ($T = 0$ K), the valence electrons remain fixed in their bonds. In this case, the crystal is unable to conduct electrical current as no free charge carriers are available.

The absolute zero temperature of $T = 0$ K (Kelvin) corresponds to a temperature of $\vartheta = -273.15$ °C.

If the temperature is now increased, then the electrons start to move due to heat oscillations. If the temperature is increased further, then individual electrons can loosen from their bonds and become available in the crystal as free electrons. In the band model, this corresponds to the case where these electrons are lifted out of the

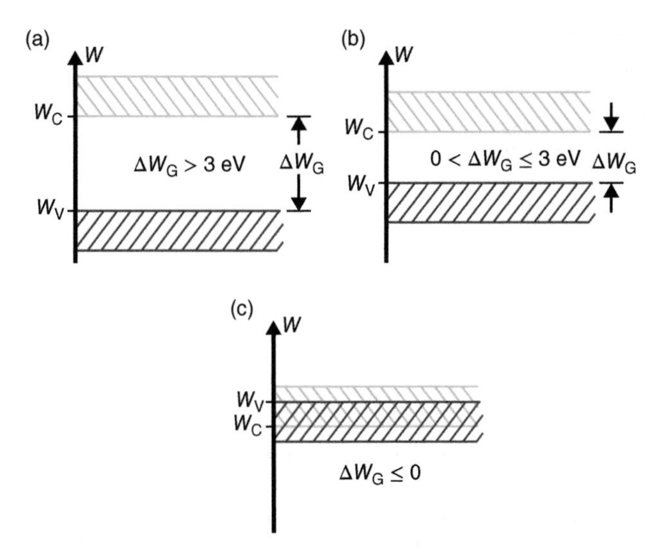

Figure 3.7 Depiction of energy bands of insulators (a), semiconductors (b), and metals (c).

valence band, overcome the forbidden zone, and arrive in the conduction band. They become conductor electrons and increase the conductivity of the crystal.

3.2.2 Differences in Isolators, Semiconductors, and Conductors

After we have made the acquaintance of semiconductors, we will expand to other materials. For this purpose, Figure 3.7 shows the comparison of the band scheme of insulators, semiconductors, and metals. In the case of insulators, the forbidden zone is very big. Insulators are typically materials whose bandgap is greater than about 3 eV. This means that almost no free electrons are available even at high temperatures. Semiconductors at low temperatures also act as insulators. At medium temperatures, however, conductivity is increased until at very high temperatures (e.g. over 200 °C) they become good conductors (thus the term *semiconductor*).

Metals are a special case. With these, we can say in simplified terms that their valence and conduction bands overlap so that they possess a high degree of conductivity even at low temperatures.

Table 3.2 shows the bandgaps of various materials.

3.2.3 Intrinsic Carrier Concentration

We will now study the processes in the semiconductor crystal more closely. Figure 3.8a shows the generation of a free electron in the crystal as well as in the band model. As soon as an electron is released from its bonds, there is a gap in the crystal called a hole. The whole process is called electron–hole pair generation. The reverse process can be seen in Figure 3.8b: The free electron falls back into the hole, and this is called electron–hole pair recombination.

The generation and recombination of electron–hole pairs occur continuously in the crystal. Depending on the semiconductor material and the current temperature, there

Table 3.2 Comparison of the bandgaps of various materials.

Material	Type of material	Bandgap ΔW_G (eV)
Diamond	Insulator	7.3
Gallium arsenide	Semiconductor	1.42
Silicon	Semiconductor	1.12
Germanium	Semiconductor	0.7

(a) Electron–hole pair generation:

Si Si Si
Si Si Si
W_C
W_V
Conduction band
Valence band

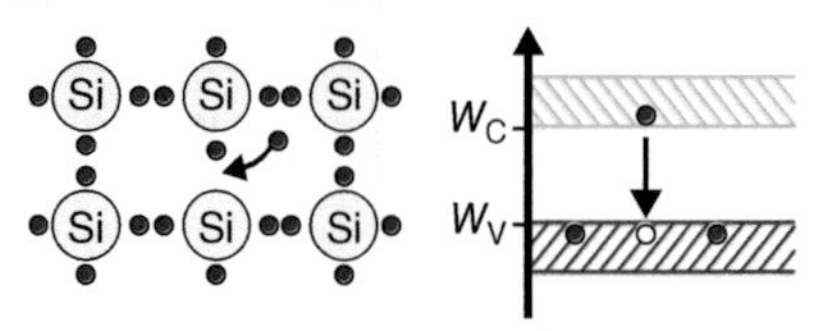

Figure 3.8 Thermal generation and recombination of electron–hole pairs: In a timed average, there is an average number of free electrons as well as holes, the intrinsic carrier concentration.

is an average number of free electrons as well as holes, known as the intrinsic carrier concentration n_i.

The index "i" stands for *intrinsic*. It indicates that this is an undoped semiconductor (see Section 3.4).

The intrinsic carrier concentration can be determined by means of the following equation:

$$n_i = N_0 \cdot e^{-\frac{\Delta W_G}{2 \cdot k \cdot T}} \tag{3.3}$$

with N_0: effective density of states; for silicon: $N_0 \approx 3 \times 10^{19}\ \text{cm}^{-3}$ and k: Boltzmann constant; $k = 1.38 \times 10^{-23}\ \text{W s K}^{-1} = 8.62 \times 10^{-5}\ \text{eV K}^{-1}$.

To a certain extent, the effective density of states N_0 gives the number of free electrons that can be generated in the extreme case (at extreme high temperature). In this, for the sake of simplicity, we have assumed that the effective density of states of the electrons is the same as one of the holes. Each generated free electron leaves a hole in the crystal lattice. Therefore, n_i describes the number of free electrons as well as the number of holes.

Example 3.3 ***Intrinsic carrier concentration***

We calculate the intrinsic carrier concentration of silicon at room temperature ($\vartheta = 25\,°\text{C}$). First, we calculate the absolute temperature T in Kelvin:

$$T = 273.15\ \text{K} + 25\ \text{K} = 298.15\ \text{K}.$$

Now, we insert all values into Equation (3.3):

$$n_i = N_0 \cdot e - \frac{1.12\ \text{eV}}{2 \cdot 8.62 \times 10^{-5}\ \text{eV K}^{-1} \cdot 298.15\ \text{K}} = 1.06 \times 10^{10}\ \text{cm}^{-3} \approx 10^{10}\ \text{cm}^{-3}.$$

■

3.3 Charge Transport in Semiconductors

3.3.1 Field Currents

Figure 3.9 shows a crystal of silicon that has an electric voltage V applied to it. As in a plate capacitor, this voltage leads to an electrical field F in the crystal:

$$F = \frac{V}{l}. \tag{3.4}$$

By means of this field, the negatively charged electrons are accelerated in the direction of the positive pole of the voltage source. Thus, there is a flow of current through the semiconductor, which is called field current (sometimes drift current). However, the electrons in the crystal repeatedly collide with the atomic nucleus, are decelerated and accelerated again by the field. In a timed average, they achieve a certain average drift velocity v_D.

The quotient from these achieved drift velocities v_D for an applied field F is called the mobility μ_N of the electrons:

$$\mu_N = \frac{v_D}{F}. \tag{3.5}$$

In Figure 3.9, the electrons are moving from left to right through the crystal. Then they flow through the outer circuit from right to left. The arrow of the current I, however, points in the opposite direction. Isn't there a mistake?

The figure is correct. The electrical current is conventionally defined as the flow of a positive charge. Thus, it has the opposite direction of the flow of the negatively charged electrons.

It is easy to understand how an increase in temperature affects the crystal: The greater the temperature, the stronger the oscillations of the crystal lattice. This increases the probability that the accelerated electrons collide with the atomic nuclei.

The average drift velocity and thus the mobility of the electrons decrease.

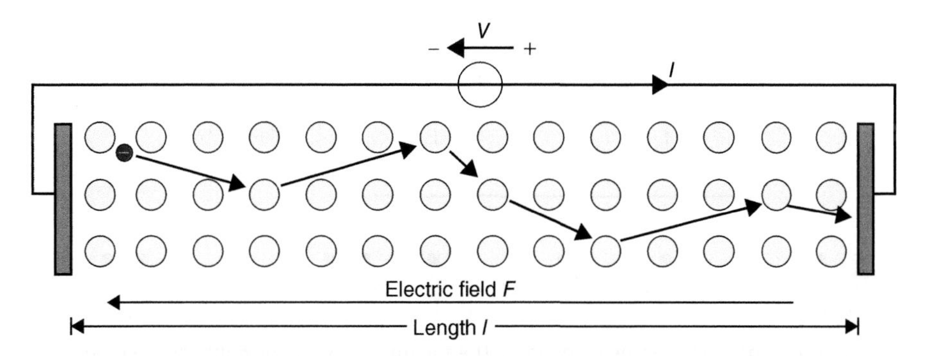

Figure 3.9 Current transport through the silicon crystal: The electrons are repeatedly decelerated by collisions with the atomic nucleus and then accelerated again.

The number of electrons in the volumes of the crystal from Figure 3.9 is N:

$$N = n \cdot \text{volume} = n \cdot A \cdot l \tag{3.6}$$

with n: electron concentration; A: cross-sectional area of the crystal; l: length of the crystal.

These electrons are pushed by the electrical field through the crystal in the time $\Delta t = l/v_\mathrm{D}$, with Equation (3.5) for the field current I_F, this results in

$$I_\mathrm{F} = \frac{\text{Charge}}{\text{Time}} = \frac{q \cdot N}{\Delta t} = q \cdot n \cdot A \cdot \frac{l}{\Delta t} = q \cdot n \cdot A \cdot v_D = q \cdot A \cdot n \cdot \mu_\mathrm{N} \cdot F \tag{3.7}$$

with n: carrier concentration.

If the current is now divided by the cross-sectional area, the result is the current density j_F:

$$j_\mathrm{F} = \frac{I_\mathrm{F}}{A} = q \cdot n \cdot \mu_\mathrm{N} \cdot F. \tag{3.8}$$

Besides the current transport by the electrons, there is also current transport through holes in the semiconductor. Consider Figure 3.10 for better understanding: Because of the applied electrical field, an electron jumps into a neighboring free space. In this way, the hole is basically moved in the opposite direction. A similar comparison, for instance, is a soccer stadium. If there is an empty seat at the end of a row and one spectator after the other moves to the next empty seat, then the empty seat "moves" in the opposite direction.

It is obvious that the hole mobility is lower than that of the electron. For the mobility of the hole, electrons must move one after the other to free spaces, which occurs much slower than the movement of a free electron in the crystal. In silicon, hole mobility μ_P with approximately 450 cm^2 (V s)$^{-1}$ is therefore only a third of the electron mobility μ_N of 1400 cm^2 (V s)$^{-1}$.

3.3.2 Diffusion Currents

Besides the field current, there is also a second type of current in semiconductors: The diffusion current. This comes from the differences in concentration in which the necessary energy is provided by the thermal lattice movement. As soon as there is an increased

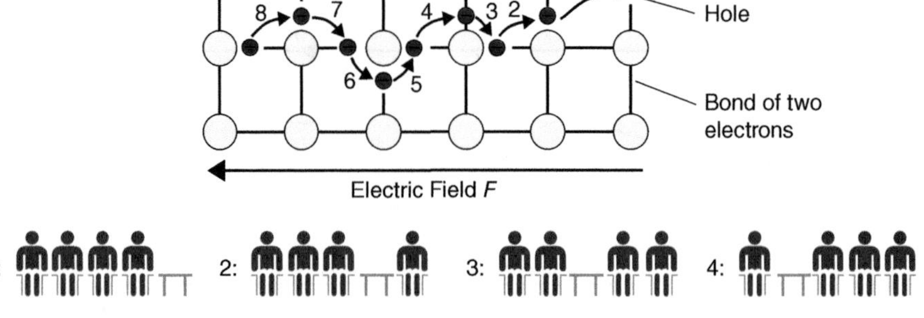

Figure 3.10 Transport of current by means of holes. The electrons move to the right one after the other. Therefore, there is a "hole movement" in the opposite direction. The situation is comparable to the movement of people along a row of seats.

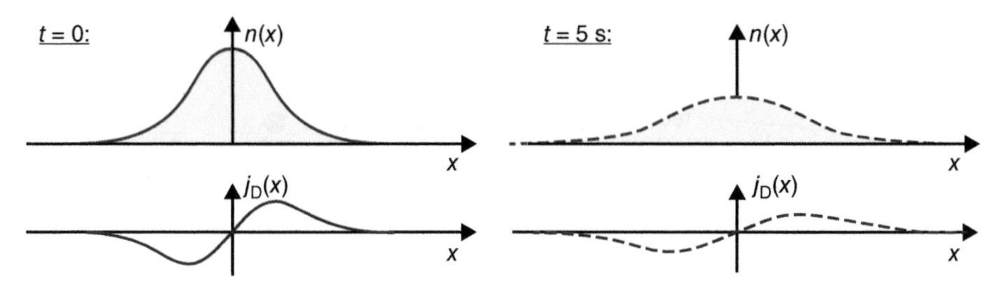

Figure 3.11 Explanation of the diffusion current on a heap of sand. If the vibrator is switched on at a time $t = 0$, then the sand heap flows apart. The highest "sand particle flows" occur at the steepest places of the heap.

concentration of charge in a place in the crystal (e.g. generated by light, see Section 3.6), then a diffusion current flows until the increased charge carrier concentration is equalized again. The magnitude of the diffusion current is proportional to the gradient (thus the equalization) of the particle concentration $n(x)$:

$$j_D = -q \cdot D \cdot \frac{dn(x)}{dx} \tag{3.9}$$

with j_D: diffusion current density; q: elementary charge; D: diffusion constant.

For better understanding, we will consider the analogy of a sand heap (Figure 3.11). This lies on a board that is shaken by a vibrator. The sand flows apart due to the vibration. The largest flows of sand will be on the steep sides, whereas the sand particles in the middle practically sink straight to the bottom.

Thus, the slope of the sand heap height $n(x)$ gives the amount of the "sand heap flow." After 5 s, the sand heap has already become wider and flatter, and consequently, the particle flows are less. The process continues until the "sand concentration differences" have been fully reduced, and the sand is therefore completely evenly distributed.

3.4 Doping of Semiconductors

As we have seen, to start off with, semiconductors are poor electrical conductors. They have achieved their particular importance in that their conductivity can be influenced in a targeted manner. For this purpose, one introduces foreign atoms into the semiconductor crystal (doping).

3.4.1 n-Doping

One speaks of n-doping when, instead of the original atom, one installs an atom from Group V (see the Periodic Table in Table 3.1). An example is phosphorous. This atom has one valence electron more than silicon and at the same time an additional proton in the nucleus. If this is inserted into a silicon lattice – as shown in Figure 3.12 – then the result is like the "Musical Chairs." Four valence electrons enter into a bond with the neighboring atom: The fifth finds no open bond. Instead, it is so weakly connected to the atomic nucleus that it is available as a free electron at room temperature. This becomes quite clear when viewing the band diagram: The doping atom generates an additional

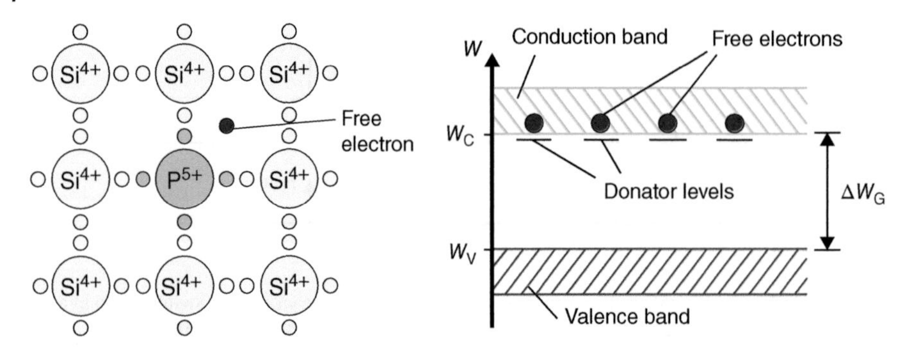

Figure 3.12 n-Doping of semiconductors; one of the five valence electrons of the phosphorous atom is not necessary for the bond and is therefore available as a free electron. Because of the doping there is a new energy level in the band diagram just below the conduction band edge.

energy level just below the conduction band edge. Only a very low energy (e.g. 1/50 eV) is required to lift the affected electron into the conduction band. The built-in foreign atom is called a donor atom from the Latin *donare*: To give or present. The donor atom more or less "presents" the crystal lattice with a free electron. As the donor atom is only bound to four electrons and at the same time has five protons at the nucleus, in total, this is a site-fixed positive charge.

Due to the n-doping, the concentration n of the free electrons rises drastically. At the same time, many of these electrons possess free bonds so that there are almost no holes left. In the case of the n-semiconductor, the electrons are therefore designated as majority carriers and the holes as minority carriers. The concentration of the free electrons in n-doped semiconductors is practically only determined by the density N_D of the donor atoms: $n \approx N_D$. These increase the conductivity of the crystal and almost turn the semiconductor into a conductor.

3.4.2 p-Doping

The second possibility of changing the conductivity of the semiconductor is by p-doping: Here, for instance, trivalent boron atoms are inserted (Figure 3.13). In this case, there are only three valence electrons available so that one bond remains

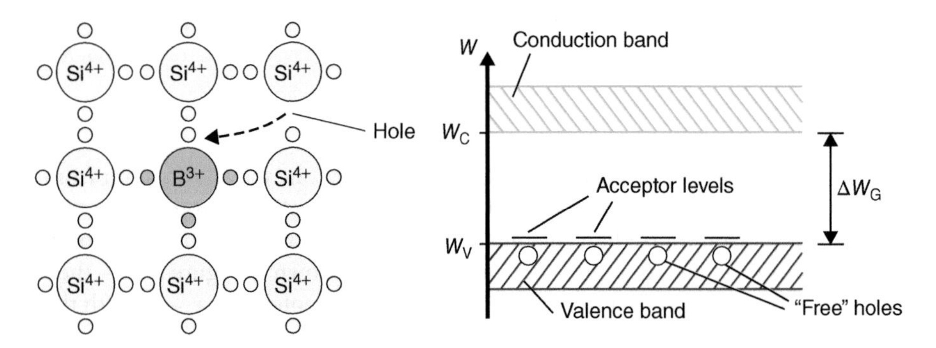

Figure 3.13 Example of p-doping of a silicon crystal with a boron atom: One of the four links remains open as the boron atom can only offer three valence electrons. A neighboring electron moves into this binding and thus "generates" a hole.

incomplete. A neighboring electron moves into this open bond so that again a noble gas configuration is available for this boron atom. Now there is a missing electron on a neighboring place, and thus a hole is formed. In this hole, conductance becomes possible. The trivalent atom is also designated as an acceptor atom, after the Latin *acceptare*: Acceptance. In a certain sense, this atom accepts an electron of the crystal. Then the acceptor atom represents a fixed negative charge as it only possesses three protons in its nucleus.

In practice, the doping densities (donor density N_D and acceptor density N_A) are very low: For instance, only every hundred-thousandth silicon atom is replaced by a doping atom. Yet the conductivity of the material can be increased by many factors of 10.

3.5 The p–n Junction

Why are we going into such details of semiconductors? They conduct electrical current so poorly and must be doped in order to conduct as well as simple metals!

In fact, the main reason for the victory march of semiconductor electronics is that the combination on n- and p-doping can create components with very special features. An important fundamental of almost all components is the p–n junction, which represents a diode in its technical realization.

3.5.1 Principle of Method of Operation

Figure 3.14 shows the principle processes in a p–n junction. The left crystal is n-doped and the right one p-doped. Both regions are electrically neutral. Thus, on the left side, the number of free electrons is equal to the number of fixed positive donor atoms. The right side corresponds to this where the positively charged holes compensate for the negative charge of the acceptor atoms.

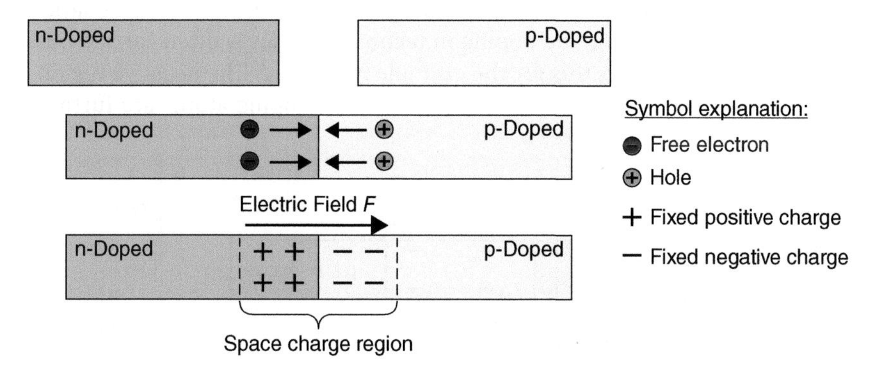

Figure 3.14 The p–n junction: Electrons flow from the n-side to the p-side where they occupy the holes. On the n-side, fixed positive charges remain behind; on the p-side, fixed negative charges are generated.

Let us now assume that both regions have just been connected. On the n-side, there is a surplus of free electrons. These diffuse due to the concentration gradient as a diffusion current to the right into the p-doped region, and there they recombine with the holes. In reverse, holes diffuse from the right to the left into the n-region where they recombine with the electrons. Thus, there are almost no free electrons and holes in the neighborhood of the junction that the fixed charges could compensate for. Because of the rising number of excess fixed charges in the junction region, an electrical field eventually comes into existence. This field again leads to the electrons being pushed to the left and the holes to the right. Finally, a new balance is built up in which diffusion and field current cancel each other and a space charge region exists at the p–n junction.

This space charge region causes a potential difference between the right and left borders of the space charge region, which is called the diffusion voltage V_D.

In Figure 3.14, we make use of a new drawing convention: Free charge carriers are shown with a border and fixed charges without a border.

We have already made the acquaintance of a good model for the diffusion current, the sand heap. If we vibrate this long enough the sand becomes fully flat. This should really be the same with the p–n junction: If we wait long enough after combining the two regions, then the electrons should distribute themselves throughout the crystal evenly, or is it not so?

Here it must be noted that there is a decisive difference between the sand particles and the electrons: Sand particles have no charge! The electrons are subjected to the influence of the field of the space charge region and are "pulled back." In this way, the flowing apart is slowed down and finally comes to a stop.

In the example shown, the left- and right-hand sides of the space charge region are the same size. This is because each electron that wanders from the left to the right side leaves a fixed positive charge behind on the left side and generates a negative fixed charge on the right (neutrality condition). The doping in technical diodes is often carried out asymmetrically. Figure 3.15 shows this for the example $N_D = 2N_A$: The negative region extends twice as far as the p-region because the individual doping atoms are further apart there. The designation n^+ denotes a strong n-doping.

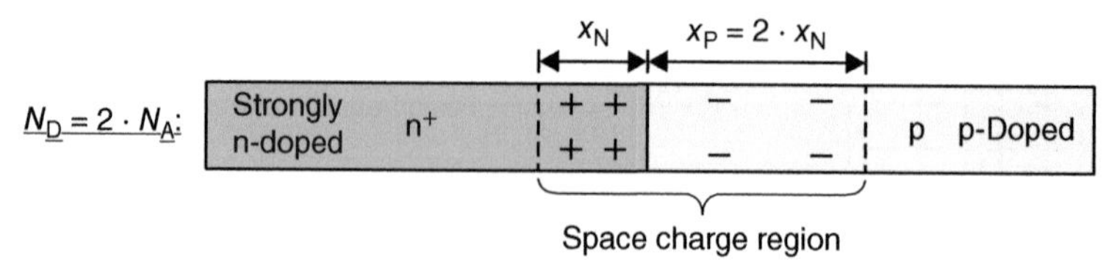

Figure 3.15 Asymmetric doping of the p–n junction: The space charge region extends mainly into the low doping region.

3.5.2 Band Diagram of the p–n Junction

We will now determine the size of the diffusion voltage. One possibility for this is through *Fermi energy*, the maximum energy occupied by an electron at 0 K, named after the Italian physicist and Nobel Prize winner Enrico Fermi (1901–1954).

Fermi energy W_F **is generally defined so that the probability for the occupation of this energy level is exactly 50%.**

Somewhat clearer (if not physically quite correct) we can describe Fermi energy such that it tells us the average energy of the electrons of a crystal. Thus, for instance, the Fermi energy of an undoped semiconductor is in the middle of the forbidden region as every electron in the conduction band generates a hole in the valence band, and the quantity of possible energy conditions in the conduction and valence bands is the same. However, as soon as the semiconductor is n-doped, then the number of electrons in the conduction band rises and with this so does the Fermi energy W_F (see Figure 3.16a).

The reverse case occurs in p-doping: There are hardly any free electrons, and most of the electrons are situated in the valence band so that W_F is just above the valence band edge.

If the p- and n-regions are now brought together, then the Fermi energy in thermal equilibrium must be at the same level in both regions. As shown in Figure 3.16b, a potential step $q \cdot V_D$ builds up that corresponds to the band distance but is reduced about both the Fermi differences ΔW_1 and ΔW_2:

$$q \cdot V_D = \Delta W_G - \Delta W_1 - \Delta W_2. \tag{3.10}$$

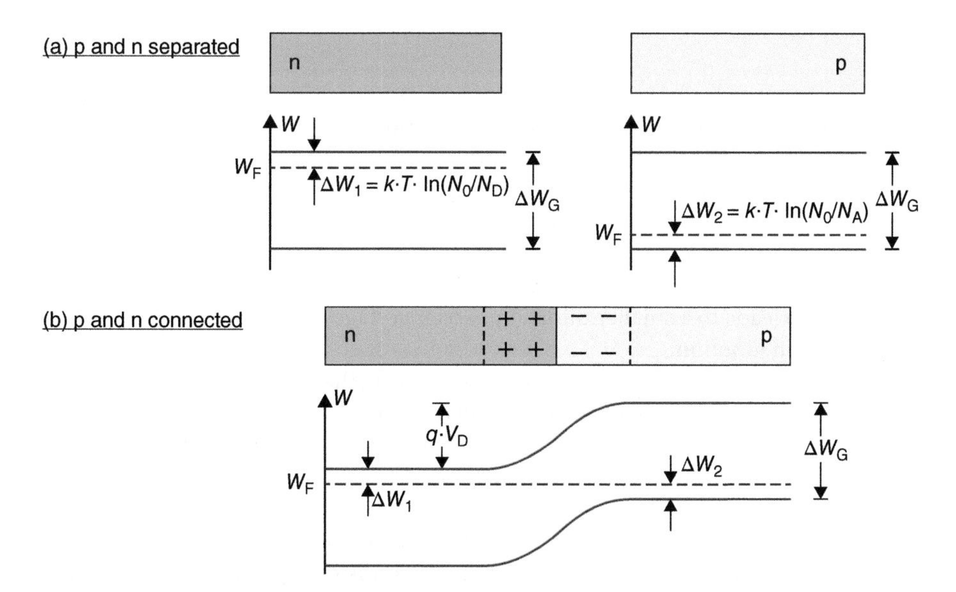

Figure 3.16 Determination of the diffusion voltage V_D of a p–n junction by means of the Fermi energies of n- and p-doped sides [24].

The Fermi differences are calculated by

$$\Delta W_1 = k \cdot T \cdot \ln\left(\frac{N_0}{N_D}\right), \tag{3.11}$$

$$\Delta W_2 = k \cdot T \cdot \ln\left(\frac{N_0}{N_A}\right). \tag{3.12}$$

Here the dimension N_0 is the effective density of states of the electrons and holes as they have already been applied in Equation (3.3).

Thus, the result is

$$q \cdot V_D = \Delta W_G - k \cdot T \cdot \ln\left(\frac{N_0}{N_D}\right) - k \cdot T \cdot \ln\left(\frac{N_0}{N_A}\right) \tag{3.13}$$

$$= \Delta W_G - k \cdot T \cdot \ln\left(\frac{N_0^2}{N_D \cdot N_A}\right),$$

$$\Rightarrow q \cdot V_D = \Delta W_G - k \cdot T \cdot \ln\left(\frac{N_0^2}{N_D \cdot N_A}\right). \tag{3.14}$$

Example 3.4 ***Diffusion voltage of a p–n junction***

We consider an asymmetric p–n junction with $N_D = 10^{17}\ \text{cm}^{-3}$ and $N_A = 10^{15}\ \text{cm}^{-3}$. The result for the diffusion voltage is

$$q \cdot V_D = 1.12\ \text{eV} - 8.62 \times 10^{-5}\ \text{eV K}^{-1} \cdot 298.15\ \text{K} \cdot \ln\left(\frac{(3 \times 10^{19}\ \text{cm}^3)^2}{10^{17}\ \text{cm}^{-3} \times 10^{15}\ \text{cm}^{-3}}\right),$$

$$= 1.12 - 0.41\ \text{eV} = 0.71\ \text{eV}.$$

Therefore, we obtain a diffusion voltage of approximately 0.7 V. ■

3.5.3 Behavior with Applied Voltage

If we apply a small positive voltage V to the p–n junction, then the electrons are driven by the voltage source into the n-region (Figure 3.17a). However, in the region of the junction, they are prevented by the field of the space charge region from moving into the p-region. They back up at the left border of the space charge region and reduce this by the neutralization of the positive fixed charges. The same occurs on the right side with the interlinkage of the holes and fixed negative charges. The reduction of the space charge region also leads to a smaller diffusion barrier voltage V_D. This results in a small current via the p–n junction.

If we increase V, then the space charge region is further reduced until finally it almost completely disappears. Now a strong current can flow, the "diode" becomes a relatively good conductor. The voltage required for this corresponds to the diffusion voltage V_D. As the diode conducts with applied positive voltage, we speak of "forward voltage."

The case of applied negative voltage ("reverse voltage") is shown in Figure 3.17b. Because of the external voltage, the space charge region is enlarged somewhat as free charge carriers are taken out at the borders. Only a minimal reverse current flows in the nA region.

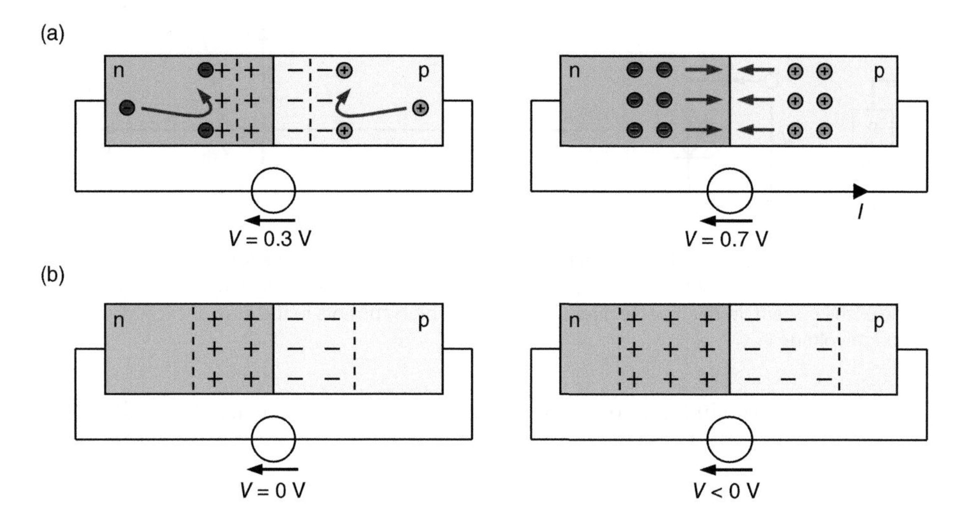

Figure 3.17 Behavior of the p–n junction with applied voltage: When V rises, the space charge region is reduced and a rising current can flow. In the case of reverse voltage, the diode blocks and the space charge region are enlarged. (a) Positive voltage (forward voltage). (b) Negative voltage (reverse voltage).

3.5.4 Diode Characteristics

For deriving the characteristics, we look at p–n junction without external circuit. In this case, the total current must be zero. Thus, the total sum of the diffusion current density j_D and the field current density j_F must be zero, This again means that in temporal average the same number of charge carriers of both types flow to the right as also to the left.

With Equations (3.8) and (3.9), we can state that

$$j_F = j_D \Rightarrow q \cdot n(x) \cdot \mu \cdot F(x) = -q \cdot D \cdot \frac{dn(x)}{dx}. \tag{3.15}$$

This equation is valid for the electrons as well as the holes. A solution for this system of differential equations can be found by means of a longer calculation and simplified assumptions (for instance, see [25]). The result is the so-called diode or Shockley equation.

$$I = I_S \cdot \left(e^{\frac{V}{V_T}} - 1\right) \tag{3.16}$$

with I_S: saturation current of the diode; V_T: thermal voltage.

The *saturation current* I_S is determined by

$$I_S = A \cdot \left(\frac{q \cdot D_N \cdot n_i^2}{L_N \cdot N_A} + \frac{q \cdot D_P \cdot n_i^2}{L_P \cdot N_D}\right) \tag{3.17}$$

with L_N, L_P: diffusion lengths of the electrons or holes.

Therefore, it depends on the concrete structure (doping, area of the junction, etc.) of the p–n junction and is typically in the nA to μA region. The diffusion lengths L_N and L_P define what distance a free particle travels on average in the foreign region until it recombines. This dimension will be discussed in greater detail in Section 4.2. The size of

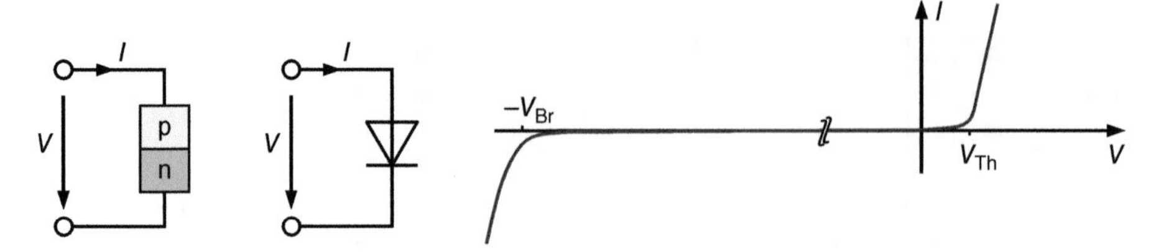

Figure 3.18 Buildup, symbol, and I/V characteristic curve of a p–n diode: In the forward direction, an exponential increase follows out of the Shockley equation. Above the threshold voltage V_{Th}, we see a strong rise of the current. In the reverse direction, there are high currents in the case of exceeding the breakthrough voltage V_{Br}.

thermal voltage V_T used in Equation (3.16) can be determined by means of the following equation:

$$V_T = \frac{k \cdot T}{q}. \tag{3.18}$$

Example 3.5 ***Thermal voltage at room temperature***

The following thermal voltage occurs at room temperature ($T \approx 300$ K):

$$V_T = \frac{k \cdot T}{q} = \frac{8.62 \times 10^{-5}\ \text{eV K}^{-1} \cdot 300\ \text{K}}{q} = 25.89\ \text{mV} = 26\ \text{mV}.$$ ■

At room temperature ($T \approx 300$ K), the thermal voltage is approximately 26 mV.

The typical characteristic curve of a p–n diode as shown in Figure 3.18 can be taken from the Shockley equation. The exponential function results in a seeming kink in the characteristic after which the current rises drastically. This occurs at the threshold voltage V_{Th} that corresponds approximately in amount to that of the diffusion voltage V_D.

If one applies a strongly negative voltage to the semiconductor diode, then the electrical field is increased in the space charge region. This accelerates the available free electrons.

If the negative voltage is increased further, then at some point the electrons reach such a high speed that they knock further electrons out of the crystal bonds. These in turn are also accelerated and increase the effect. Thus, we speak of an avalanche effect or avalanche breakthrough. At a later point, we will return to our knowledge of the p–n junction when dealing with solar cells.

3.6 Interaction of Light and Semiconductors

3.6.1 Phenomenon of Light Absorption

We learned of the effect of light absorption on individual atoms when discussing Bohr's atomic model (see Section 3.1.1). The behavior of the semiconductor is similar. In place of the individual energy levels, however, the bandgap ΔW_G is decisive here for the absorption behavior.

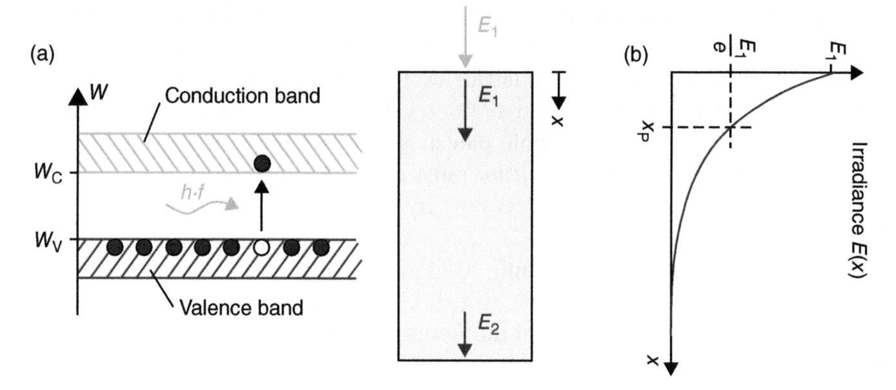

Figure 3.19 Principle of light absorption in the semiconductor. (a) The photon is absorbed only with sufficient light energy and an electron is raised into the conduction band. (b) Incidental light radiation into a semiconductor crystal: Due to absorption in the material, the light intensity sinks with increasing penetration depth.

3.6.1.1 Absorption Coefficient

Figure 3.19 shows the effect of light absorption in the semiconductor crystal. Incident light photons lift individual electrons from the valence into the conduction band. In order to trigger this effect, the energy W_{Ph} of the photons must be greater than the bandgap:

$$W_{Ph} = h \cdot f = \Delta W_G. \tag{3.19}$$

What is the effect of the light absorption in the solid-state body? Figure 3.19 shows the incidence of a light ray in a semiconductor crystal. In passing through the crystal, the irradiance E sinks continuously due to absorption. The course of the irradiance in the absorbing material can be described by means of a decaying exponential function:

$$E(x) = E_1 \cdot e^{-\alpha \cdot x} \tag{3.20}$$

with E_1: irradiance at $x = 0$; α: absorption coefficient.

The absorption coefficient α indicates the absorption "ability" of the respective material. Occasionally use is made of the penetration depth x_p. This describes according to which light path the intensity has decayed by 1/e times (thus approximately 37%). The connection between the two quantities is given by the following equation (see Exercise 3.3):

$$x_P = \frac{1}{\alpha}. \tag{3.21}$$

Example 3.6 ***Penetration depth of silicon***
In the visible spectrum ($\lambda = 600\,\text{nm}$), crystalline silicon (c-Si) has an absorption coefficient of approximately $4000\,\text{cm}^{-1}$. With Equation (3.21), the result from this is a penetration depth of 2.5 μm. ■

3.6.1.2 Direct and Indirect Semiconductors

In order to understand why different materials possess greatly different absorption coefficients, we must study the interactions between light and the semiconductor crystal more closely. A semiconductor crystal is a system of coupled oscillating lattice particles and for this reason the energy of the lattice oscillations cannot take on every desired state. Similar to a photon, these lattice oscillations can also be allocated to a particle character. Such a particle is called a phonon. With this model, we can describe the optical generation of an electron–hole pair as an impact process for which both energy as well *as conservation of momentum* must apply. An incidental photon has a relatively high energy but only a low momentum. In contrast, a phonon has low energy but at the same time high momentum.

Semiconductor materials are divided into two groups: Indirect and direct semiconductors. With an indirect semiconductor (e.g. Si), the minimum of the conduction band edge is situated at a different crystal momentum to the maximum of the valence band edge (Figure 3.20).

This means that an electron–hole pair can only be formed with the participation of a phonon. In the case of a direct semiconductor, however, no phonon is necessary as here the minimum of the conduction band and the maximum of the valence band exist with the same crystal momentum (Figure 3.21).

In order to really understand these processes, we should go into the details of solid-state physics (see, for instance, [26]). Instead, we will select a simple mechanical model for description. Let us view Figure 3.20 again. The x axis now shows the

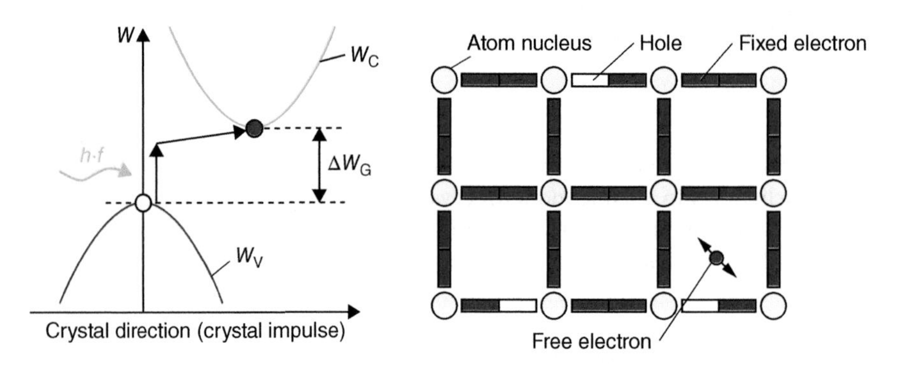

Figure 3.20 Simple model for understanding an indirect semiconductor: In order for it to absorb the photon, the electron must change its energy as well as its direction of oscillation.

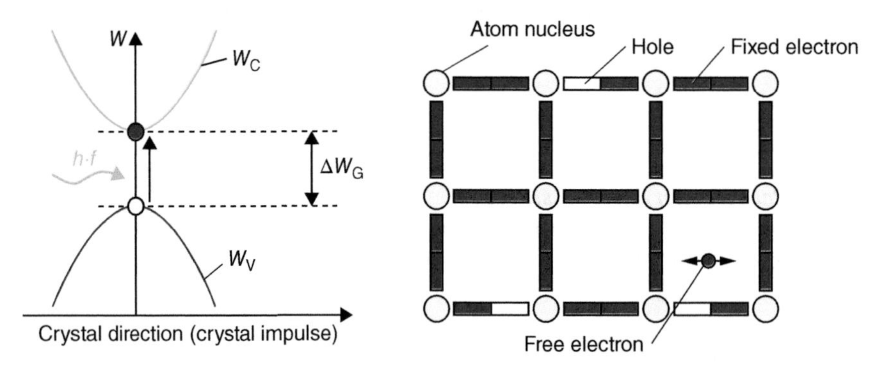

Figure 3.21 Situation in the direct semiconductor: A free electron can be generated by means of absorption in that it changes only its energy but not its direction of oscillation.

oscillation direction of an electron. We see that the minimum of the conduction band is positioned in a different oscillation direction to the maximum of the valence band. If a photon with energy $W_{Ph} \geq \Delta W_G$ is to lift an electron into the conduction band, then the electron must not only change its energy but also its oscillation direction at the same time.

In the crystal model, on the one hand, we show this by depicting the holes not as circles but as slots. On the other hand, we represent a free electron as a sphere swinging to-and-fro, which can only exist when it achieves a diagonal oscillation direction. This only occurs when the electron collides with the nucleus of an atom (more or less when it takes up an additional momentum from the lattice).

As this coincidence is relatively unlikely, a photon moves comparatively far into an indirect semiconductor crystal before it is absorbed. For this reason, indirect semiconductors like silicon or germanium only possess a low absorption coefficient.

The position is different with a direct semiconductor (Figure 3.21). The absorption of a photon can take place in that the electron is torn out of its bond without the direction of oscillation being changed. The result is that absorption is relatively probable and thus a high absorption coefficient is given. Table 3.3 lists the absorption properties of various direct and indirect semiconductors.

The strongly differing absorption behavior of individual semiconductors is shown even more clearly in Figure 3.22. In the case of direct semiconductors, the absorption coefficient rises steeply above the bandgap energy. In crystalline silicon, however, the rise is much more moderate, resulting overall in a relatively low absorption coefficient.

3.6.2 Light Reflection on Surfaces

3.6.2.1 Reflection Factor

We will consider two materials with different *refractive indices* n_1 and n_2 (see Figure 3.23a). The refractive index n of a material indicates by which factor the speed of light is reduced compared to its speed in a vacuum: $n = c_0/c$. If the ray of light falls on an interface between two materials, then *reflection* results. The strength of the reflection is given by the reflection factor R [30]:

$$R = \frac{E_R}{E_0} \tag{3.22}$$

with E_0: incident irradiance; E_R: reflected irradiance.

Table 3.3 Comparison of the absorption coefficients of different materials for light of wavelength 600 nm [27–29].

Material	Type	Bandgap ΔW_G (eV)	Absorption coefficient α (cm^{-1})	Penetration depth x_p (μm)
c-Si	Indirect	1.12	4000	2.5
a-Si	Direct	1.7	40 000	0.25
CdTe	Direct	1.45	37 000	0.3
GaAs	Direct	1.42	47 000	0.2

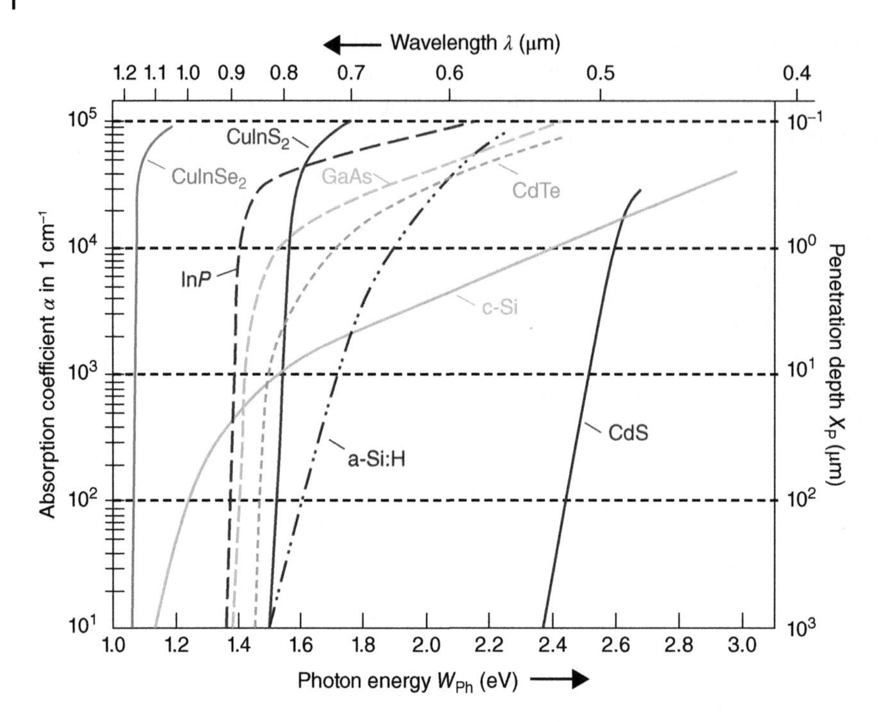

Figure 3.22 Absorption coefficient of different semiconductor materials versus photon energy: The direct semiconductors show a steep rise in absorption above the bandgap energy [27–29].

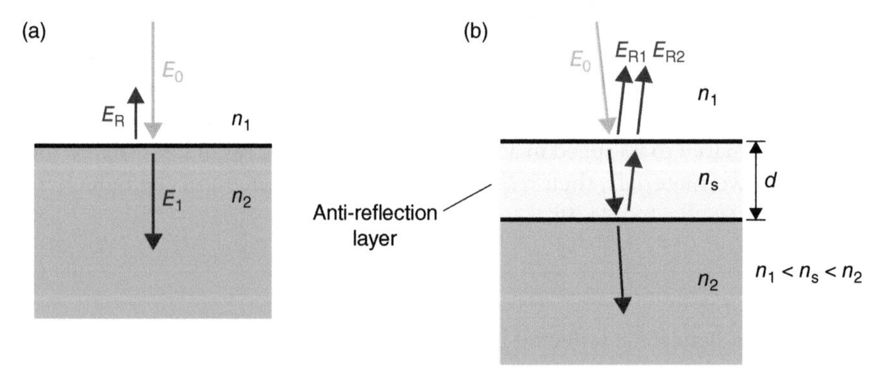

Figure 3.23 Reflection of light on the interface between two media: The reflection can be reduced with the use of an antireflection layer (b).

For vertical incidental radiation, the reflection factor is calculated according to the following equation:

$$R = \left(\frac{n_1 - n_2}{n_1 + n_2}\right)^2. \tag{3.23}$$

Example 3.7 ***Reflection on a silicon surface***

In the case of silicon, the refractive index is in the visible spectrum at approximately $n = 3.9$ [31]. If a ray of light ($n = 1$) impinges vertically on the silicon surface, this results in a reflection factor of

$$R = \left(\frac{1-3.9}{1+3.9}\right)^2 = \left(\frac{-2.9}{4.9}\right)^2 = 0.35 = 35\%.$$

Thus, approximately one-third of the light is reflected. ■

If, instead of a vertical, one selects a flat incidence of radiation, then the reflection factor rises further. This effect can be seen well on a pane of glass: The more we turn a pane of glass to look at it as flat, the more it acts as a mirror.

3.6.2.2 Antireflection Coating

The incidental reflection must be reduced in order to achieve a high degree of efficiency in a solar cell. A standard means of doing this is antireflection coating. Figure 3.23b shows the principle: A material of thickness d is inserted between the two media. At the interface, there is a reflective ray E_{R1}. There is also a reflection at the transition from n_S to n_2 that appears at the surface with the strength E_{R2}. The trick is to make the layer d so thick that the ray 2 is displaced by 180° phase compared to ray 1 so that the reflective radiations cancel each other out due to interference.

I can just imagine that the two reflecting rays could cancel each other out. This doesn't bring anything useful, as in this way no more light penetrates the semiconductor, or is this not the case?

First of all, we must admit that the right-hand sketch of Figure 3.23 does not mirror the actual case accurately enough. In fact, the ray reflected at the border layer n_S/n_2 is partly reflected again at the front border layer n_S/n_1. This results in an infinite number of continuously weakening to-and-fro reflections. The rays moving downward have a phase displacement of 360° to each other and superimpose themselves constructively (for details see [30]). As a result, in the optimum case actually the whole of the incidental light penetrates into the semiconductor.

The conditions for the optimal layer thickness can be determined as follows:

$$d = \frac{\lambda}{4 \cdot n_S} \cdot (2 \cdot m + 1) \quad (3.24)$$

with m: 0, 1, 2, 3, …

In matter, the speed of light c and thus the wavelength compared to that in a vacuum is reduced: $\lambda_{Mat} = \lambda / n_S$. The layer thickness must therefore correspond to an uneven multiple of a quarter wavelength:

$$d = \frac{\lambda_{Mat}}{4} \cdot (2 \cdot m + 1). \quad (3.25)$$

The remaining reflection factor can now be calculated according to the Fresnel equations [30]:

$$R = \left(\frac{n_S^2 - n_1 \cdot n_2}{n_S^2 + n_1 \cdot n_2}\right)^2. \quad (3.26)$$

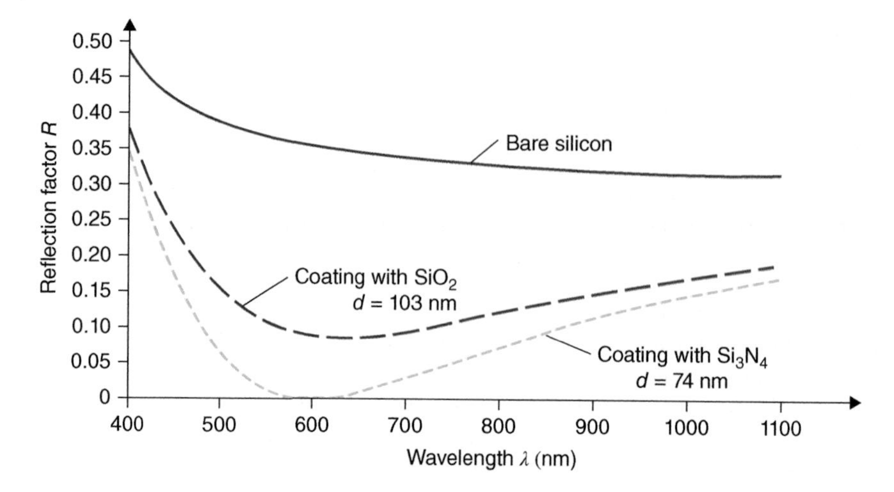

Figure 3.24 Spectral progression of reflection for uncoated and, for silicon oxide or silicon nitride, coated silicon: For both materials, the reflection factor can be clearly reduced compared with bare silicon.

From Equation (3.26), it can be seen that the reflection factor even reaches zero, when the refractive index n_S is at the geometric average of the two other indices:

$$n_S = \sqrt{n_1 \cdot n_2}. \tag{3.27}$$

In reality, however, a complete prevention of reflection is not achievable. Thus, Equation (3.27) cannot be fulfilled optimally as only a few suitable materials are available for antireflection coatings.

Example 3.8 ***Antireflection coating with SiO_2***

In the case of silicon, the refractive index of $n_S = \sqrt{3.9} = 1.97$ is optimum. The easy-to-use material silicon oxide (SiO_2) has a refractive index of 1.46. With Equation (3.26), this gives a reflection factor of $R = \left(\frac{1.46^2 - 1 \cdot 3.9}{1.46^2 + 1 \cdot 3.9}\right)^2 = 8.6\%$ ■

Silicon nitride (Si_3N_4) is much better and is used today as the standard material for antireflective coatings of solar cells. It possesses a refractive index of 2.0, which again leads to a remaining reflection factor of less than 1%.

However, it has to be kept in mind that a selected layer thickness can always function only as the optimum for one single wavelength. But with solar cells, we want to utilize the largest possible region of the Sun's spectrum. Mostly, the layer thickness is defined for minimal reflection at $\lambda = 600$ nm. Figure 3.24 shows the result: With the use of a Si_3N_4 layer, the reflection disappears almost completely at 600 nm and rises at the borders of the viewed spectral region by up to 34%.

A third limitation must be noted, that the considerations made here apply only to vertical incidence. In a flat incidence, the path difference changes between the two reflected rays so that an increased reflection also applies here. In Chapter 4, we will discuss further possibilities for reducing the reflection.

4

Structure and Method of Operation of Solar Cells

The basis of photovoltaic power generation is solar cell. For this reason, this chapter will deal with its structure and function in greater detail. We will pay special attention to the question of how to achieve a higher degree of efficiency and present the current efficiency records of solar cells.

4.1 Consideration of the Photodiode

A good foundation for understanding the solar cell is the photodiode.

4.1.1 Structure and Characteristics

We can represent a photodiode in the simplest case as a p–n junction that is illuminated from the side (Figure 4.1).

Penetrating photons are absorbed, and they generate free electron–hole pairs. These are separated again by the electrical field prevailing in the space charge region and "brought home": The electrons to the n-side and the holes to the p-side. There they are majority carriers, which reduce the probability of undesired recombinations. Now the generated power can be drawn off at the contacts. As it is generated by photons, it is called the photocurrent I_{Ph}.

We will assume that every absorbed photon also leads to an electron–hole pair and, therefore, makes a contribution to the photocurrent. Thus, photocurrent I_{Ph} is proportional to the irradiance E:

$$I_{Ph} = \text{const} \cdot E. \tag{4.1}$$

The resulting characteristic curve is shown in Figure 4.2.

Provided that no light shines on the photodiode, it behaves like a normal p–n junction. With a reverse voltage, only a small reverse current flows, which is called the dark current. As soon as light shines on the diode, a photocurrent that is independent of the voltage V is added to the diode characteristic curve. Because it flows in the reverse direction, it displaces the depicted I/V curve downwards. The use of the photodiode in quadrant III is called photodiode operation as photodiodes are typically operated with applied reverse voltage in order, for instance, to serve as detectors for optical data receivers. In quadrant IV, the photodiode is operated as a solar cell – with positively

Photovoltaics – Fundamentals, Technology, and Practice, Second Edition. Konrad Mertens.

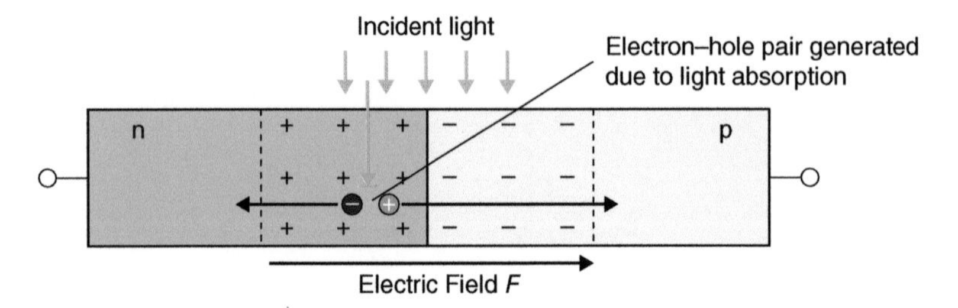

Figure 4.1 Illuminated p–n junction: The free electrons and holes generated by light absorption are separated from the field of the space charge region and "brought home."

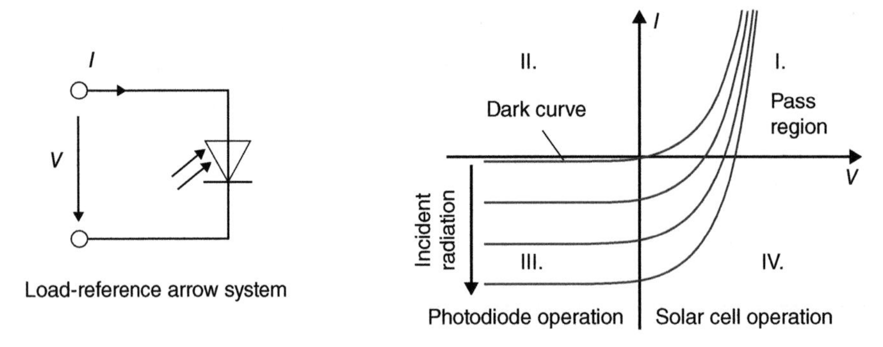

Figure 4.2 Symbol and characteristic curves for a photodiode.

applied voltage, the result is a negative current. In the depicted load reference-arrow system, this means that energy is not used from the device but that energy is generated.

In the load reference-arrow system, the voltage V is applied to the device and then the current I flowing from the voltage source to the component is counted as positive.

In principle, one could also dispense with the p–n junction in the diode. Also in a silicon crystal without a junction, the absorption of light would lead to the generation of electron–hole pairs. Could one also use this external generation of electrical energy?

Of course, electron–hole pairs are also generated in an illuminated semiconductor without a junction. This increases the carrier concentration and, therefore, the conductivity of the crystal. Thus, we have a light-dependent resistor (LDR). Although this can be used for the measurement of light (e.g. in a twilight switch), it cannot be used for the generation of electrical energy as no voltage is established due to the lack of a p–n junction.

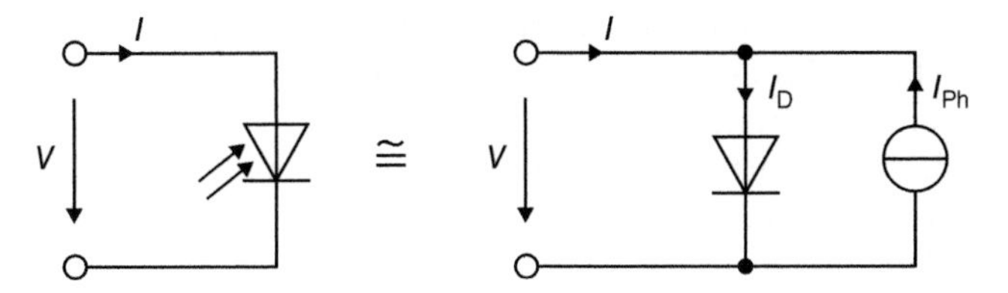

Figure 4.3 Equivalent circuit of the photodiode.

4.1.2 Equivalent Circuit

The electrical behavior of the photodiode can be expressed by the Shockley equation (3.16) in combination with the photocurrent:

$$I = I_D - I_{Ph} = I_S \cdot \left(e^{\frac{V}{V_T}} - 1\right) - I_{Ph}, \tag{4.2}$$

where the dimension I_S is the saturation current mentioned in Chapter 3:

$$I_S = A \cdot q \cdot n_i^2 \cdot \left(\frac{D_N}{L_N \cdot N_A} + \frac{D_P}{L_P \cdot N_D}\right). \tag{4.3}$$

Equation (4.2) can be illustrated by an electrical equivalent circuit (Figure 4.3), where a current source with the strength I_{Ph} is combined with a passive diode. We will return to this equivalent circuit when considering the solar cell.

4.2 Method of Function of the Solar Cell

4.2.1 Principle of the Structure

What is the structure of a solar cell? Figure 4.4 provides information on this. Basically, like the photodiode, it consists of a p–n junction. This is asymmetrically doped: At the bottom is the p-base and at the top the heavily doped n^+-emitter. The terms *base* and *emitter* come from the starting times of the bipolar transistors and have been adopted for solar cells. If light penetrates the cell, then every absorbed photon generates an

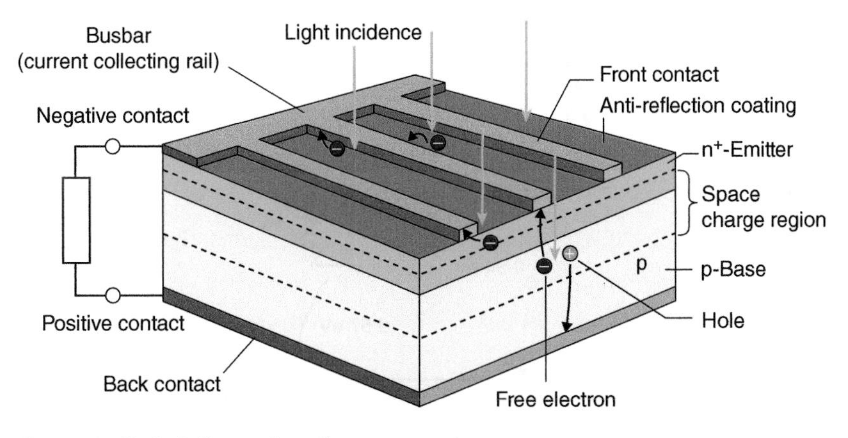

Figure 4.4 Typical silicon solar cell.

electron–hole pair. The particles are separated from the field of the space charge region and moved to the contacts: The holes through the base to the bottom back contact and the electrons through the emitter to the front contacts.

These are small metal strips that transport the generated electrons to the current collector rail (busbar). If a load is connected to the two poles of the solar cell, then this can draw off the generated electrical energy.

4.2.2 Recombination and Diffusion Length

Before we consider the method of operation of the cell in greater detail, we must acquaint ourselves with the behavior of the minority charge carriers. With incident light, electron–hole pairs are generated by the absorption of photons, and these are then available as "surplus" charge carriers. When the source of light is switched off, the particles recombine after a short time in order to re-create the starting condition.

The mechanism of greatest importance to us is defect recombination. It occurs when the theoretically ideal crystal is impure because of foreign atoms, crystal structure errors, and so on. In this case, the forbidden zone is no longer empty but has additional levels (Figure 4.5). Thus, an iron atom in a silicon crystal, for instance, leads to a level in the middle of the forbidden zone, but a sulfur atom is situated only 0.18 eV below the conduction band. For an electron, additional levels represent something like step levels over which a descent into the valence band becomes simpler and thus more probable. The level of the sulfur atom is easy for the electron to reach from the conduction band, but then it must still bridge almost the whole of the bandgap of $1.12 - 0.18\ \text{eV} = 0.94\ \text{eV}$. With an iron atom, however, the step height is reduced to 0.56 eV so that the recombination probability is extremely large. The recombination centers caused by the foreign atoms are also called traps. In addition to foreign atoms, crystal errors such as empty lattice places or crystal displacements also lead to increased recombination.

A crystal surface is also a disturbance of the ideal, infinitely extended crystal. The electrons of the outer atoms do not find bonding partners and remain as open bonds. These then lead to undesired surface recombinations.

From these considerations, it becomes clear that a crystal used for solar cells should be single crystalline as far as possible and of a high degree of purity. In order to compare

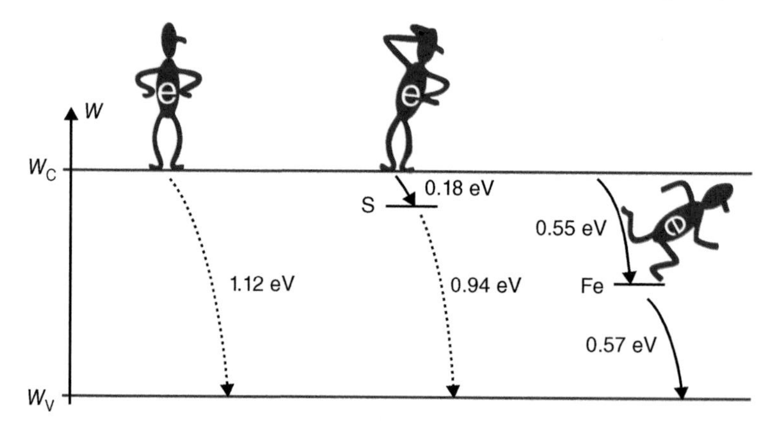

Figure 4.5 Recombination of electron–hole pairs in the case of impurities: The energy levels of the foreign atoms form "step levels," which increase the probability of an electron transferring from the conduction to the valence band.

various materials, one measures the carrier lifetime of a particular material sample. The carrier lifetime τ_N defines how long a generated electron exists on average until it recombines again. Depending on the quality of the silicon and the doping concentrations, it falls within the milliseconds to microseconds range.

A more useful parameter is the diffusion length L_N, which describes the distance that a generated electron travels in the semiconductor until it recombines again. It can be calculated from the lifetime of the carrier:

$$L_N = \sqrt{D_N \cdot \tau_N}, \tag{4.4}$$

where D_N is the diffusion constant of the electron; $D_N = 35\ \text{cm}^2\ \text{s}^{-1}$ for c-Si.

Typical numerical values for silicon, for instance, are between 50 and 500 μm.

4.2.3 What Happens in the Individual Cell Regions?

The situation in the interior of the solar cells is shown in greater detail in Figure 4.6. As we have already seen in Section 3.6.1, light is absorbed differently for different wavelengths. Blue light has the highest absorption coefficient with penetration depths of less than 1 μm; infrared light, in comparison, has penetration depths of more than 100 μm. For this reason, we will look more closely at the generation of photocurrent at different depths in the cell.

4.2.3.1 Absorption in the Emitter

Let us consider Photon ①. It is absorbed in the highly doped emitter. Because of the high degree of doping, the diffusion length is extremely short so that the generated hole probably recombines before reaching the space charge region. The particularly highly

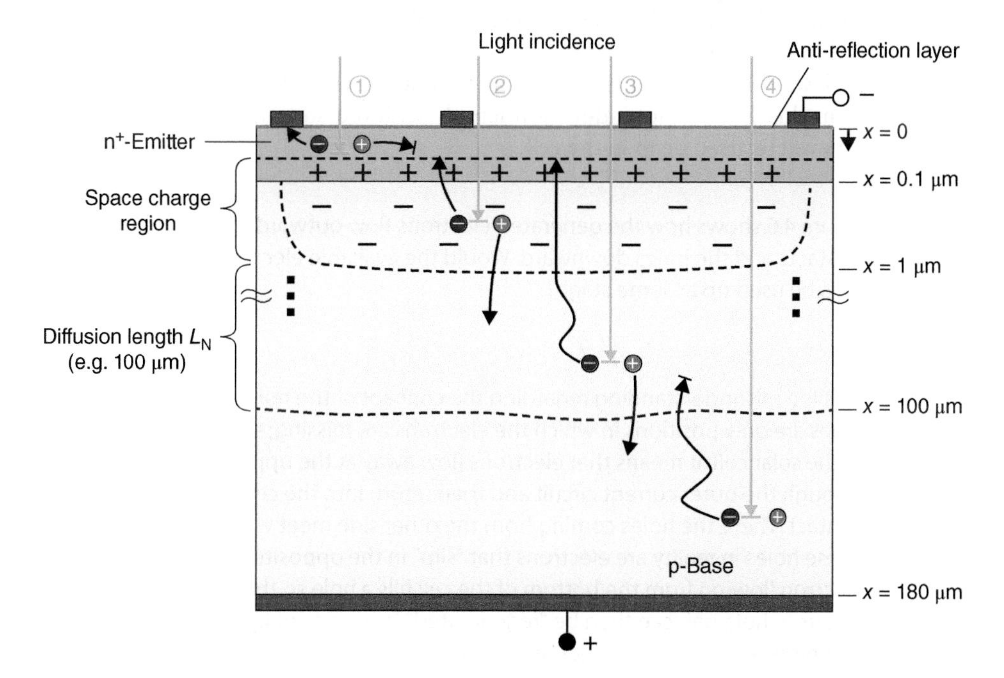

Figure 4.6 Cross-section of a solar cell: Each individually generated electron–hole pair has a different chance of making a contribution to the photocurrent.

doped upper edge of the emitter is also, occasionally, called the dead layer in order to emphasize that this is where the highest recombination probability is situated.

4.2.3.2 Absorption in the Space Charge Region

What happens to Photon ②? Absorption takes place within the space charge region. The field prevailing in the space charge region separates the generated electron–hole pair and drives the two charge carriers into different directions. The electron is moved to the n-region and from there, further to the minus contact of the solar cell. The hole is moved in the opposite direction. It must travel a relatively long way through the base to the plus contact. As it is in the p-region during this movement, the probability of recombination is slight. Thus, practically all generated electron–hole pairs can be used for the photocurrent.

4.2.3.3 Absorption Within the Diffusion Length of the Electrons

Photon ③ is absorbed only deep in the solar cell. The generated electron is not situated in an electrical field but diffuses as a minority charge carrier with little motivation throughout the crystal. If, by chance, it arrives at the edge of the space charge region, then it is pulled to the n-side by the prevailing field, where it can flow as a majority carrier to the contact. As the electron was still generated within the diffusion length, the probability that it can maintain itself up to the space charge region is relatively high.

4.2.3.4 Absorption Outside the Diffusion Length of the Electrons

Photon ④ is a true loser and is absorbed only in the lowest region of the solar cell. Although the electron diffuses through the p-base, it recombines with a hole before it can reach the space charge region. Thus, although an electron–hole pair is formed due to light absorption, an electron and a hole are "eliminated" afterward. Thus, the absorbed photon ④ has contributed nothing to the photocurrent. As no electric energy was produced in this case, the crystal has only become a bit warmer due to the energy conservation law. This confirms the importance of good crystal quality for high efficiency. Only in this way is a long diffusion length achieved so that absorbed infrared light rays can even be used deep in the cell.

Figure 4.6 shows how the generated electrons flow outward via the upper contacts and the holes downward. Would the available electrons and holes in cells be used up at some stage?

This is a misunderstanding regarding the concept of the holes. In fact, the holes are only positions in which the electrons are missing; see Section 3.3.1. In the solar cell, it means that electrons flow away at the upper contact through the outer current circuit and then return into the cell at the lower contact. There the holes coming from the other side meet with the electrons. These holes in reality are electrons that "slip" in the opposite direction. An electron flowing from the bottom of the cell fills a hole so that this electron–hole pair can then be "regenerated" by means of light absorption. Therefore, electrons and holes are not "used up."

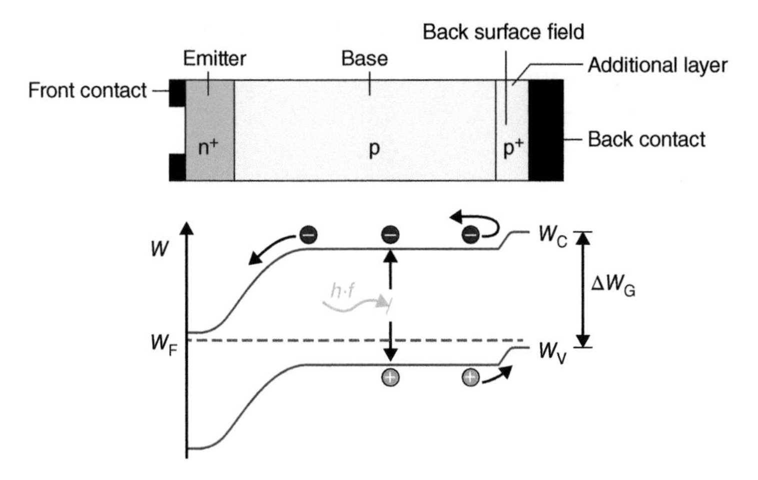

Figure 4.7 Effect of the back-surface field: The generated electrons are stopped by the layer-forming potential step at the pp+ interface and driven back into the p-region [14].

4.2.4 Back-surface Field

A special danger for the electrons that are generated in the lower region of the solar cell is the metal semiconductor interface, as massive surface recombinations can occur there. A common trick for avoiding this danger is the application of a highly doped p+-layer between the metal and the semiconductor. This is achieved, for instance, by doping with boron or aluminum atoms.

How does this trick work? Because of the concentration gradient, holes flow out of this highly doped p^+-layer into the p-region and leave site-fixed negatively charged acceptor atoms behind. The generated electrical field is called *back-surface field* (BSF). It acts as an electric mirror that returns the electrons generated by means of absorption in the direction of the space charge region. The probability of an undesired recombination at the rear of the cell is thus greatly reduced.

To gain a better understanding of this effect, an alternative way of viewing it is shown in the band diagram in Figure 4.7. Analogous to Figure 3.16, we first consider what Fermi energy the individual regions of the cells possess. After the merger, the result is the depicted band process. At the transition of the p+ to the p-layer, the resulting small potential step that prevents the electrons from moving further up to the back contact can be seen. In addition to being an electric mirror, the BSF function has a further advantage. The whole of the voltage drop at the cell is now divided into the potential level at the actual p–n junction and the additional level at the pp^+ interface. The resulting reduced voltage at the p–n junction leads to a reduced dark current and then finally to an increase in the open-circuit voltage V_{OC} [32].

4.3 Photocurrent

On the one hand, the size of the photocurrent depends on the number of incident photons that are absorbed by the solar cell. On the other hand, the electron–hole pairs

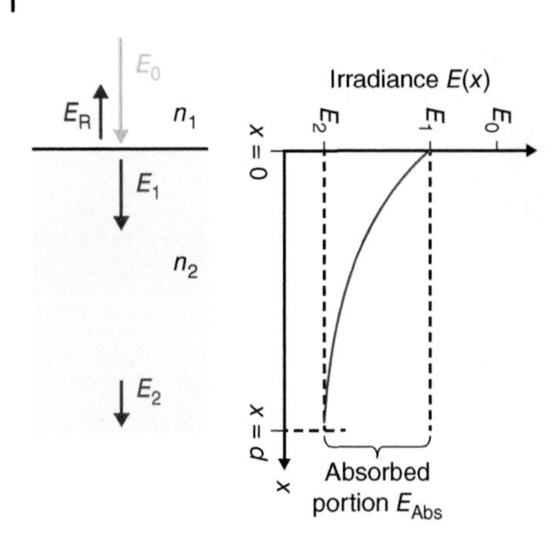

Figure 4.8 Light absorption in the solar cell.

generated by light absorption must be separated and brought home safely. We will look at the conditions for this more closely.

4.3.1 Absorption Efficiency

Figure 4.8 shows the light absorption of a solar cell similar to that in Figure 3.19. One part E_R of the overall irradiance E_0 is reflected at the surface (see Section 3.6.2). Thus the portion $E_1 = (1 - R) \cdot E_0$ penetrates into the cell. The intensity of the light is now weakened by absorption by passing through the cell according to Equation (3.20). At the bottom end, $E_2 = E(x = d) = E_1\, e^{-\alpha \cdot d}$ still remains. The difference $E_{Abs} = E_1 - E_2$ gives the proportion of light absorbed in the cell.

We define the absorption efficiency η_{Abs} as the relationship between the number of absorbed photons and the number of photons incident from outside:

$$\eta_{Abs} = \frac{\text{Number of absorbed photons}}{\text{Number of incident photons}} = \frac{N_{Ph\_Abs}}{N_{Ph}} = \frac{E_{Abs}}{E_0} = \frac{E_1 - E_2}{E_0}. \tag{4.5}$$

After inserting the above equations, we obtain

$$\eta_{Abs} = (1 - R) \cdot (1 - e^{-\alpha \cdot d}), \tag{4.6}$$

where R = reflection factor; α = absorption coefficient.

η_{Abs} can reach values of almost 100%. For this, on the one hand, the reflection on the surface must be reduced (e.g. by means of an anti-reflection layer; see Section 3.6.2). On the other hand, the cell should be made thick enough that for $x = d$ almost no photons are left over. The problem here is that the absorption coefficient is strongly dependent on the wavelength. Light in the near-infrared region is absorbed relatively weakly.

Example 4.1 ***Absorption of infrared light***

Infrared light with a wavelength $\lambda = 1000$ nm has an absorption coefficient of approximately 50 cm^{-1}; according to Equation (3.21), this corresponds to a penetration depth of 200 μm. For a cell thickness of 200 μm, a proportion of $1/e \approx 37\%$ is lost as unused (assume $R = 0$). If one wishes to make the cell thick enough that a maximum of 1% of the light is lost, then the minimum thickness according to Equation (4.6) is

$$0.99 = 1 - e^{-\alpha \cdot d} \rightarrow e^{-\alpha \cdot d} = 1 - 0.99 \rightarrow d = \frac{\ln(1 - 0.99)}{-\alpha} = 921\ \mu\text{m}.$$ ■

The cell in the example would, therefore, be approximately 900 μm thick. In addition to the high cost of producing such a cell, there would still be the problem that the electrons generated in the depth of the cell would recombine on the way to the space charge region. A better solution is, for instance, to provide the back of the cell with an optical reflector. In this way, the optical path distance can be doubled for the same thickness of cell.

4.3.2 Quantum Efficiency

Even if it were successful in driving the absorption efficiency up to 100%, not all electron–hole pairs generated would contribute to the photocurrent. For this reason, the external quantum efficiency η_{Ext} is defined as the relationship between the electron–hole pairs usable for the photocurrent and the overall incident photons:

$$\eta = \frac{\text{Number of usable electron–hole pairs}}{\text{Number of impinging photons}} = \frac{N_{\text{EHP}}}{N_{\text{Ph}}}. \tag{4.7}$$

In addition to the external, we also define the internal quantum efficiency η_{Int}, where the losses caused by reflections are not considered:

$$\eta_{\text{Int}} = \frac{\eta_{\text{Ext}}}{1 - R}. \tag{4.8}$$

Naturally, its value is always greater than that of the external quantum efficiency.

4.3.3 Spectral Sensitivity

The spectral sensitivity $S(\lambda)$ shows which photocurrent is generated with the incidence of a particular optical power:

$$S(\lambda) = \frac{I_{\text{Ph}}}{P_{\text{Opt}}}. \tag{4.9}$$

The connection between the two quantities can easily be found when the current is interpreted as a charge Q per unit time and the optical power as optical energy W_{Opt} per unit time:

$$S(\lambda) = \frac{I_{\text{Ph}}}{P_{\text{Opt}}} = \frac{\frac{Q}{\Delta t}}{\frac{W_{\text{Opt}}}{\Delta t}} = \frac{N_{\text{EHP}} \cdot q}{N_{\text{Ph}} \cdot (h \cdot f)} = \frac{N_{\text{EHP}}}{N_{\text{Ph}}} \cdot \frac{q}{\frac{h \cdot c}{\lambda}} = \frac{q}{h \cdot c} \cdot \lambda \cdot \eta_{\text{Ext}}(\lambda). \tag{4.10}$$

Here we have kept in mind that the external quantum efficiency is dependent on the wavelength. This follows, on the one hand, out of the wavelength dependence of the absorption coefficient and, on the other, from the fact that the refractive index of the semiconductor changes with the wavelength.

The prefactor $q/(h \cdot c)$ consists only of natural constants and can be combined to give

$$\frac{q}{h \cdot c} = \frac{1.6 \times 10^{-19}\,\text{As}}{3.6 \times 10^{-34}\ \text{W s} \cdot 3 \times 10^{8}\ \text{m s}^{-1}} = 0.808 \cdot \frac{\text{A}}{\text{W}\ \mu\text{m}} = \frac{1}{1.24\ \mu\text{m}} \cdot \frac{\text{A}}{\text{W}}. \tag{4.11}$$

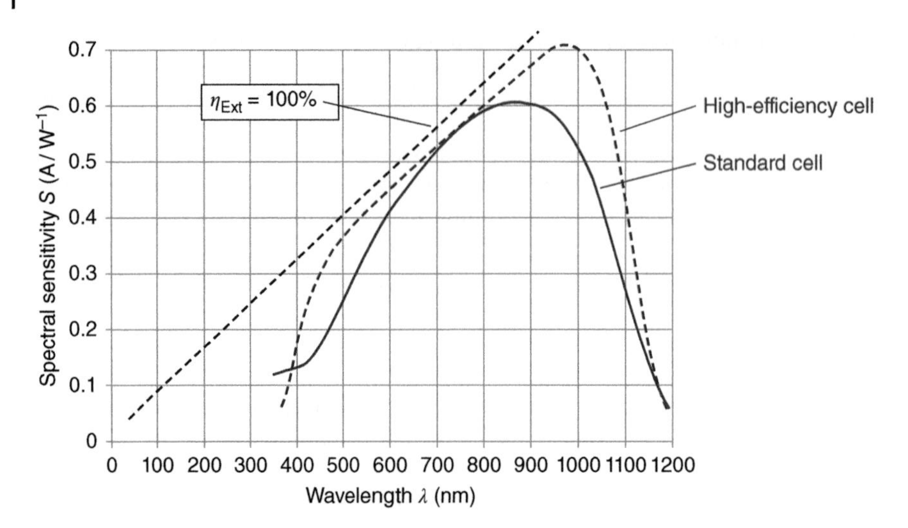

Figure 4.9 Spectral sensitivity of two solar cells: In the blue and infrared regions, the measured curves deviate especially strongly from the ideal line [33, 34].

Thus, finally, for spectral sensitivity we have

$$S(\lambda) = \frac{\lambda}{1.24\ \mu\text{m}} \cdot \frac{\text{A}}{\text{W}} \cdot \eta_{\text{Ext}}(\lambda). \tag{4.12}$$

Figure 4.9 shows the measured curve of the spectral sensitivity of a c-Si standard solar cell and also a high-efficiency cell (see Section 4.7). The ideal curve for the case of $\eta_{\text{Ext}} = 100\%$ is also shown. It is noticeable that the quantum efficiency in the region of blue light (400–500 nm) is relatively poor. This is because blue light is absorbed mostly in the n^+-emitter and a large proportion of the holes generated recombine there without contributing to the photocurrent. In the infrared region, η_{Ext} is reduced again as the absorption occurs only in the lower region of the solar cell. Above 1100 nm, the energy of the light photons becomes too small to overcome the bandgap of the silicon and for this reason $S(\lambda)$ collapses relatively suddenly.

4.4 Characteristic Curve and Characteristic Parameters

The characteristic curve of a solar cell corresponds to the principle of a photodiode. However, with a solar cell, the generator reference-arrow system is mostly selected (see Figure 4.10).

In the generator reference-arrow system, the voltage *V* is measured at the energy source and counts the current *I* flowing from the energy source to the load as positive.

Compared with Figure 4.2, the voltage is maintained and only the prefix of the current is reversed. The generation of energy now takes place in the first quadrant, and mostly for this reason, only the characteristic curve of the first quadrant of the solar cell

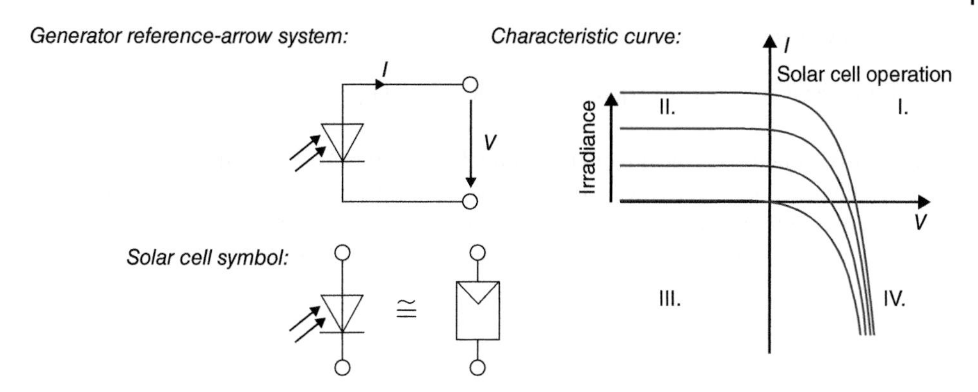

Figure 4.10 Characteristic curves for a solar cell in the generator reference-arrow system.

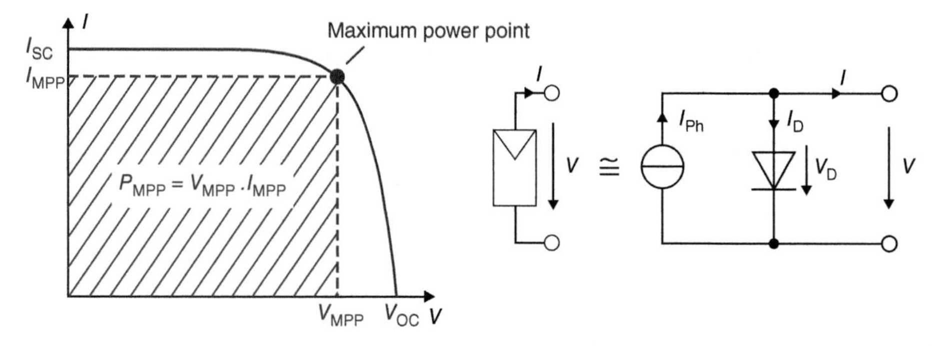

Figure 4.11 Characteristic curve for a solar cell and its associated simplified equivalent circuit.

is shown. Instead of the symbol for the photodiode, the special solar cell symbol has become standard (see Figure 4.10), and we will use it in the following.

A typical solar cell characteristic curve is shown in Figure 4.11, including the equivalent circuit that we have become acquainted with from the photodiode. We will call this the simplified equivalent circuit as it describes the behavior of real solar cells only approximately (see Section 4.5).

The characteristic curve equation, similar to Equation (4.2), is

$$I = I_{Ph} - I_D = I_{Ph} - I_S \cdot \left(e^{\frac{V}{m \cdot V_T}} - 1\right). \tag{4.13}$$

However, we have introduced an ideality factor m into the exponent that permits us to model real solar cell curves better. The ideality factor is usually between 1 and 2.

We will consider the individual points of the characteristic curve in Figure 4.11 in more detail in order to derive various parameters of solar cells from them.

4.4.1 Short-circuit Current I_{SC}

The short-circuit current I_{SC} is delivered by the solar cells when it is short circuited at its connections; the voltage V is thus 0. From Equation (4.13) this results in

$$I_{SC} = I(V = 0) = I_{Ph} - I_S \cdot (e^0 - 1) = I_{Ph}. \tag{4.14}$$

We can thus define:

The short-circuit current I_{SC} is equal to the photocurrent I_{Ph}.

This is also immediately clear from the equivalent circuit: An external short circuit also short-circuits the internal diode so that $I_D = 0$ applies. In this way, the whole photocurrent I_{Ph} can be removed outside. From Equation (4.1), we already know that the photon current is proportional to the irradiance E. Therefore, we can immediately derive:

The short-circuit current I_{SC} of a solar cell is proportional to the irradiance E.

4.4.2 Open-circuit Voltage V_{OC}

The second extreme case occurs when the current becomes zero. In this case, the resulting voltage is called the open-circuit voltage V_{OC}.

In order to determine the open-circuit voltage, we resolve Equation (4.13) according to V and set $I = 0$. The result is $I_{Ph} = I_{SC}$:

$$V_{OC} = V(I = 0) = m \cdot V_T \cdot \ln\left(\frac{I_{SC}}{I_S} + 1\right). \tag{4.15}$$

With very small currents, the value of 1 for I_{SC}/I_S can be ignored so that, in a simplified manner, the equation becomes

$$V_{OC} = m \cdot V_T \cdot \ln\left(\frac{I_{SC}}{I_S}\right). \tag{4.16}$$

The dependence of the open-circuit voltage is thus much lower than that of the short-circuit current:

The open-circuit voltage V_{OC} of a solar cell varies only with the natural logarithm of the irradiance E.

4.4.3 Maximum Power Point (MPP)

The solar cell provides different capacities depending on the actual working point in which it is operated. The operating point at which the maximum power is provided is called the *maximum power point* (MPP). As the power of a working point always corresponds to the surface area $V \cdot I$, this area must be the maximum in the case of the MPP. This case is shown in Figure 4.11. The current and voltage values associated with the MPP are I_{MPP} and V_{MPP}. In Figure 4.12, also the power curve $P = V \cdot I$ is depicted. It reaches its maximum at the MPP.

4.4.4 Fill Factor (FF)

The *fill factor* (*FF*) describes the relationship between MPP power and the product of the open-circuit voltage and short-circuit current (see Figure 4.12). As depicted, FF shows the size of the area under the MPP working point compared with the area $V_{OC} \cdot I_{SC}$:

$$\text{FF} = \frac{V_{MPP} \cdot I_{MPP}}{V_{OC} \cdot I_{SC}} = \frac{P_{MPP}}{V_{OC} \cdot I_{SC}}. \tag{4.17}$$

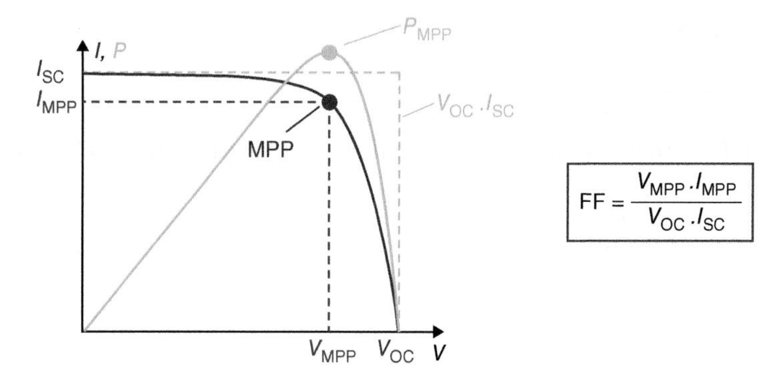

Figure 4.12 The fill factor describes the relationship of the red bordered to the green bordered area.

The FF is a measure of the quality of a cell; typical values for silicon cells are between 0.75 and 0.85 and between 0.6 and 0.75 in the region of thin-film materials.

For the FF, an approximation equation for the function of the open-circuit voltage is used ("idealized FF") [35]:

$$\mathrm{FF} = \frac{1 + \ln\left(\frac{V_{\mathrm{OC}}}{V_{\mathrm{T}}} + 0.72\right)}{\frac{V_{\mathrm{OC}}}{V_{\mathrm{T}}} + 1}. \tag{4.18}$$

4.4.5 Efficiency η

The efficiency of a solar cell describes what proportion of the optical power P_{Opt} incident on the cell is output again as electrical energy P_{MPP}:

$$\eta = \frac{P_{\mathrm{MPP}}}{P_{\mathrm{Opt}}} = \frac{P_{\mathrm{MPP}}}{E \cdot A} = \frac{\mathrm{FF} \cdot V_{\mathrm{OC}} \cdot I_{\mathrm{SC}}}{E \cdot A}, \tag{4.19}$$

where A = cell area.

Typical efficiencies of crystalline silicon cells are between 15% and 22%. The calculation of the efficiency is described in greater detail in Section 4.6.

4.4.6 Temperature Dependence of Solar Cells

A rise in temperature of a semiconductor causes an increase in the thermal movement of the electrons built into the crystal lattice. From Section 3.2, we know that in this case more and more electrons are torn from their bonds and move into the conduction band and that therefore the intrinsic carrier concentration n_{i} increases.

Higher intrinsic carrier concentrations again lead to an increase in saturation current I_{S}.

How does this affect the solar cell? According to Equation (4.16), the increased saturation current leads to a reduction in open-circuit voltage. For calculating this, we insert Equations (3.3) and (4.3) in Equation (4.16), resulting in

$$V_{\mathrm{OC}} = m \cdot V_{\mathrm{T}} \cdot \ln\left(\frac{I_{\mathrm{SC}}}{I_{\mathrm{S}}}\right) = m \cdot V_{\mathrm{T}} \cdot \ln\left(\frac{I_{\mathrm{SC}}}{B}\right) + m \cdot \frac{\Delta W_{\mathrm{G}}}{q}. \tag{4.20}$$

The constant B is expressed by

$$B = A \cdot q \cdot N_0^2 \cdot \left(\frac{D_N}{L_N \cdot N_A} + \frac{D_P}{L_P \cdot N_D} \right). \tag{4.21}$$

If Equation (4.20) is differentiated with respect to T, then, applying Equation (4.20) again, the result is

$$\frac{V_{OC}}{dT} = \frac{m \cdot k}{q} \cdot \ln \left(\frac{I_{SC}}{B} \right) = \frac{V_{OC} - m \cdot \Delta W_G / q}{T}. \tag{4.22}$$

For a typical solar cell, we obtain ($m = 1$)

$$\frac{\Delta V_{OC}}{\Delta \vartheta} = \frac{0.6\ \text{V} - 1.12\ \text{V}}{300\ \text{K}} = 1.7\ \text{mV K}^{-1}. \tag{4.23}$$

In this derivation, we have not taken into account that the bandgap and the intrinsic carrier concentration of the semiconductor are also temperature-dependent. A more accurate derivation gives [36]

$$\frac{\Delta V_{OC}}{\Delta \vartheta} = \frac{V_{OC} - \Delta W_{G0}/q - \gamma \cdot V_T}{T}, \tag{4.24}$$

where ΔW_{G0} = Bandgap at $T = 0$; for silicon: $\Delta W_{G0} = 1.17$ eV; γ = temperature parameter; typically $\gamma = 1$–4.

Thus, for the Si cell we obtain ($\gamma = 3$)

$$\frac{\Delta V_{OC}}{\Delta \vartheta} = -2.3\ \text{mV K}^{-1}. \tag{4.25}$$

For a typical open-circuit voltage of 600 mV, the temperature coefficient TC(V_{OC}) is thus approximately –0.4% K^{-1}.

The open-circuit voltage V_{OC} of a Si solar cell is reduced by 2.3 mV K^{-1}, which corresponds to a temperature coefficient of approximately –0.4% K^{-1}.

The position is different in the case of a short-circuit current I_{SC}. The reduction of the bandgap has the effect that even energy-poor photons still have enough energy to be absorbed and to generate an electron–hole pair. For this reason, the short-circuit current I_{SC} increases slightly with increase in temperature, for instance, by 0.06% K^{-1}.

Figure 4.13 shows the temperature dependence of a monocrystalline cell as an example of the Bosch M-3BB solar cell. With an increase in temperature, the open-circuit voltage and thus also the MPP are clearly displaced to the left.

How can I imagine in a descriptive way why the bandgap of the semiconductor is reduced with increase in temperature?

The crystal expands with a rise in temperature. This also increases the mean spacing between the atoms (lattice constant). As a result, the attractive force of the positively charged atom nuclei on the negatively charged electrons is reduced, which in practice corresponds to a reduction of the bandgap. Although this explanation does not quite cover the physical causes, it is still fairly descriptive.

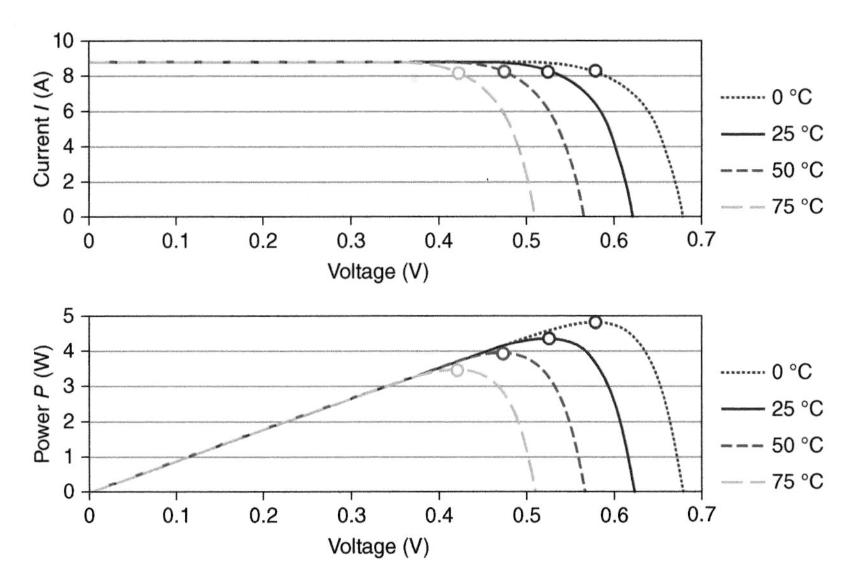

Figure 4.13 Temperature dependence of an Si solar cell on the example of the Bosch M-3BB solar cell: The circles indicate the position of the MPP [37].

The question of how the temperature changes the power is of interest especially for the user of solar cells and solar modules. As the decay of the open-circuit voltage is much greater than the slight rise in I_{SC}, the power P_{MPP} also decreases. Added to this is the fact that according to Equation (4.18), the FF is dependent on V_{OC}/V_T: an increase in temperature increases V_T and thus reduces the FF. All the three effects finally result in a temperature coefficient TC_P of the power of

$$\mathrm{TC}(P_{\mathrm{MPP}}) = \frac{\Delta P_{\mathrm{MPP}}}{\Delta\vartheta \cdot P_{\mathrm{MPP}}} = -0.\dot{4} \text{ to } -0.5\% \ \mathrm{K}^{-1}. \tag{4.26}$$

The power of an Si solar cell decreases by 0.4–0.5% per kelvin temperature change.

Roughly, one can say that the power of a solar cell is reduced by approximately 5% with an increase of temperature of 10 K. As solar modules with full sunlight can easily reach a temperature of 60 °C, this means a clear loss of power compared with the assumed 25 °C normally assumed in the data sheets of solar modules.

4.5 Electrical Description of Real Solar Cells

4.5.1 Simplified Model

This model (see Figure 4.14a) is already known from Figure 4.11 and Equation (4.13):

$$I = I_{\mathrm{Ph}} - I_{\mathrm{D}} = I_{\mathrm{Ph}} - I_{\mathrm{S}} \cdot \left(e^{\frac{V}{m \cdot V_{\mathrm{T}}}} - 1\right). \tag{4.27}$$

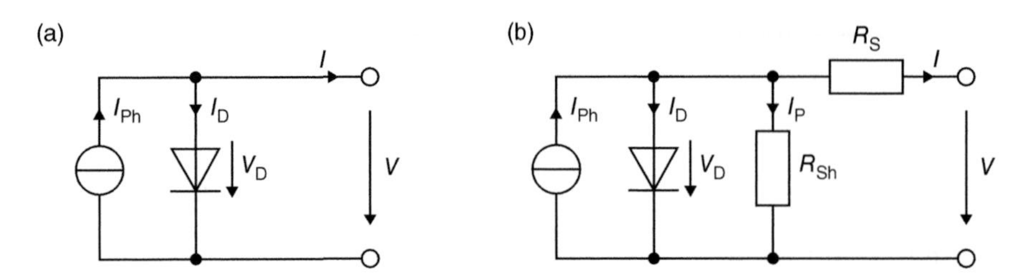

Figure 4.14 (a) Simplified and (b) standard equivalent circuits for electrical description of solar cells and solar modules.

4.5.2 Standard Model (Single-diode Model)

The standard model, also called the single-diode model, goes deeper into electrical losses in the solar cell (Figure 4.14b). The series resistance, R_S, describes especially the ohmic losses in the front contacts of the solar cell and at the metal–semiconductor interface. In contrast, leak currents at the edges of the solar cell and also any point that short circuits of the p–n junction are modeled by the shunt resistance, R_{Sh}.

For deriving the characteristic curves of the standard model, the current I becomes $I = I_{Ph} - I_D - I_{Sh}$, and we find I_{Sh} as

$$I_{Sh} = \frac{V_D}{R_{Sh}} = \frac{V + I \cdot R_S}{R_{Sh}}. \tag{4.28}$$

This gives the characteristic curve equation of the standard model:

$$I = I_{Ph} - I_S \cdot \left(e^{\frac{V+I \cdot R_S}{m \cdot V_T}} - 1\right) - \frac{V + I \cdot R_S}{R_{Sh}}. \tag{4.29}$$

This equation can only be solved numerically as the current I appears on both the left- and the right-hand sides.

The influence of the series resistance on the I/V characteristic curve is shown in Figure 4.15a. With increase in R_S, the curve flattens and the FF decreases significantly. The situation is similar in the case of falling values of the shunt resistance R_{Sh} (Figure 4.15b). Here even the open-circuit voltage is affected as the rising shunt current I_P causes the diode voltage V_D to drop.

4.5.3 Two-diode Model

In the derivation of the Shockley Equation (3.16), it was assumed for the sake of simplicity that there would be no recombination in the space charge region. Especially for semiconductors with larger bandgaps, this leads to deviations between actual and simulated characteristic curves.

In these cases, one makes use of the two-diode model in which the diffusion current is modeled by means of a diode with an ideality factor of 1 and a recombination current through an additional diode with an ideality factor of 2 (Figure 4.16).

The characteristic curve equation can be determined in a similar manner to Equation (4.29):

$$I = I_{Ph} - I_{S1} \cdot \left(e^{\frac{V+I \cdot R_S}{V_T}} - 1\right) - I_{S2} \cdot \left(e^{\frac{V+I \cdot R_S}{2 \cdot V_T}} - 1\right) - \frac{V + I \cdot R_S}{R_{Sh}}. \tag{4.30}$$

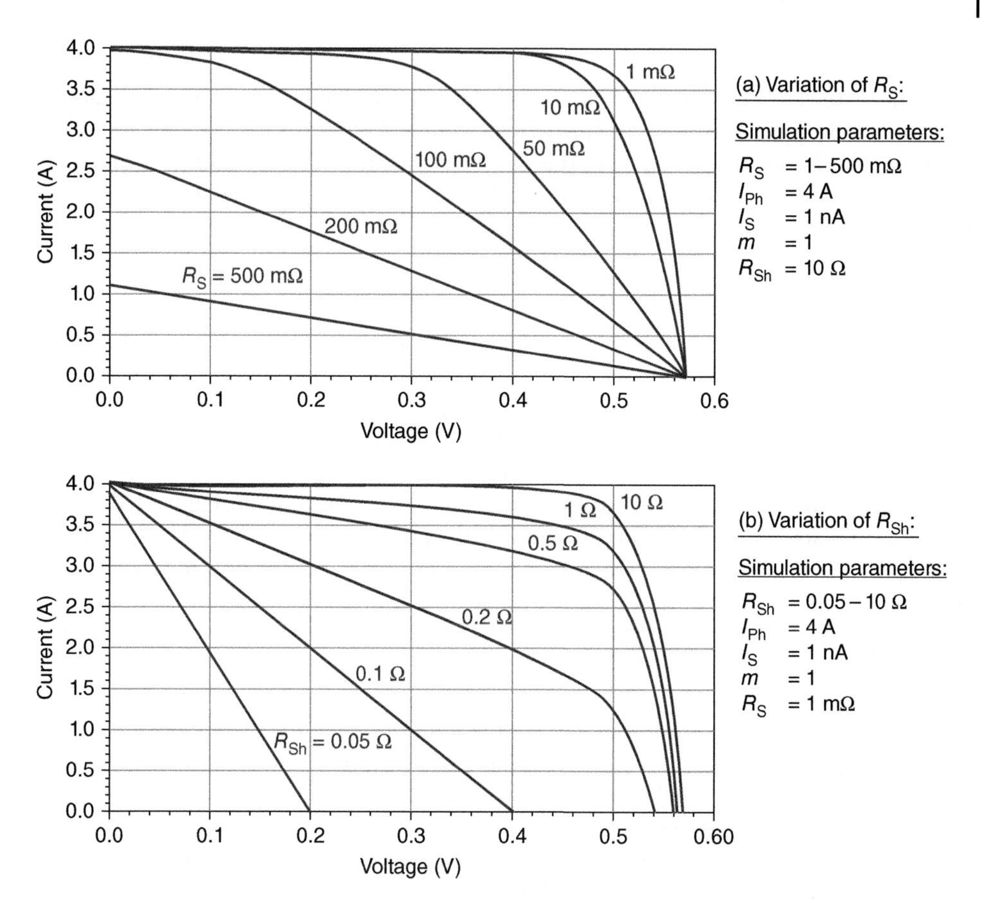

Figure 4.15 Influence of series resistance, R_S (a) and shunt resistance, R_{Sh} (b), on the solar cell characteristic curve: The fill factor decreases significantly with an increase in R_S and a decrease in R_{Sh}.

Figure 4.16 The two-diode model for possible exact modeling of the solar cell characteristic curve.

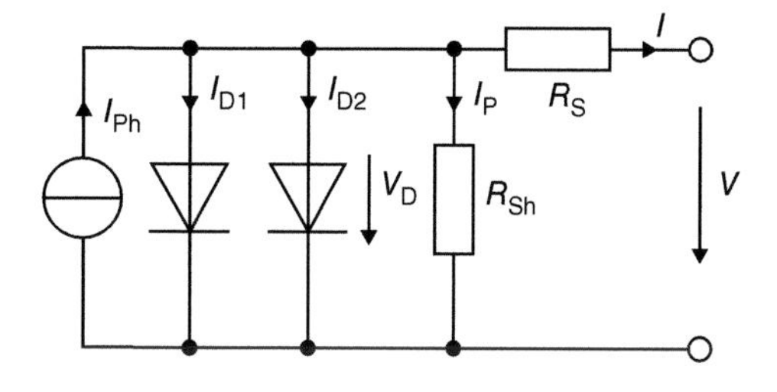

In addition to the three equivalent circuits presented, use is also occasionally made of the effective characteristic curve model. This corresponds to the standard model but without the use of shunt resistance R_{Sh}. Also, negative values are permitted for the variable of the series resistance in this model in order to be able to achieve a good approximation quality of measured curves. Details of this can be found in [14].

4.5.4 Determining the Parameters of the Equivalent Circuit

If the measured I/V characteristic curve of a solar cell is available, then the parameters of the simplified equivalent circuit can be derived from it. As described in Section 4.4.1, the photocurrent I_{Ph} may be set equal to the short-circuit current I_{SC}. In addition, one first takes the diode ideality factor m as 1. The saturation current I_S can then be determined from Equation (4.16):

$$I_S = I_{SC} \cdot e^{-V_{OC}/V_T}. \tag{4.31}$$

However, the agreement of the curve calculated from these parameters with the original measured curve is usually poor, the reason being that the ideality factor of real solar cells is greater than 1. Here the only help is a simulation of the progression of the curve (e.g. with Excel, Mathematica, etc.) with variations of the two parameters m and I_S until the best possible agreement between simulation and measured curve is reached. Figure 4.17 shows an example of the measured curve of a solar module. Figure 4.17a also shows a simulated curve based on the simplified equivalent circuit. Here, there is still a clear deviation between measurement and simulation even after the optimization of the parameters.

The approximation quality with the use of the standard equivalent circuit is much better. In the example in Figure 4.17b, there is practically no longer any deviation between measurement and simulation.

The really good starting values for R_S and R_{Sh} are obtained from the curve gradient in the short-circuit and open-circuit points. For this purpose, we will now consider the short-circuit point: here the largest part of I_{Ph} flows to the outside so that the voltage V_D in Figure 4.14b becomes small.

The current I_D over the diode can therefore be ignored. The remaining characteristic curve equation compared with Equation (4.29) is

$$I = I_{Ph} - \frac{V + I \cdot R_S}{R_{Sh}}. \tag{4.32}$$

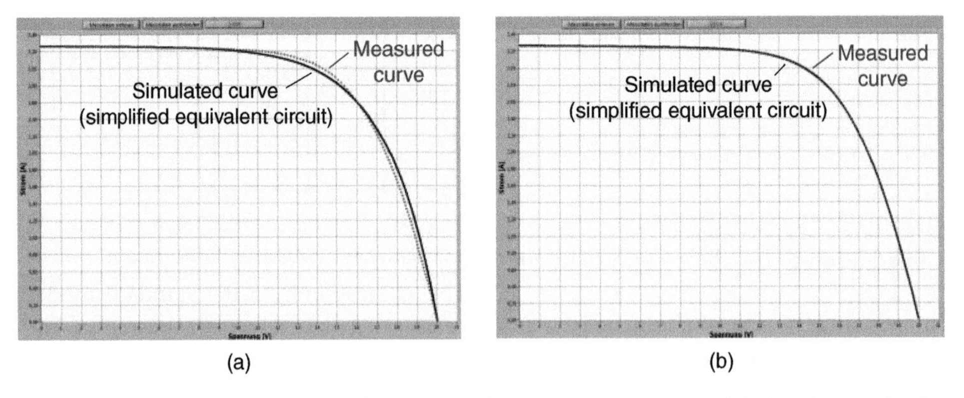

Figure 4.17 Simulation of the curve of a solar module with *PV-Teach:* The simplified equivalent circuit achieves an inadequate agreement with the measured curve, whereas the standard equivalent circuit shows an almost perfect fit.

The slope of the curve is found from the differentiation

$$\frac{\mathrm{d}I}{\mathrm{d}V} = 0 - \frac{1}{R_{\mathrm{Sh}}} - \frac{R_{\mathrm{S}}}{R_{\mathrm{Sh}}} \cdot \frac{\mathrm{d}I}{\mathrm{d}V}. \tag{4.33}$$

Resolving the equation according to $\mathrm{d}I/\mathrm{d}V$ gives

$$\frac{\mathrm{d}I}{\mathrm{d}V} = \frac{1}{R_{\mathrm{S}} + R_{\mathrm{Sh}}}. \tag{4.34}$$

In general, $R_{\mathrm{S}} \ll R_{\mathrm{Sh}}$ applies so that we finally write

$$R_{\mathrm{Sh}} = -\left.\frac{\mathrm{d}V}{\mathrm{d}I}\right|_{V=0}. \tag{4.35}$$

Thus, the shunt resistance R_{Sh} can be determined directly from the slope of the tangent of the short-circuit point (Figure 4.18).

A similar treatment is applied for the open-circuit case: Here the voltage V_{D} is large, but the diode becomes very low resistant so that in Figure 4.14b the current I_{D} can be ignored compared with I_{Sh}. The remaining equation from Equation (4.29) is now

$$I = I_{\mathrm{Ph}} - I_{\mathrm{S}} \cdot \left(\mathrm{e}^{\frac{V+I \cdot R_{\mathrm{S}}}{m \cdot V_{\mathrm{T}}}} - 1\right). \tag{4.36}$$

We differentiate this with respect to the current and rearrange it for $\mathrm{d}V/\mathrm{d}I$. Here we take into account that V is dependent on the current I:

$$\frac{\mathrm{d}(I)}{\mathrm{d}I} = 1 = 0 - I_{\mathrm{S}} \cdot \mathrm{e}^{\frac{V+I \cdot R_{\mathrm{S}}}{m \cdot V_{\mathrm{T}}}} \cdot \frac{1}{m \cdot V_{\mathrm{T}}} \cdot \left(\frac{\mathrm{d}V}{\mathrm{d}I} + R_{\mathrm{S}}\right), \tag{4.37}$$

$$\frac{\mathrm{d}V}{\mathrm{d}I} = -R_{\mathrm{S}} - \frac{m \cdot V_{\mathrm{T}}}{I_{\mathrm{S}}} \cdot \mathrm{e}^{-\frac{V+I \cdot R_{\mathrm{S}}}{m \cdot V_{\mathrm{T}}}}. \tag{4.38}$$

At the open-circuit point, $V = V_{\mathrm{OC}}$ and $I = 0$ apply, hence the equation simplifies to

$$\left.\frac{\mathrm{d}V}{\mathrm{d}I}\right|_{V=V_{\mathrm{OC}}} = R_{\mathrm{S}} + \frac{m \cdot V_{\mathrm{T}}}{I_{\mathrm{S}}} \cdot \mathrm{e}^{-\frac{V_{\mathrm{OC}}}{m \cdot V_{\mathrm{T}}}} = R_{\mathrm{S}}. \tag{4.39}$$

The second term of the sum has been ignored here. This represents the forward resistance of the diode, which at the open-circuit point is typically significantly lower than R_{S}. The series resistance can thus be determined in that the gradient of the open-circuit

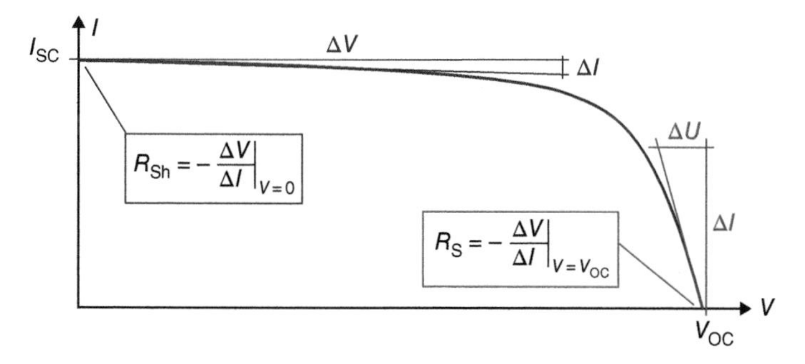

Figure 4.18 Determination of R_{S} and R_{Sh} from the solar cell characteristic curve: The two resistances can be determined from the gradient of the short-circuit or open-circuit point.

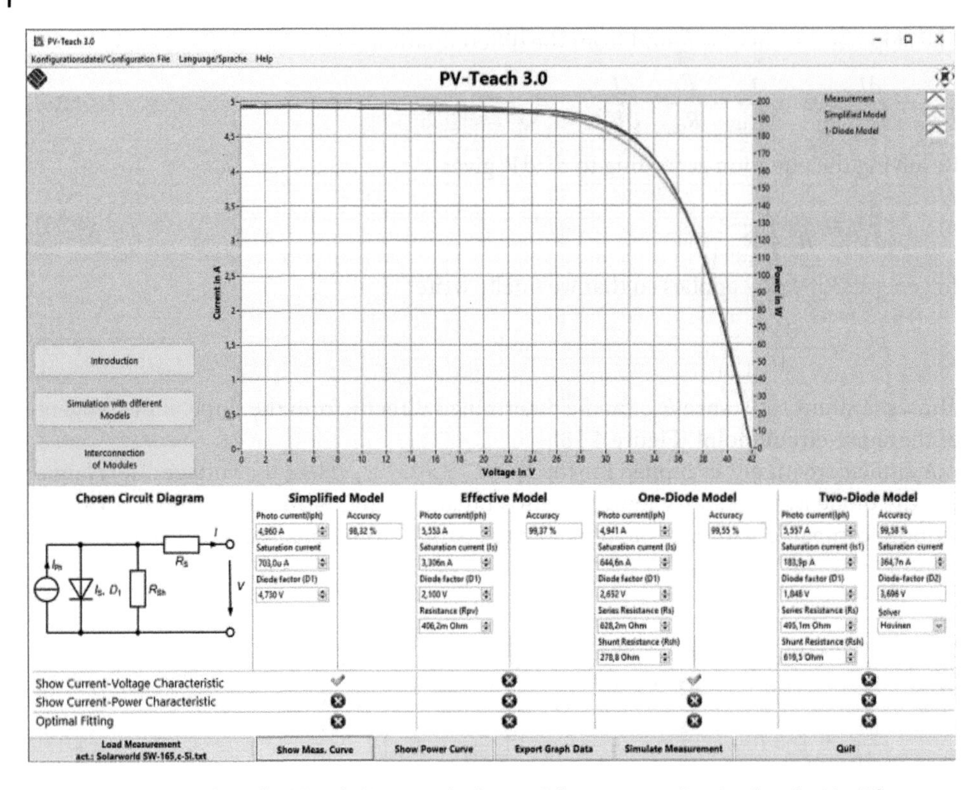

Figure 4.19 Screenshot of PV-Teach: Measured solar module curves can be simulated with different equivalent circuits.

point is measured (see Figure 4.18). In Section 8.2, we will find a somewhat more accurate method of determining R_S.

Instead of the ideality factor, a second saturation current as a second unknown variable will be used in the case of the two-diode equivalent circuit. A practical proposal for determining the parameters of the equivalent circuit is given in [38].

An aid to understanding and comparing the different equivalent circuits is provided by the software *PV-Teach* (see Figure 4.19). It offers the possibility of loading measured I–V curves of solar cells and solar modules and to approximate them with different models. Moreover, different modules can be defined with data sheet values to compare them with the simulated curves. In addition to the manual input of the different equivalent circuit parameters, there exists an option for automatic optimization of the parameters to achieve the best possible approximation between measured and simulated curves.

PV-Teach is offered as a free download at *www.textbook-pv.org*.

4.6 Considering Efficiency

The efficiency of solar cells is a deciding parameter for using solar energy efficiently and economically. We will now, on the one hand, consider what upper limits physics places on the efficiency and, on the other, learn about techniques for approaching these upper limits.

4.6.1 Spectral Efficiency

A fundamental limit of the efficiency of a solar cell is the fact that every semiconductor material has a bandgap ΔW_G. The wavelength at which light is just absorbed is called the bandgap wavelength λ_G:

$$\lambda_G = \frac{h \cdot c}{\Delta W_G}. \tag{4.40}$$

The portion of the solar spectrum that lies above λ_G thus cannot be used for providing electrical energy. We call this portion transmission losses (see Figure 4.20).

On the other hand, the radiation below λ_G represents photon energies that are larger than the bandgap necessary for absorption. This surplus energy is given by impacts on the crystal lattice; we call these thermalization losses.

It is now interesting to find out what electrical energy can theoretically be won from the solar spectrum with a semiconductor of bandgap ΔW_G. First, we will consider the maximum possible current density j_{Max} that an ideal solar cell with a radiation of an AM 1.5 (air mass 1.5) spectrum can generate.

N_{Ph} is the number of photons that can impinge within a time interval Δt on an area A. It can be determined from the spectral irradiance $E_\lambda(\lambda)$ in that we divide the optical energy W_λ of the radiation at a given wavelength by the energy of a single photon W_{Ph} of this wavelength. Then we have to integrate over all wavelengths:

$$N_{Ph} = \int_0^\infty \frac{W_\lambda(\lambda)}{W_{Ph}(\lambda)} \cdot d\lambda = \int_0^\infty \frac{A \cdot E_\lambda(\lambda) \cdot \Delta t}{\frac{h \cdot c}{\lambda}} \cdot d\lambda = \frac{A \cdot \Delta t}{h \cdot c} \int_0^\infty E_\lambda(\lambda) \cdot \lambda \cdot d\lambda. \tag{4.41}$$

We assume as an idealized case that every photon in the cell is absorbed and an electron–hole pair is generated that contributes to the current density. This, however, does not apply to photons whose energy is less than the bandgap so that we only integrate up to the bandgap wavelength λ_G:

$$j_{Max} = \frac{\text{Charge}}{\Delta t \cdot A} = \frac{q \cdot N_{Ph}}{\Delta t \cdot A} = \frac{q}{h \cdot c} \int_0^{\lambda_G} E_\lambda(\lambda) \cdot \lambda \cdot d\lambda. \tag{4.42}$$

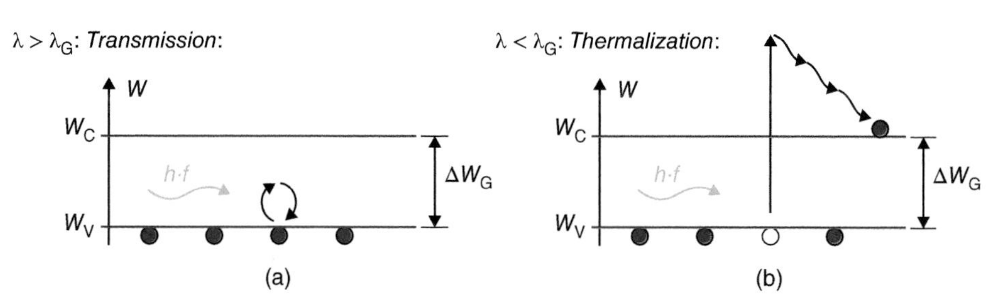

Figure 4.20 Loss mechanisms due to unsuitable energy of the photons: In the case of too little photon energy, the electron cannot be raised to the conduction band. If the energy is too great, then a portion of it is given up to the lattice as heat energy.

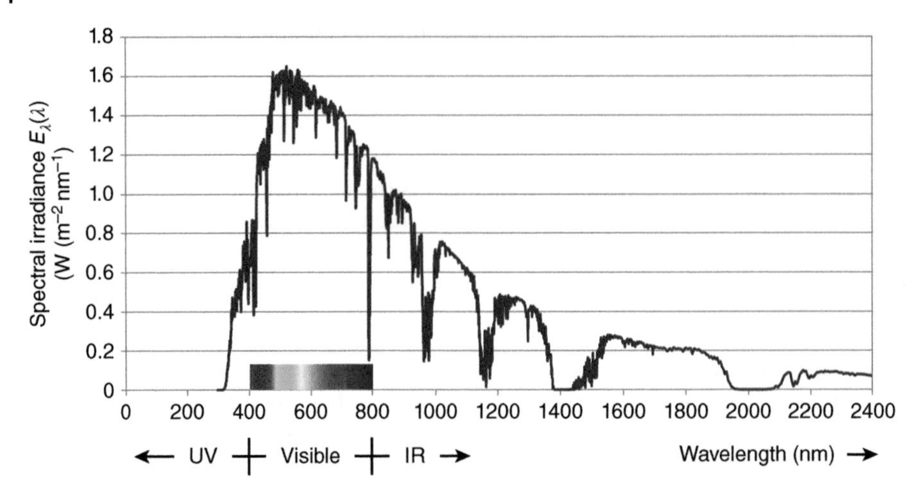

Figure 4.21 Standard spectrum with an irradiance of 1000 W m^{-2} based on STC [39].

As a standard AM 1.5 spectrum, we use a spectrum that is enhanced by a factor of about 1000/850 = 1.1976 with respect to the curve shown in Figure 2.2 (see Figure 4.21). Thus it possesses an overall power density of 1000 W m^{-2} as required for STC conditions.

Figure 4.22 shows a diagram[1] created with Equation (4.42). It shows the maximum possible current density as a function of the bandgap ΔW_G for the two spectra AM 0

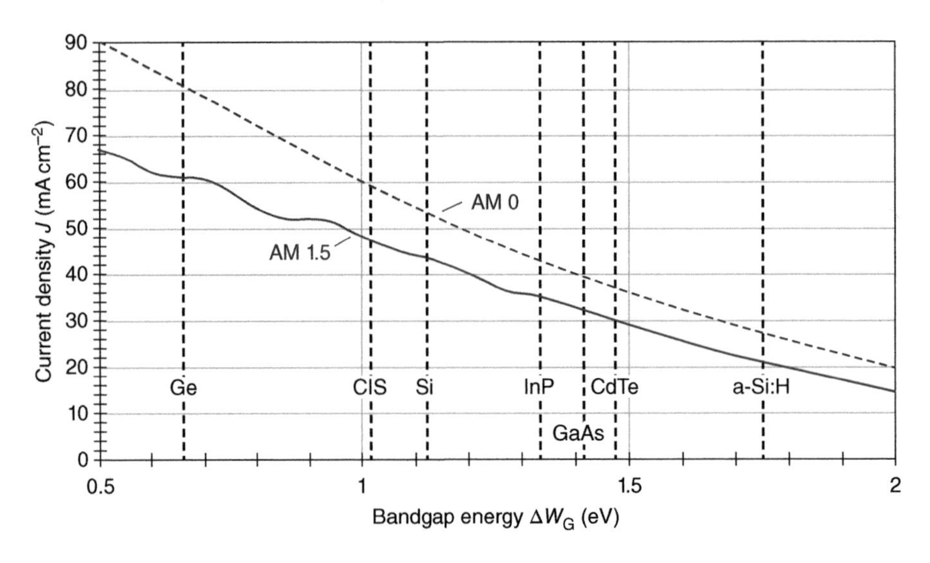

Figure 4.22 Maximum possible current density j_{Max} depending on the bandgap.

1 The calculations for this and the following diagrams were carried out using the finely resolved solar spectra of the American standard ASTM-G173-03 [ASTM]. These correspond to the international standard Norm IEC 60904-3, Edition 2, of 2008.

and AM 1.5. Naturally, j_{Max} increases for semiconductors with small bandgaps as these can also make use of light in the deep infrared region.

It is noticeable that the AM 1.5 curve has kinks and flat parts in some places. This is due to the irregular course of the AM 1.5 spectrum in which whole wavelength regions are filtered out in the atmosphere. For silicon, there is a maximum current density of $j_{Max} = 44.1\ \text{mA cm}^{-2}$ for an AM 1.5 spectrum.

After establishing what the maximum current is, we need to determine the dimension of the maximum possible voltage as a function of the bandgap. We assume that our ideal solar cell manages to give up the full energy of each photon to the outer electric circuit. The maximum possible voltage is then

$$V_{Max} = \Delta W_G / q. \tag{4.43}$$

The maximum electrical power P_{El} of the cell is

$$P_{El} = V_{Max} \cdot I_{Max} = V_{Max} \cdot j_{Max} \cdot A. \tag{4.44}$$

With this, we are in a position to calculate the so-called *spectral efficiency* η_S of the ideal solar cell [40]:

$$\eta_S = \frac{P_{El}}{P_{Opt}} = \frac{V_{Max} \cdot j_{Max}}{E}. \tag{4.45}$$

With Equation (4.42), this results in

$$\eta_S = \frac{\Delta W_G}{E} \cdot \frac{1}{h \cdot c} \cdot \int_0^{\lambda_G} E_\lambda(\lambda) \cdot \lambda \cdot d\lambda. \tag{4.46}$$

Figure 4.23 shows the spectral efficiency for AM 0 and AM 1.5 calculated with this equation. For decreasing values of ΔW_G, we can again see a rise produced by the increasing current density from Figure 4.22. However, below 1 eV this rise is overcompensated

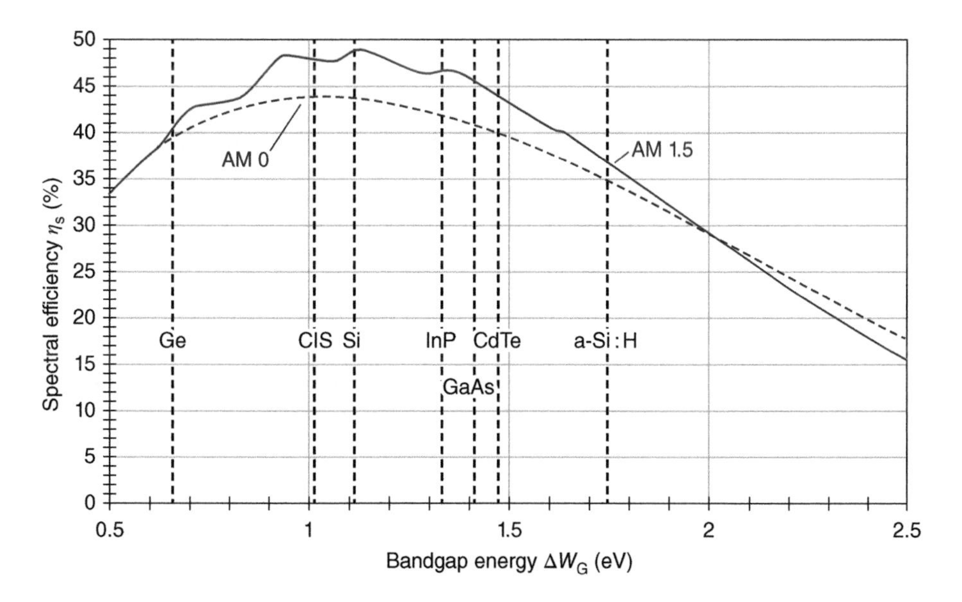

Figure 4.23 Spectral efficiency of the ideal solar cell.

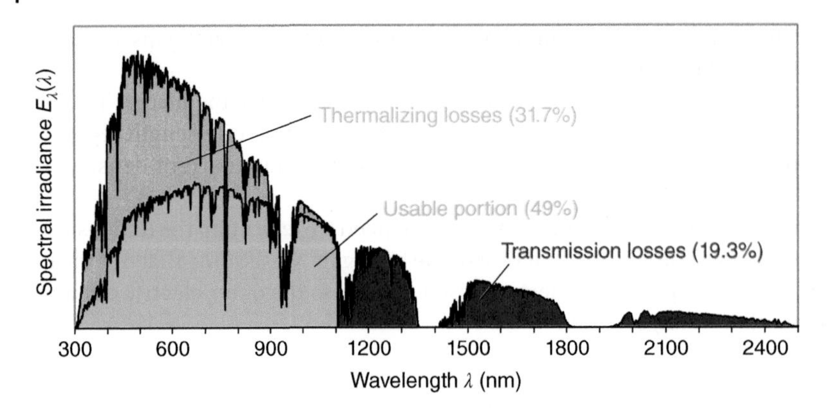

Figure 4.24 Spectral losses in a c-Si solar cell.

by the decreasing voltage $\Delta W_G/q$. We thus have an optimum that reaches a value of almost $\eta_S = 49\%$ for the AM 1.5 spectrum. With a bandgap of 1.12 eV, silicon is placed almost exactly at this optimum.

For the sake of clarity, Figure 4.24 shows the losses due to transmission and thermalization in an ideal Si solar cell. Photons above 1120 nm have too little energy to be absorbed. Because the AM spectrum above this wavelength has an irradiance of 193 W m^{-2}, there are transmission losses of 19.3%. In the short-wavelength region, things are different: Here, only a maximum energy in the amount of the bandgap can be used by the energy-rich photons. The calculation shows losses due to thermalization of 31.7%. The sum of both types of losses is 51%, hence a maximum of 49% of the solar radiation can be used. This corresponds exactly to the previously calculated spectral efficiency.

4.6.2 Theoretical Efficiency

There are two things that have yet to be considered in the discussion on efficiency:

1. In a real solar cell, it is not possible to use the full voltage $V_{Max} = \Delta W_G/q$.
2. Because of a FF, that is, <100%, the current I_{MPP} is smaller than I_{SC} and the voltage V_{MPP} is smaller than V_{OC} (see Figure 4.12).

Both limitations refer to the fact that a real solar cell also has a p–n junction. All other properties of the cell are meant to remain ideal (especially, each incident photon with $W_{Ph} > \Delta W_G$ is absorbed and contributes to the photocurrent). Under these conditions, we define the theoretical efficiency η_T as [40]

$$\eta_T = \frac{P_{MPP}}{E \cdot A}. \tag{4.47}$$

With Equation (4.17), this results in

$$\eta_T = \frac{FF \cdot V_{OC} \cdot I_{SC}}{E \cdot A} = FF \cdot \frac{V_{OC}}{V_{Max}} \cdot \frac{V_{Max}}{E} \cdot \frac{I_{SC}}{A} = FF \cdot \frac{V_{OC}}{V_{Max}} \cdot \frac{V_{Max}}{E} \cdot j_{Max}. \tag{4.48}$$

With the use of Equation (4.45), we can directly determine a connection with the already calculated spectral efficiency η_S.

$$\eta_T = FF \cdot \frac{V_{OC}}{V_{Max}} \cdot \eta_S. \tag{4.49}$$

We, therefore, require a high open-circuit voltage and a large FF. By way of Equation (4.16), the open-circuit voltage depends on the saturation current I_S; the smaller this is, the higher is the achievable open-circuit voltage. The saturation current again depends on the bandgap. This can easily be proved when we consider Equation (4.3): The determining dimension is the intrinsic carrier concentration n_i, which, according to Equation (3.3), is determined exponentially on the bandgap. The result is the dependence

$$j_S = K_S \cdot e^{-\frac{\Delta W_G}{k \cdot T}}, \tag{4.50}$$

where j_S is the saturation current density.

In the literature, for the lower limit of the constant K_S a value of 40 000 A cm^{-2} is given [40, 41].

Figure 4.25 shows the theoretical efficiency calculated with Equations (4.18), (4.49), and (4.50) as a function of the bandgap energy. A noticeable deterioration of efficiency can be seen in comparison with Figure 4.23. A maximum value of $\eta_T = 30.02\%$ is given for an energy of 1.38 eV; thus InP and GaAs are very near to optimum. For silicon, there is still a good value of 28.6%.

The theoretical efficiency of 28.6% is the upper limit of the achievable efficiency of a cell of crystalline silicon (assumption: only one p–n junction).

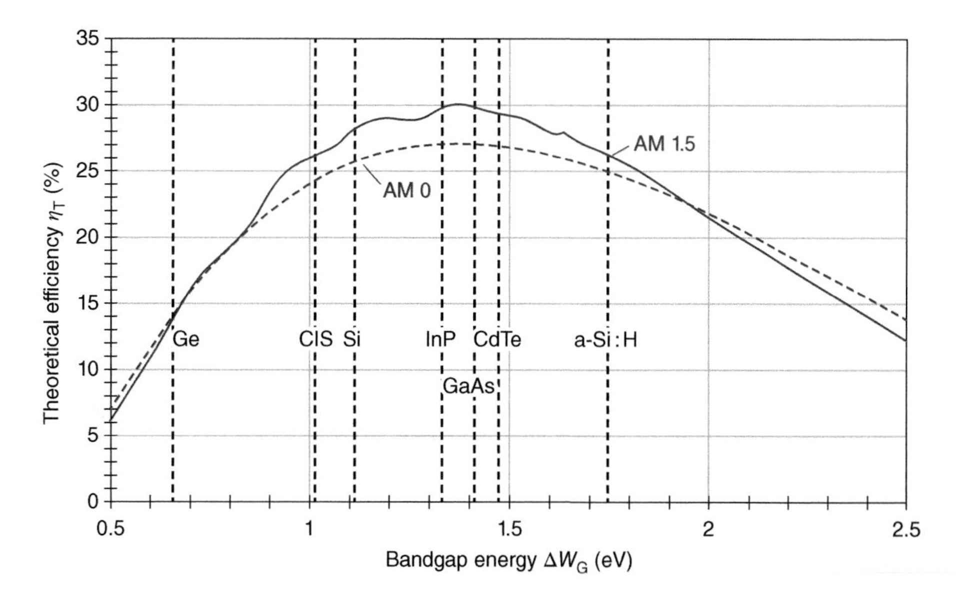

Figure 4.25 Theoretical efficiency as a function of the bandgap energy.

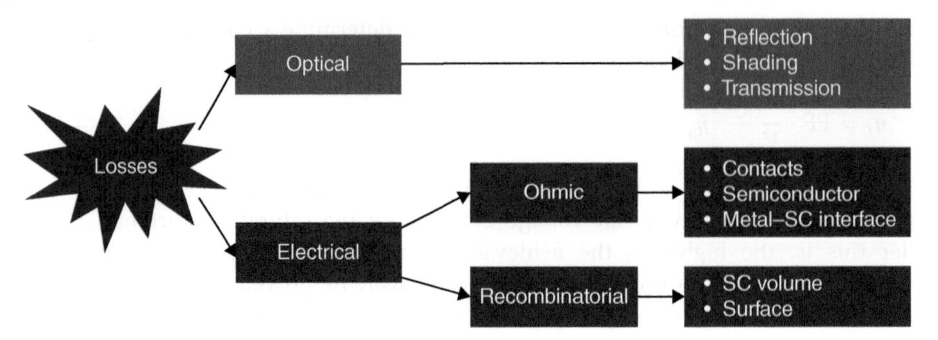

Figure 4.26 Types of losses in a solar cell.

4.6.3 Losses in Real Solar Cells

After having determined the theoretical limits of the maximum achievable efficiency discussed in the previous section, we will now find out how to come as near as possible to these limits. For this purpose, we will first consider the optical and electrical losses that occur. An overview of this is shown in Figure 4.26.

4.6.3.1 Optical Losses, Reflection on the Surface

As we have already seen in Chapter 3, the refractive index step of air on silicon causes a reflection of approximately 35%. Help is obtained with an anti-reflective coating that lowers the average reflection of an AM 1.5 spectrum to around 10%. A further measure is texturing of the cell surface: The surface is etched with an acid in order to roughen it. In the case of monocrystalline silicon, with anisotropic etching processes (e.g. with potassium hydroxide, KOH) pyramid structures can also be fabricated. The results are pyramids with an angle at the top of 70.5° (Figure 4.27).

What does this texturization yield? Figure 4.27 shows how incident rays partly penetrate the cell and are partly reflected. According to the Fresnel equations, the strength of the reflection factor R can be determined from the angle of incidence α_1 [30]:

$$R(\alpha) = \left(\frac{\sin(\alpha_1 - \alpha_2)}{\sin(\alpha_1 + \alpha_2)}\right)^2, \tag{4.51}$$

where the exit angle α_2 can be determined by the *law of refraction*:

$$n_1 \cdot \sin\ \alpha_1 = n_2 \cdot \sin\ \alpha_2. \tag{4.52}$$

However, the reflected ray is not lost; it impinges on the surface of the cell where once again part of the light is reflected. The light is, therefore, given a "second chance" in a way.

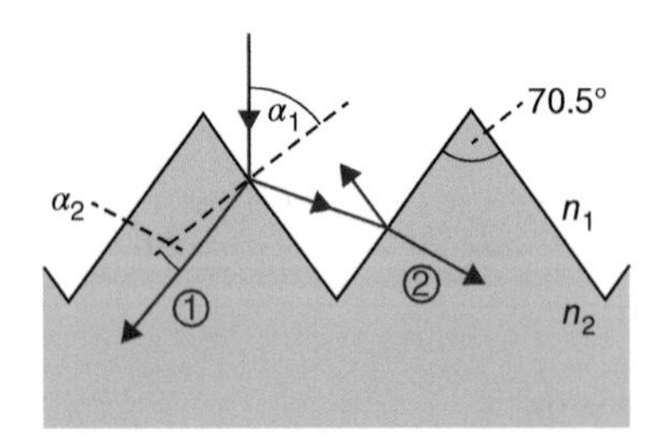

Figure 4.27 Reduction of overall reflection by means of texturing: Giving light "a second chance."

Table 4.1 Remaining reflection losses for antireflection layer and texturization [24, 36].

Antireflection coating	Texturization	Average reflection factor (%)
No	No	30
Yes	No	10
No	Yes	10
Yes	Yes	3

Overall, more light penetrates the cell, and in the case shown, the improvement is more than 20% compared with the simple vertical incidence. As a result, the short-circuit current is increased and thus the efficiency of the cell. Table 4.1 shows the remaining reflection factor losses for the combination of anti-reflection layer and texturization.

Shading by Means of Contact Finger The current generated by the solar cell must be led via the contact fingers to the connection wires. The cross-section must not be too small in order to ensure that they possess a low ohmic resistance.

The shading losses increase with greater finger width; the usual widths are 100–200 μm. Broader strips serve as current collectors, the so-called *busbars*. These are tapered at the ends, because that is where the current density is at its lowest (see Figure 4.28).

A further possibility for optimization is to make the contact fingers narrow and high instead of broad and flat. The contacts are "buried" in the semiconductor material so as not to create additional shadows in case of an oblique incidence. To accommodate these buried contacts, small grooves are first cut in the cell surface by means of lasers and these are then filled with an Ni–Cu mixture (Figure 4.29). The shading losses can be reduced by approximately 30% [35].

Losses Through Transmission Long-wavelength light possesses little absorption. Thus, for instance, the penetration depth of light with a wavelength of 1000 nm already lies in the

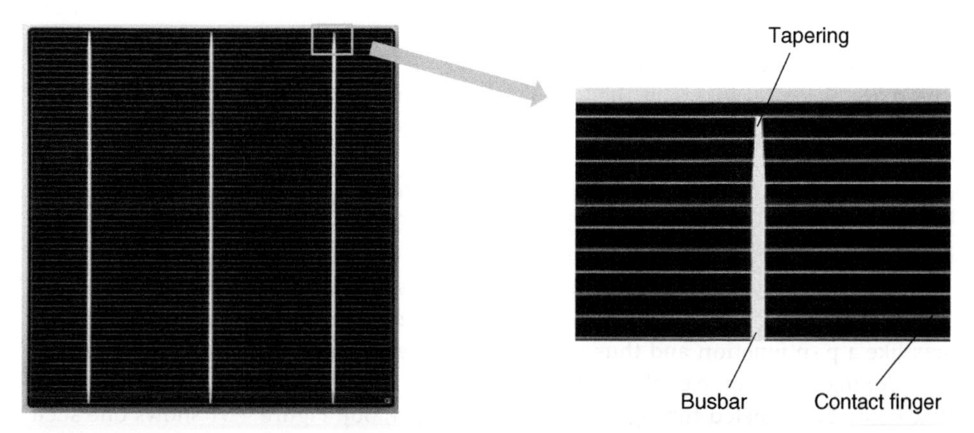

Figure 4.28 Front contacts of a solar cell with contact fingers and current collector rails (busbars). Source: *Q*-cells.

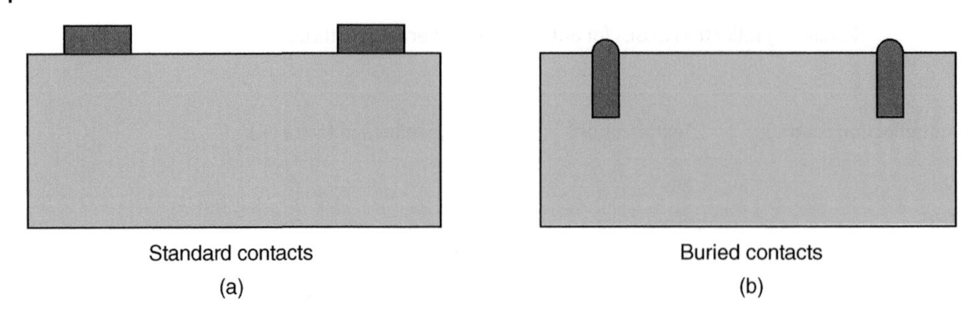

Figure 4.29 Comparison of standard contacts with the buried contact technology: The shading losses can be significantly reduced.

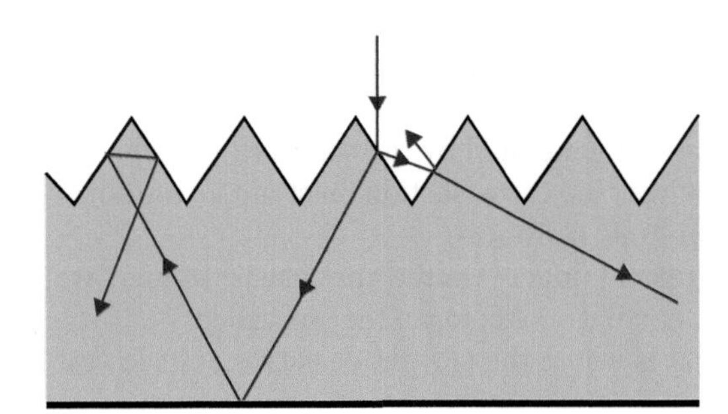

Figure 4.30 Depiction of light trapping: The obliquely refracted rays are reflected on the rear side and thus travel a longer path through the semiconductor.

range of the currently normal cell thickness of 150–200 μm. Without further measures, this leads to transmission losses.

A further improvement can be achieved by means of the texturization shown in Figure 4.27. As can be seen from Figure 4.30, the vertical incident light rays are refracted in an oblique path to the bottom and thus travel a longer way through the cell. An additional improvement would be to use a reflecting material at the bottom of the cell for which the normal aluminum bottom contact layer is well suited as it provides a reflection factor of more than 80% [42].

4.6.3.2 Electrical Losses and Ohmic Losses

There are electrical losses in the contact fingers on the top side of the cell. Narrow and high contacts (in the optimum case as buried contacts) help in this. In addition, ohmic losses can occur in the semiconductor material as the conductivity of the doping material is limited. High currents must be led to the front contacts, especially with thin emitters. An increase in the n-doping brings an improvement but also leads to stronger recombination in the doped area. Finally, there are also losses at the metal–semiconductor junction. The reason for this is that in bringing the metal and the semiconductor together, a potential step is caused (so-called *Schottky contact*). This acts like a p–n junction and thus reduces the achievable cell voltage. Here extremely high doping is of help (e.g. $N_D = 10^{20}$ cm^{-3}) leaving such a narrow space charge region that it can be tunneled through by the electrons [24, 36]. Figure 4.31 shows one such structure. The n^{++}-doping is only applied in the immediate surroundings of the metal contact in order to reduce recombination losses.

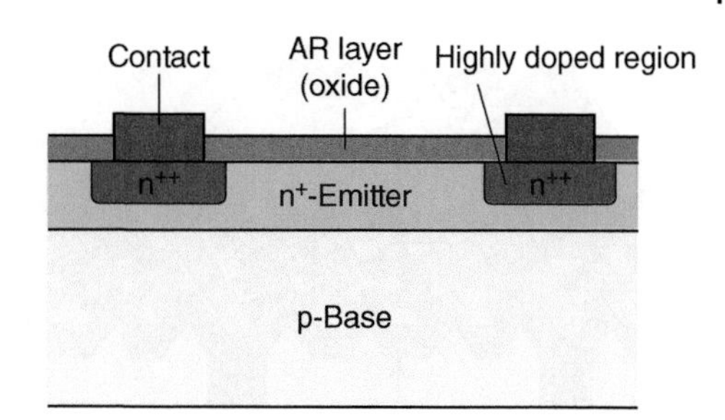

Figure 4.31 Structure for preventing a Schottky contact: The high doping permits tunneling of the electrons from the semiconductor to the metal contact.

Recombination Losses The various reasons for recombination of generated charge carriers in the semiconductor volume have already been discussed in Section 4.2.2. Added to this in real cells are recombinations at surfaces that are created by the open bonds at the border of the crystal lattice. A means of reducing the recombination at the bottom is the BSF discussed in Section 4.2.4. At the top, one attempts to cover the largest possible area with an oxide that saturates the open bonds and thus passifies them. As is shown in Figure 4.31, the antireflection layer (e.g. of Si_3N_4) is used for this. At the same time, the n^{++}–n^{+}-layer at the front contact leads to a "front surface field" that keeps the holes away from the contacts.

4.7 High-efficiency Cells

In the following, we will consider some examples of current high-efficiency cells.

4.7.1 Buried-contact Cell

The best-known high-efficiency cell is the buried-contact cell that was developed in the 1980s by Professor Martin Green at the University of New South Wales (UNSW) in Australia. Martin Green is a true luminary in the field of cell development and has repeatedly achieved world records for efficiency. Figure 4.32 shows the structure of the buried-contact cell. The pits cut by a laser in which the front contacts have been inserted are clearly visible. Around the contacts the n^{++}-regions for preventing the Schottky contact can be seen. At the bottom is the p^{+}-layer for forming the BSF. A feature is that the cell was also texturized at the bottom. The light rays arriving there are thus reflected back and travel a particularly long way through the cell (light trapping).

The cell concept was developed by BP Solar under license and subsequently released for mass production under the name Saturn Cell. This possesses an efficiency of 17.5% on a surface area of 150 cm^2 [43].

4.7.2 Point-contact Cell (IBC Cell)

Figure 4.33 shows a point-contact cell developed at Stanford University.

A noticeable feature is that both the negative and positive contacts are positioned on the rear side of the cell, hence no shading occurs. This is possible as good-quality silicon is used. Thus, the diffusion length is so long that almost all charge carriers find their way to the rear side without recombining. As the whole of the front side is passivated, there is hardly any recombination there.

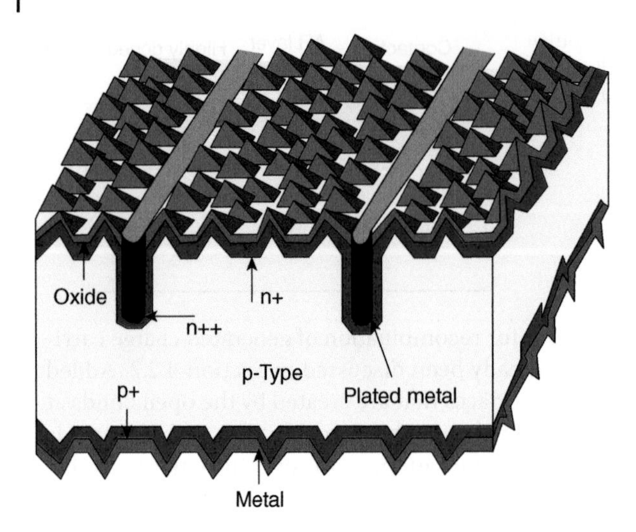

Figure 4.32 Buried-contact cell (more accurately: Laser grooved buried contact: LGBC cell).Source: Reprinted with kind permission of Martin Green.

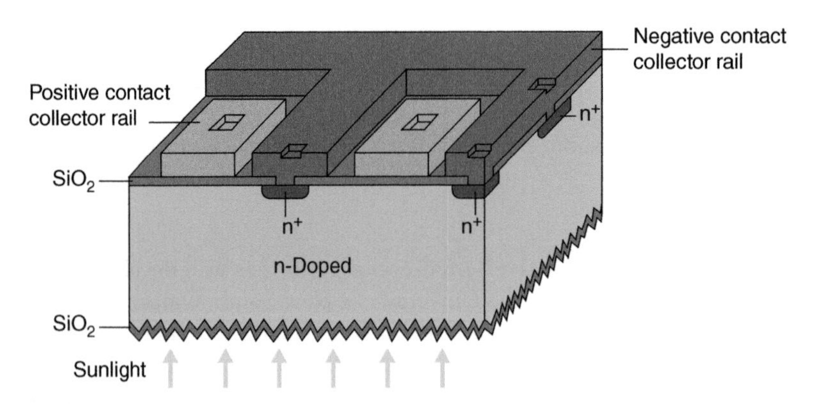

Figure 4.33 View of the point-contact cell: All contacts are positioned on the rear side of the cell and can thus be made as thick as desired [44].

A special trick is used on the rear side: An oxide layer is also inserted between the contacts and the silicon. This is drilled through by laser only at certain points and then a local emitter is diffused in. The surface recombination is, therefore, reduced to a minimum.

A similar trick can be used for the plus contacts. Here a p^+ island is diffused through the drilling hole (not to be seen in Figure 4.33). Between the respective n^+ and p^+ regions, a space charge region is established, which guarantees charge separation of the produced electron–hole pairs.

This technology has been marketed for several years by SunPower Corporation, first under the name A-300 and now as an improved version called the Maxeon cell. The cells with a dimension of 13 × 13 cm achieve an efficiency of 24%!

In addition to the term "point-contact cell" for back-contact cells in general, the name IBC-cell (interdigitated back contact) has also been established, as the plus and minus contacts mutually mesh into each other.

4.7.3 PERL and PERC Cell

The current efficiency champion again comes from the UNSW in Australia. As shown in Figure 4.34, the passivated emitter rear locally diffused (PERL) cell also uses the principle of the rear-side point contacts. The front side features regular texturization in the form of inverted pyramids – a very effective measure of good light trapping. The upper passivating layer is made up of two layers of silicon oxide and silicon nitride and acts as an anti-reflective double layer.

The cell achieves a short-circuit current of 42 mA cm^{-2} and an open-circuit voltage of 714 mV. With the FF of 83%, this results in a record efficiency of 25% [45, 46]. However, this cell is only a laboratory cell, with an area of merely 2 cm^2.

Approximately 100 process steps were required for the manufacture of the record cell, which is an unacceptable effort for industrial production. Suntech produces a much simplified cell on the principle of the PERL concept under the name "Pluto." The upper side is structured similarly to the cell in Figure 4.34. This leads to a remarkably small reflectance factor of only 1%, although only a single antireflection layer is used. At first, the cells were produced with a "normal" backside (totally covered with aluminum with BSF). In that format, efficiency at least of 19% was achieved. Since then, further development of the PERL concept (backside passivation with local through-contacts) with cells with an area of 155 cm^2 has led to efficiencies of more than 20% [46, 47].

Nowadays, numerous other companies have developed a related concept with the name passivated emitter and rear cell (PERC). In this case, instead of the local point-shaped contacts, stripe-shaped contacts are applied (see Figure 4.35). After the production of grooves in the backside passivating layers with a laser, aluminum paste can be applied on all of the backside so that the grooves are also filled. With the subsequent "firing" of the cell, a local BSF is established (see Section 5.1.3). In addition to the passivation, the SiO_x/SiN_x layers also offer relatively good reflection of the infrared light arriving at the bottom so that it gains a "second chance" of being absorbed.

The German company SolarWorld has already achieved cell efficiencies of 21% and declares that efficiencies of up to 24% will be possible in the future.

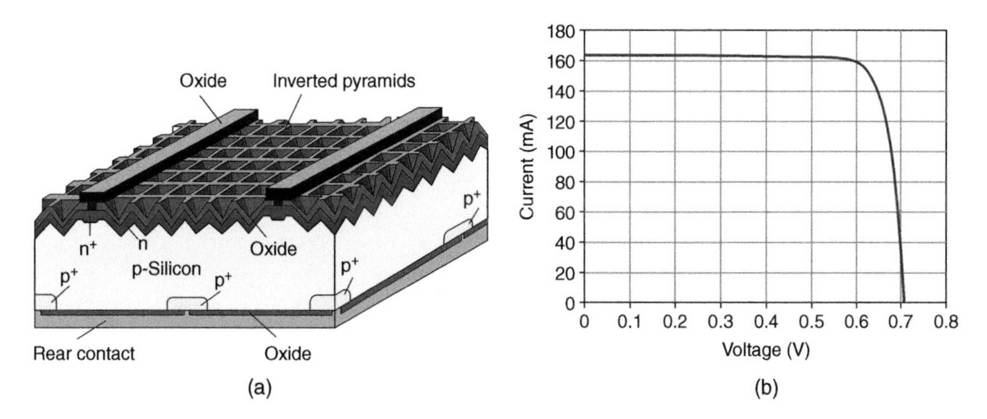

Figure 4.34 PERL cell together with *I*/*V* characteristic curve. Source: After [35]. Reprinted with kind permission of Martin Green.

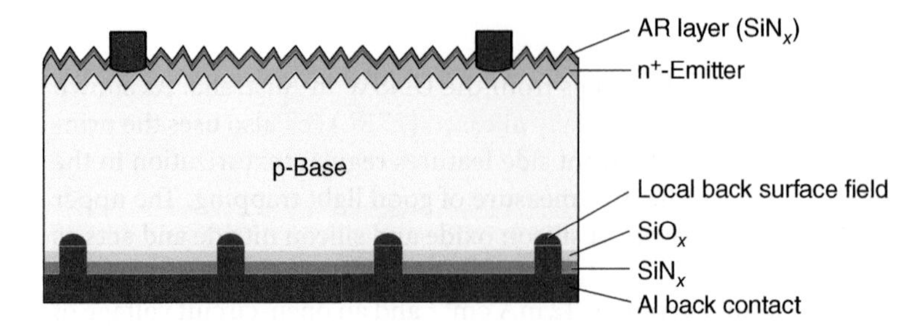

Figure 4.35 Construction of the PERC cell: As the back contact only reaches the base at a few spots, the result is only minor surface recombination and thus a high efficiency of the cell. Source: After [48].

Finally, the spectral sensitivity of a high-efficiency cell should be discussed (similarly to a PERL cell) (dashed line in Figure 4.9). In a wide wavelength region, it achieves almost the ideal value that corresponds to an external quantum efficiency of 100%. In the short-wavelength region, this is much better than the standard cell as there is no dead layer due to the surface passivation and very small n^{++} areas. Above 800 nm, it works especially effectively as the infrared light travels through the cell several times by means of light trapping and is thus also absorbed.

5

Cell Technologies

Now that we have become experts in the method of operation of solar cells, we will look in more detail at how solar cells are produced out of crystalline silicon. Then we will make the acquaintance of cells of alternative materials such as amorphous silicon or gallium arsenide. Finally, ecological aspects of individual technologies will be handled.

5.1 Production of Crystalline Silicon Cells

The workhorse of photovoltaics is the silicon solar cell. For this reason, we will deal with its production from sand via the silicon, the wafer, and cell processing up to the finished solar module in detail.

5.1.1 From Sand to Silicon

The first step is the conversion of quartz sand into high-grade silicon for the production of wafers.

5.1.1.1 Production of Polysilicon

The starting point of the solar cell is silicon (from the Latin *silicia*: gravel earth). After oxygen, it is the second most common element on Earth. However, it almost never comes in its pure form in nature, but mostly in the form of silicon oxide (quartz sand). Therefore, in the truest sense, silicon is like the sand at the beach.

First, the silicon is reduced in an electric arc furnace with the addition of coal and electrical energy at approximately 1800 °C (Figure 5.1):

$$SiO_2 + 2C \rightarrow Si + 2CO.$$

Thus, we obtain metallurgical silicon (Metallurgical Grade: MG-SI) with a purity of approximately 98%. The designation is because this type of silicon is also used in the production of steel. For use in solar cells, the MG-Si must still go through complex cleaning. In the so-called silane process, the finely ground silicon is mixed in a fluidized bed reactor with hydrochloric acid (hydrogen chloride, HCl). In an exothermic reaction this results in trichlorosilane ($SiHCl_3$) and hydrogen.

$$Si + 3HCl \rightarrow SiHCl_3 + Si + H_2.$$

Photovoltaics – Fundamentals, Technology, and Practice, Second Edition. Konrad Mertens.

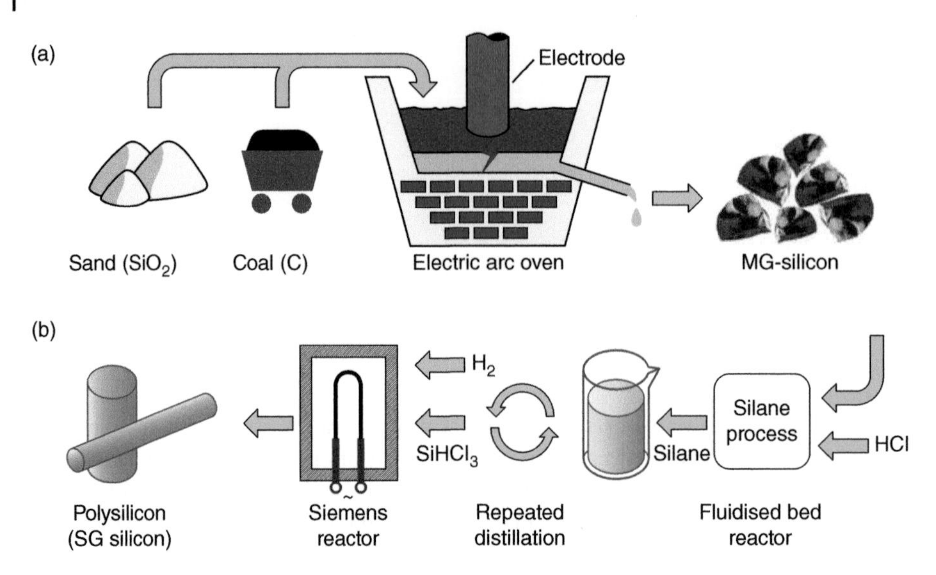

Figure 5.1 Production of polysilicon from quartz sand. (a) Production of metallurgical silicon. (b) Processing to highly purified polysilicon (Solar Grade).

Now the trichlorosilane can be further cleaned by means of repeated distillation. Fortunately, the boiling point is only at 31.8 °C. The reclamation of the silicon takes place in a reactor (Siemens reactor) in which the gaseous trichlorosilane with hydrogen is fed past a 1350 °C hot thin silicon rod. The silicon separates out at the rod as highly purified polysilicon. This results in rods, for instance, of length 2 m with a diameter of approximately 30 cm (Figure 5.1).

The polysilicon should have a purity of at least 99.999% (5 nines, designation 5N) in order to be called Solar Grade Silicon (SG-Si). However, for a normal semiconductor technology used in the production of computer chips, and so on, this degree of purity would be insufficient; here a purity of 99.9999999% (9N, Electronic Grade: EG-Si) is normal.

As the Siemens process is relatively energy-intensive, the search has been on for years for alternatives to cleaning silicon. One possibility is the use of fluidized bed reactors (FBR) in which the purest silicon is continually separated. This is achieved by blowing small dust-shaped silicon seed crystals into the reactor instead of using seed rods. These then grow with the help of trichlorosilane and hydrogen to small silicon spheres. The FBR have higher production rates and a 70% lesser energy usage than the Siemens reactor [49]. However, processing is difficult and requires much skill and experience.

Very promising is the production of directly purified silicon (Upgraded Metallurgical Grade: UMG-Si), for which there are now different versions. For instance, the process of the 6N-Silicon Company consists in that MG-silicon is melted in liquid aluminum. This is already possible in practice at 800 °C in contrast to the much higher "normal" melting point of silicon at 1414 °C. Then one allows the melt to cool so that silicon crystals are formed. The foreign atoms such as boron or phosphorous are taken up by the aluminum. However, after cooling, the silicon is contaminated with aluminum that must be extracted in further steps. The production of UMG silicon requires only approximately half the energy effort of the Siemens process. However, the desired purity does not yet approach that attained by the Siemens process [50].

The structure of the polycrystalline silicon is too poor for it to be used directly in solar cells. Therefore, in the following, we will consider the two most important processes for improving the properties of the crystal.

5.1.1.2 Production of Monocrystalline Silicon

The *Czochralski process* (CZ process) is the process that has been used most for production of monocrystalline silicon (see also Section 1.6.1). For this purpose, pieces of polysilicon are melted in a crucible at 1450 °C, and a seed crystal, fixed to a metal rod, is dipped into the melt from above. Then, with light rotation, it is slowly withdrawn upward whereby fluid silicon attaches to it and crystallizes (Figure 5.2). Thus, eventually, a monocrystalline silicon rod (ingot) is formed, whose thickness can be adjusted by the variation of temperature and withdrawal speed. Rods with a diameter of up to 30 cm and a length of up to 2 m can be produced with this method. For photovoltaics, the diameter is typically 5–6 in (12.5–15 cm).

If the crystal quality still needs to be improved, then the Float-Zone (FZ or Zone Melt) process can be used instead of the CZ process. Here, a seed crystal is placed under a vertical hanging polysilicon rod (Figure 5.3). Then an induction coil is slowly pushed from below upward over the rod. In this, the silicon is only melted in the induction zone so that the monocrystal forms from the bottom upward. Because any existing contaminants are driven upward with the melt in the crystallization, the zone process achieves

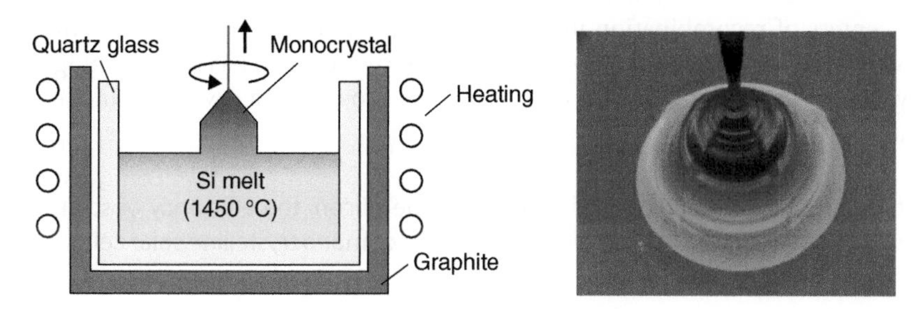

Figure 5.2 Production of monocrystalline silicon rods by means of the Czochralski process. Photo source: PVA Crystal Growing Systems GmbH.

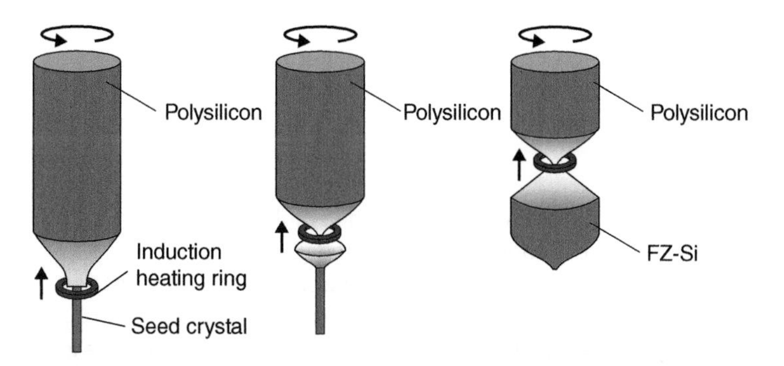

Figure 5.3 Principle of the float-zone process: The upward moving heating ring melts the polysilicon only locally so that impurities are driven upward during crystallization.

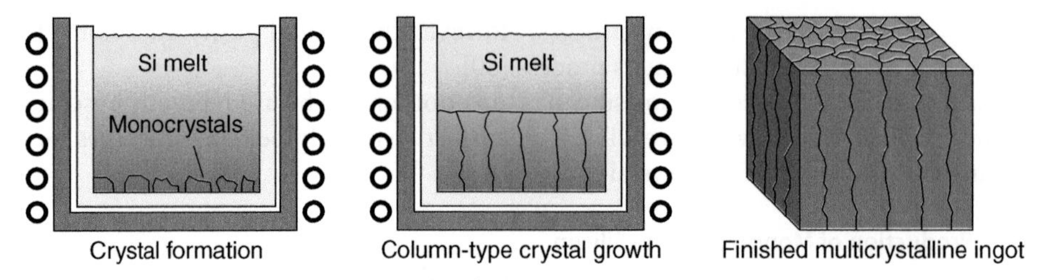

Figure 5.4 Production of multicrystalline ingots.

a particularly high crystal quality. However, FZ-Si is significantly more expensive than CZ-Si, and for this reason, it is used for photovoltaics only in exceptional circumstances. The PERL world record cell mentioned in Section 4.7, for instance, was established with FZ silicon.

5.1.1.3 Production of Multicrystalline Silicon

The production of multicrystalline silicon is much simpler. Figure 5.4 shows the principle: Pieces of polysilicon are poured into a graphite crucible and brought to a melt, for instance, using induction heating. Then the crucible is allowed to cool from the bottom by the heating ring slowly pulled upward. At various places on the bottom of the crucible, small monocrystals are formed that grow sideways until they touch each other. With the vertical cooling process, the crystals grow upward in a column (columnar growth). At the boundary layers, crystal displacements are formed that later become centers of recombination in the cell. For this reason, one tries to let the monocrystals become as large as possible. The column structure also has the advantage that minority carriers generated by light do not have to cross over a crystal boundary in the vertical direction.

Because of the poorer material quality of multicrystalline silicon, the efficiency of solar cells made from this material is typically 2–3% below that of monocrystalline solar cells.

After the crystallization of the whole melt, the silicon block (ingot) is divided into cubes (bricks) of 5 or 6 in along the edges.

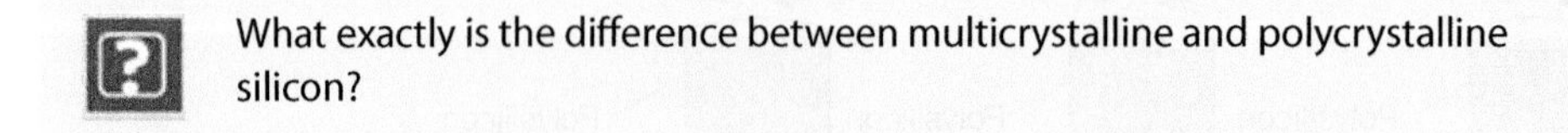

What exactly is the difference between multicrystalline and polycrystalline silicon?

Polycrystalline silicon is of poorer crystal quality than multicrystalline material; the diameter of the monocrystals contained is in the micro to millimeter region. With multicrystalline material, one speaks of monocrystals in the order of millimeters to 10 cm [51]. If the monocrystals are larger than 10 cm, then monocrystalline silicon is present. However, this clear difference is not always reflected in the literature.

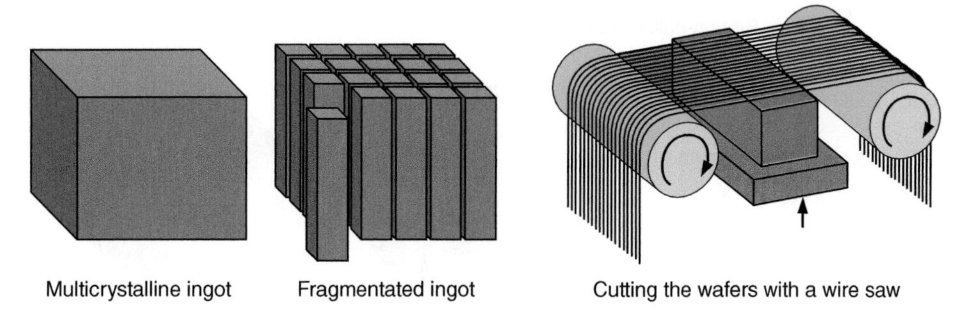

Figure 5.5 Production of multicrystalline wafers: After the fragmentation of the ingots into individual bricks, they are cut into wafers with the wire saw.

5.1.2 From Silicon to Wafer

5.1.2.1 Wafer Production

After production, the ingots must be sawed into individual sheets (wafers). This is mostly done with wire saws that remind one of an egg cutter (Figure 5.5). A wire with the thickness of 100–140 μm moves at high speed through a paste (slurry) of glycol and extremely hard silicon-carbide particles and carries these with it into the saw gap of the silicon. This is more a grinding or lapping process rather than sawing. The saw gap is at least 120 μm.

Unfortunately, the silicon chips cannot be recycled with sufficient purity. With the current wafer thicknesses of 180 μm, there are saw losses (often called kerf losses) that are almost as large as the used parts. First, producers use saw wires encrusted with diamond particles in order to refrain from the use of silicon-carbide particles. In this case, it should be possible to clean the silicon chips and to use them for new wafers [52].

5.1.2.2 Wafers from Ribbon Silicon

If pulling the wafers directly from the melt is successful, then the saw losses can be completely prevented. This is the idea of the ribbon silicon. In the so-called Edge-Defined-Film-Fed-Growth process (EFG process) from the Schott Solar Company, a former of graphite is dipped into the silicon melt (Figure 5.6). The fluid silicon rises in the gap by means of capillary force and can be "docked" on a longitudinal seed crystal and pulled upward as a thin sheet.

In the real process, an octagon-shaped gap is used and eight-cornered tubes of 6 m with a diameter of 30 cm and silver-thin wall thickness of only 300 μm are pulled. The production of a tube takes about 5 h. These tubes are then cut into individual wafers of 12.5 cm by means of lasers.

Although the EFG process with the elimination of the saw losses promises clear advantages, the Schott Solar decided in 2009 to shut down the production of EFG wafers. Apparently, the company could not follow the industry trend to larger and thinner wafers. Added to this was a relatively slow pulling speed in order to achieve a sufficient crystal quality.

A second process for producing silicon film has been developed by the Evergreen Solar Company under the name String-Ribbon that resembles the principle of the soap bubble. Two parallel heated wires are pulled upward through the Si melt. In doing so,

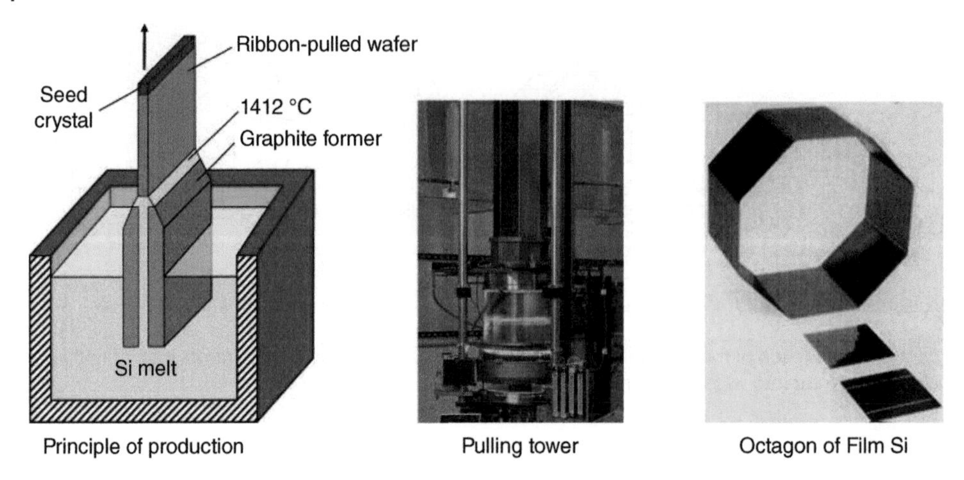

Figure 5.6 Production of wafers according to the EFG process: The wafers are pulled without sawing directly from the melt. Photos – Schott Solar AG.

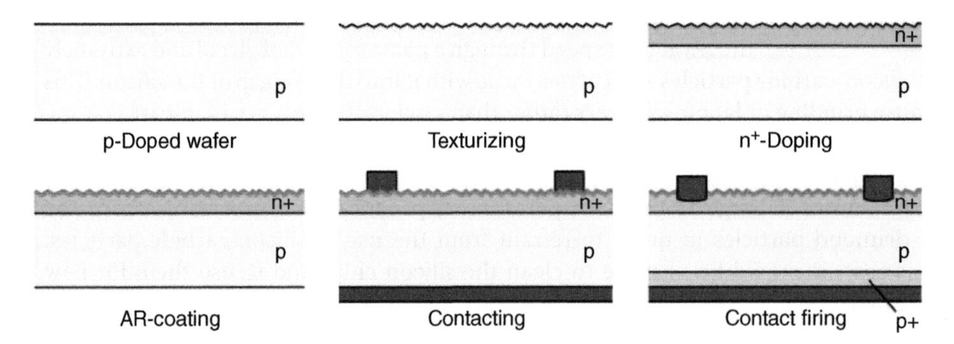

Figure 5.7 Process steps for producing standard cells.

a "soap skin" of silicon forms between the two and hardens as multicrystalline silicon in the air. Evergreen states that their process has the lowest use of energy of all wafer production processes.

5.1.3 Production of Standard Solar Cells

In the following, we will consider the typical steps for producing a modern silicon cell (see Figure 5.7). First, the already-doped wafers are dipped into an etching bath in order to remove the contaminants or the crystal damage on the surface (damage-etching). Then follows texturizing of the surface (e.g. by means of etching with potassium solvent). The formation of the p–n junction is then achieved by the formation of the n^+-emitter by means of phosphorous diffusion. This is a relatively energy-intensive process as temperatures of 800–900 °C are required. In the next step, the deposition of the antireflection coating of silicon nitride (Si_3N_4) is carried out, which causes a passivation of the surface at the same time.

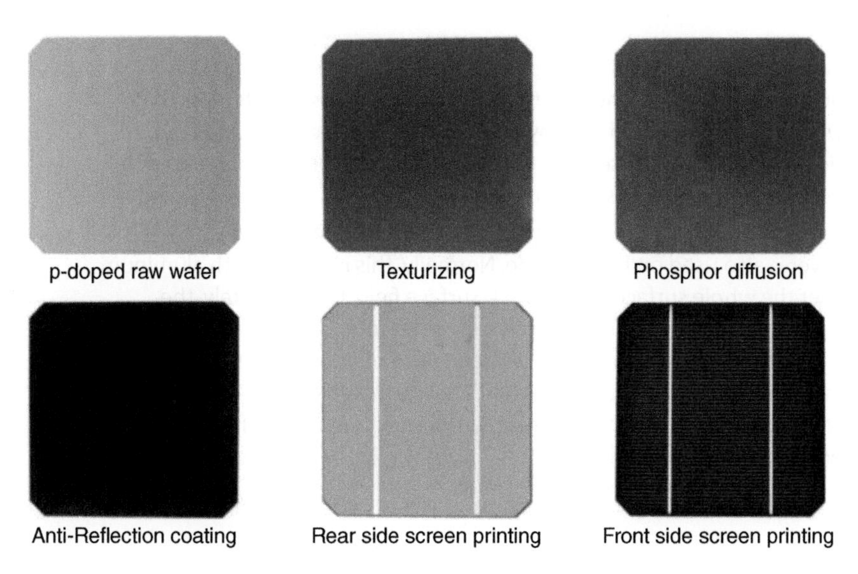

Figure 5.8 View of a monocrystalline cell after the respective production steps.

The application of the contacts occurs in the screen printing process. For this purpose, a mask with slits is placed on the cell, and metal paste is brushed on. In this way, it is placed on the wafer only at particular positions. The formation of the rear-side contacts occurs in two steps. First, the soldering contact surfaces of silver paste are applied in order later to solder the connection wires to them. Then the rest of the rear side is fully covered with aluminum. The front-side contacts are applied in the next step. Here, too a silver paste is used in order to achieve a low series resistance.

The subsequent contact firing of the cells (~800 °C) ensures the hardening of the pastes and the "firing" of the antireflection layer between the front contact and emitter. Besides this, the firing achieves a diffusion of the Al atoms from the rear contact into the base in order to generate the required p^+-layer for the back-surface field (see Section 4.2.4). Because of the phosphorous diffusion, the border regions of the cell are also n-doped whereby the p–n junction is basically short-circuited. Thus, as a last step, an edge insulation of the cell is carried out (etching or laser cutting process). The production process of the solar cell is completed with the measurement of the I/V characteristic curve under standard test conditions in order to allocate the cell to a quality class. Figure 5.8 shows the views of a monocrystalline cell after the respective production steps.

The use of silver for solar cell contacts is certainly quite expensive. Are there alternatives to this?

In fact, the cost of the silver is a real hurdle in the further reduction of production costs. Meanwhile, there are many promising attempts to do without silver. One of the many attempts by producers is to use copper. However, copper diffuses into the cell already at room temperature and generates traps there (see also Section 4.2.2). Attempts are now being made to use nickel, which acts as a barrier between copper and the Si emitter [53].
Silver is also used on the rear side. Normally, this is coated with aluminum over the whole surface for the back-surface field. Unfortunately, the zinc-covered cell-connector strips cannot be soldered directly to aluminum, which is the reason why silver has been used up to now. A new process makes use of zinc contact strips that are connected directly with aluminum. Thus, besides the silver paste, a screen-printing step also falls away in the production [54].

The solar cell in the earlier figures always has a p-doped base and an n-doped emitter. Could one not do this just as well the other way around?

Nowadays, it is common practice that cell producers purchase wafers, which are entirely doped with boron and then dope them in a well-known process with phosphor for the emitter. However, these p-type cells have an important disadvantage: together with contaminants (e.g. chrome or iron), boron forms complexes that again form additional recombination centers. Added to this is the problem of the oxygen that enters the melt during the production of the wafer. The metastable boron–oxygen complexes that form from this lead to additional recombination centers with the incidence of light. In this way, depending on the quality of the material, the efficiency is reduced by about 7%, for example, within a month of operation (degradation) [55].
Meanwhile, an increasing number of producers are changing to n-type cells, as these problems do not occur there. In this case, the base is typically doped with phosphor, and aluminum or boron can be used as the doping material for the emitter. However, special measures must be taken in order to passivate the surfaces [56]. For instance, cells that use n-type wafers are the point-contact cell (see Section 4.7.2) and the HIT cell (see Section 5.4.1).

The next production step is the integration of the solar cells into the solar module and we will consider this in the following sections.

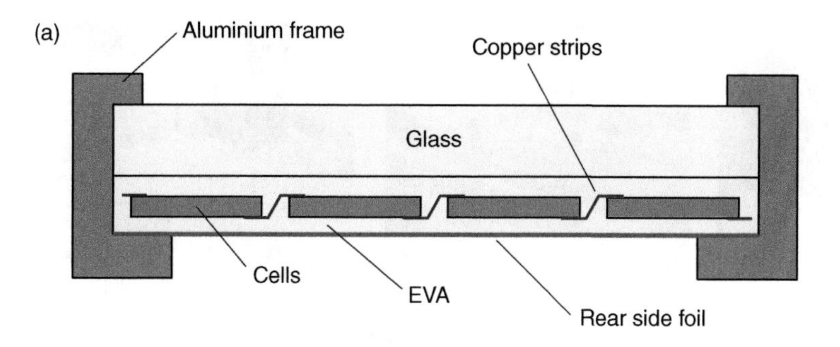

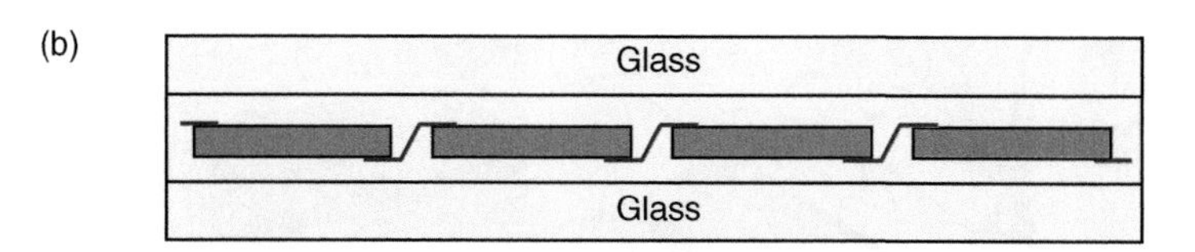

Figure 5.9 Structure of a glass–foil (a) as well as a glass–glass module (b).

5.1.4 Production of Solar Modules

In order to make solar cells manageable for power supply, they are integrated into solar modules. Figure 5.9 shows the principle of the structure of a glass-foil solar module. The individual cells are connected electrically in series into a cell string by means of galvanized copper strips. This string is bedded between two sheets of Ethyl-Vinyl-Acetate (EVA), a transparent plastic. To finish off, a glass sheet of 4 mm thickness is placed on the front side and a rear-side foil on the rear side. This sandwich is then heated in a laminator under vacuum up to 150 °C. The EVA material softens and flows around the cells and then hardens again.

The rear-side foil is for protection from moisture and is also an electric insulator. It is made up of a layered film of polyvinyl fluoride and polyester and is mostly designated a Tedlar foil, a trade name of the DuPont Company. The edge of the module must be sealed (e.g. by means of adhesive tape) before being inserted into the aluminum module frame.

An alternative to the glass-foil module is the glass–glass module (Figure 5.9). For architectural reasons, this is often used on facades or for integration in roofs. The second sheet of glass is for increasing mechanical stability, as there is no metal frame. In the last years, there is a new trend to glass–glass modules. This is due to the fact that, meanwhile, very stable glass sheets of only 2 mm thickness are available. Thus, a glass–glass module is not heavier than a conventional one with 4 mm front glass. At the same time, these modules offer a whole slew of advantages. The back glass provides a vapor proof and insensitive sealing of the module. As the cells are positioned symmetrically between the two glass panes, bending leads only seldom to cell cracks. Some manufactures are convinced in such a way of their glass–glass modules that they offer a power guarantee of up to 30 years.

Figure 5.10 shows the individual steps in the production of a solar module.

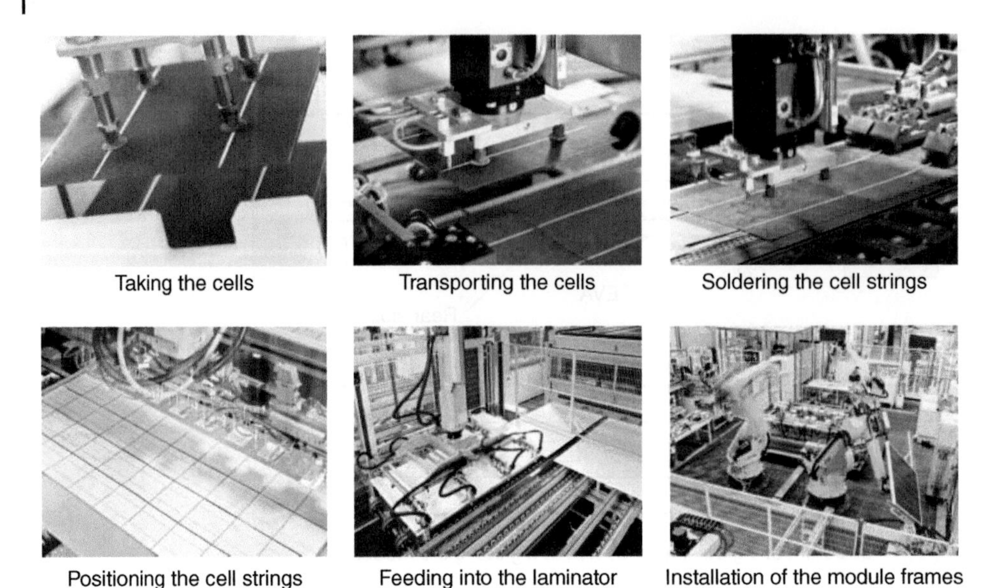

Figure 5.10 Steps for producing a solar module. Photos – Solar-Fabrik AG.

The features of solar modules are described in Chapter 6, and questions on optimum cell circuitry are discussed in greater detail. In the following, we will look at alternative cell technologies.

5.2 Cells of Amorphous Silicon

As we have learned in Chapter 3, direct semiconductors possess an extremely high absorption coefficient. With them, it is possible to absorb sunlight in a "thin film cell" of one micrometer. The best-known thin film material is amorphous silicon, which we will now examine.

5.2.1 Properties of Amorphous Silicon

If one deposits silicon out of the gas phase onto a carrier material, then an extremely irregular structure of silicon atoms is formed (*amorph*: Greek: without structure). It consists of a multitude of open bonds that are called dangling bonds. They form recombination centers for electron–hole pairs and make the material unsuitable for solar cells. The trick is to add hydrogen for passivation during the deposition in order to saturate the dangling bonds.

The structure of the material designated as a-Si:H is shown in Figure 5.11a. Unfortunately, not all bonds can be saturated as this would require the hydrogen portion to be increased to such an extent that the optical properties of the material would be impaired [58]. Depending on the hydrogen portion, the crystal structure of a-Si:H possesses a direct bandgap in the region of $\Delta W_G = 1.7–1.8$ eV [59]. As can be seen in Figure 3.22, the absorption coefficient is one or two factors above that of c-Si. At a wavelength

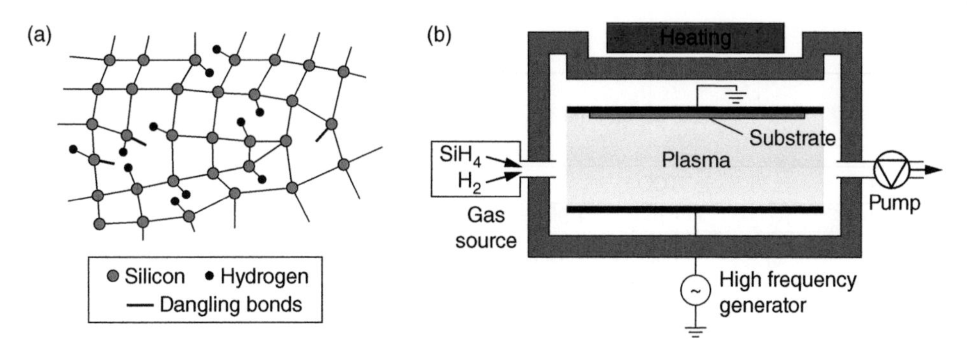

Figure 5.11 Structure of the a-Si lattice and depiction of the Plasma-Enhanced Chemical Vapor Deposition (PECVD) for producing a-Si thin film cells [57].

of 600 nm, the penetration depth is only 0.25 μm. Thus, cell thicknesses of 0.5 μm are sufficient to absorb a large part of sunlight.

5.2.2 Production Process

For the production of a-Si thin film cells, use is mainly made of Plasma-Enhanced Chemical Vapor Deposition (PECVD), see Figure 5.11b. The starting gases, silane (SiH_4) and hydrogen (H_2), flow into the approximately 200 °C hot process chamber, and there enter into a strong high-frequency field. This accelerates individual electrons that in their turn separate the molecules of the starting gases by means of impact ionization into their constituent parts (SiH_3^+, etc.). The charged particles form plasma that contains highly reactive ions, which react with the substrate surface and settle there. This layer of a-Si:H continues to grow with further addition of the two process gases. The process would also function without the plasma enhancement (normal gas phase deposition – CVD) but then one would require temperatures of more than 450 °C for breaking up the starting gases, and this would strongly limit the selection of substrate materials.

Typical deposition rates are in the region of 0.2 nm s^{-1}. The production of a 0.5 μm thick a-Si-H layer takes about 40 min. This time is actually too long for mass production; desirable would be a reduction by a factor of 10. Many promising new processes are available for this (for instance, Very-High-Frequency-PECVD as well as Hot-Wire-CVD) with high deposition rates, which, however, often increase the number of defects in the a-Si:H [58].

5.2.3 Structure of the Pin Cell

The typical structure of an a-Si thin film cell produced with the PECVD process in shown in Figure 5.12.

A glass sheet is coated over its whole area with a transparent electrode of conducting oxide called Transparent Conducting Oxide – TCO; a technology that is also used in the production of flat screens. Typical materials are indium-tin oxide (ITO) or zinc oxide (ZnO). Connected to this is a sandwich of p-doped, intrinsic (undoped), and n-doped amorphous silicon. The final coat is a thin rear contact of aluminum or silver. The surprising thing is the low material usage for the cell: The layers applied to the glass have

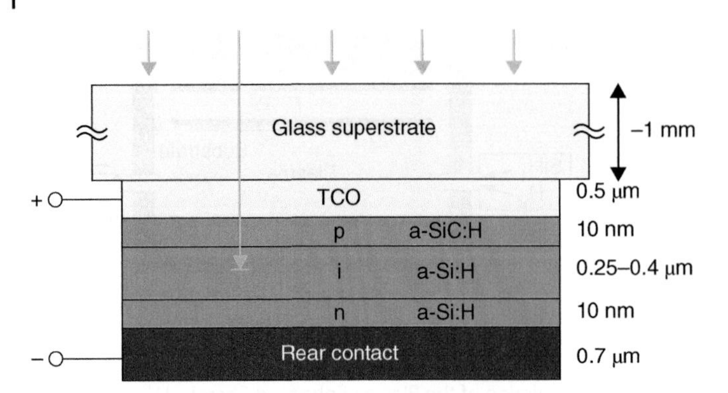

Figure 5.12 Structure of the thin film cell: The overall thickness of the deposed material is less than 2 μm.

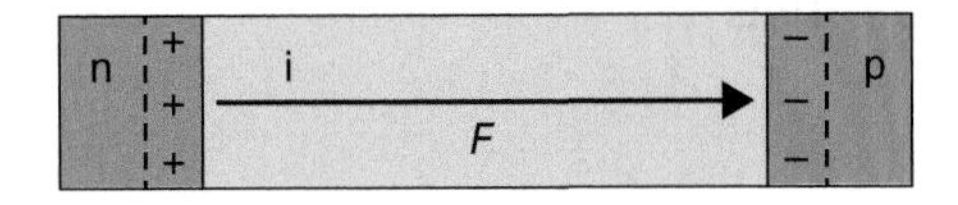

Figure 5.13 Results of the build-up as a pin cell: The space charge region extends practically over the whole of the intrinsic region.

a combined thickness of less than 2 μm. The structure shown in Figure 5.12 is called a *superstrate cell*, as the glass sheet on which the layers are deposited lie above the rest of the cell in sunlight (*super*: Latin for over).

As far as possible, the light absorption should occur in the intrinsic layer as the electron–hole pairs in doped materials recombine within a few nanometers. For this reason, one adds carbon to the p-region, and this increases the bandgap to approximately 2 eV so that the a-SiC:H layer is almost transparent.

The probability of recombination is also very large in the undoped a-Si. The generated minority carriers must be "brought home" as quickly as possible (see Section 4.2). This is achieved by building up pin cells that generate a large electric field, which separates the particles as soon as they are generated and transports them to their home area.

For a better understanding of the operating method of the pin cell, Figure 5.13 shows the space charge region (compare with Figure 3.14). Analogous to the normal p–n junction, the electrons diffuse out of the n-region into the intrinsic region and leave positive donator atoms behind.

As they can find no holes for recombination there, they "roll" further until they drop into holes in the p-region, and there they form negative space charges. The result is the formation of a constant electric field over the whole i-area.

The thin film cells are also called drift cells or field cells as the optically generated minority carriers flow here as pure field current. However, we had named the c-Si cells diffusion cells: There the particles diffuse into the space charge region where they then flow as field current into the home region.

Could one interchange the layer sequence of an a-Si cell just as well? A nip instead of a pin?

Basically, yes. However, in a-Si:H, the mobility of the holes, and therefore the drift velocity, is smaller by one order of magnitude than the electrons [60]. After generation, they take longer to arrive in the p-region and are in special danger of recombination. This is the reason for the advantage of the pin structure. As the absorption of light occurs mainly in the upper half of the i-layer (see Figure 4.8), the generated holes so do not have far to go to the p-layer.

5.2.4 Staebler–Wronski Effect

The large bandgap of 1.75 eV has the result that light above approximately 700 nm can no longer be absorbed. Because of these transmission losses, the theoretically possible efficiency is 26% (see Figure 4.25). In fact, the record efficiency of real cells is only around 10%. Standard cells achieve approximately 7–8%. Besides the high concentration of defects of the a-Si and the shunt resistance of the TCO layer, an important reason is that newly produced cells degrade under the influence of light. Figure 5.14 shows this in the example of two pin cells that were exposed to full sun radiation for more than 10 000 h at AM 1.5 (*light soaking*) after production. After approximately 3000 h, the power reduction was 29%, then 37%, and then stabilized.

The cause of the degradation is the Staebler–Wronski effect, named after the two scientists who first described it in detail in Ref. [62]. The reason is in the strained Si—Si crystal bonds that are caused by the irregular crystal structure. In the recombination of the electron–hole pairs generated by light, these weak bonds are "split open" and, as new dangling bonds, form new recombination centers for the minority carriers (see Figure 5.15). The split-open bonds also represent additional space charge regions that can weaken the built-in field in the pin cells.

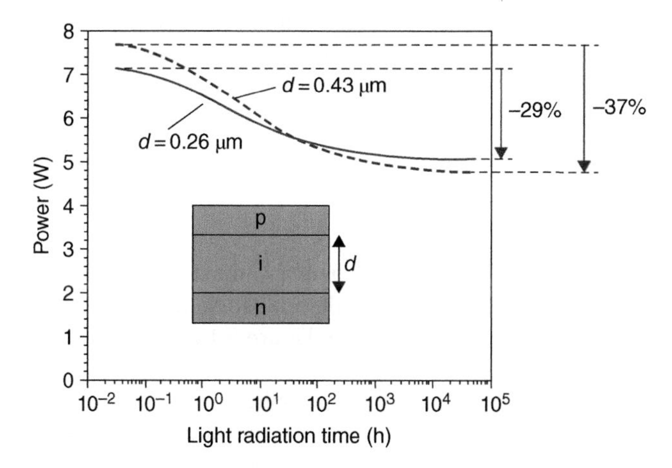

Figure 5.14 Light degradation measurement of two s-Si pin cells on different thicknesses: Within 3000 h with full-light radiation, the power is reduced by up to 37% [61].

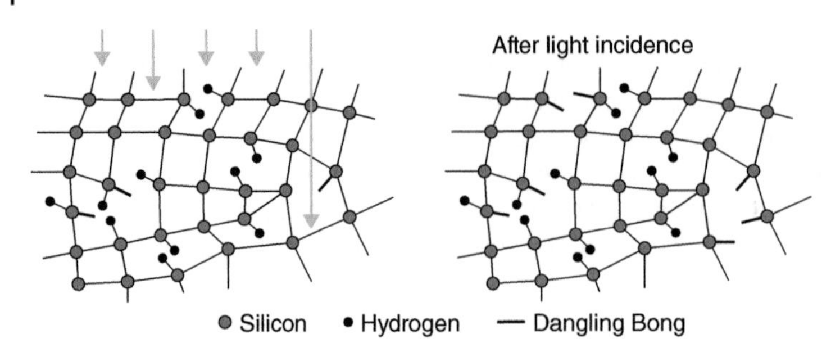

Figure 5.15 Depiction of the Staebler–Wronski effect: With incident light, the weak bonds in the crystal are split open [58].

After a certain light radiation time, all the weak bonds are split open so that the efficiency of the cell stabilizes itself.

The exact method of operation of the effect is not yet fully understood. However, a method for reducing this has been found. Thus, the additional crystal defects can be healed by means of tempering (from 150 °C) and the Staebler–Wronski effect can be reversed [63]. In addition, Figure 5.14 shows that thinner i-layers are subject to less degradation than thick layers. The reason for this is in the larger electric field that brings the electron–hole pairs safely home even in the case of the existence of disturbing space charge regions [57].

When the layers are made ever thinner, don't you think that then not enough sunlight would be absorbed? Or do you think otherwise?

Of course, transmission losses rise with thinner i-layers. However, there is a trick we have seen in Chapter 4 that helps light trapping by means of texturizing. For this purpose, both the TCO layers and the rear-side electrode are roughed so that the light path is significantly lengthened and layer thicknesses of 250 nm become possible.

5.2.5 Stacked Cells

In order to significantly increase the efficiency, the Sun's spectrum must be used more effectively. For this purpose, two pin cells of materials of different bandgaps are stacked into a tandem cell optimized to a particular spectral range. Figure 5.16 shows this by means of two examples.

In the case of the tandem cell, an additional a-SiGe layer is applied on top of the a-Si absorber layer. Depending on the portion of germanium atoms, this alloy can possess a bandgap between 1.4 and 1.7 eV. Also depending on the wavelength, incident light is absorbed at different depths: Short-wavelength light ("blue") above 1.7 eV only manages to reach the first pin cell. The upper pin cell is transparent for long-wavelength ("red") light; it is, therefore, only absorbed in the lower cell. As both cells are switched in series,

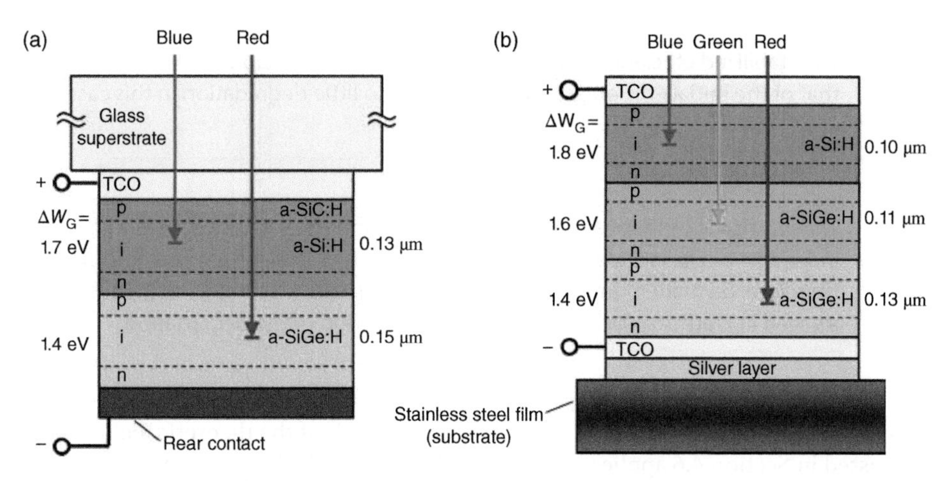

Figure 5.16 Examples of superstrate tandem and substrate triple cells [57, 64]. (a) Tandem cell. (b) Triple cell.

the weaker cell determines the overall current. For this reason, the thickness of the individual absorber layers must be selected so that the two cells achieve approximately the same current (current matching).

Even better is the light separation in the case of the triple cell: Here a further a-SiGe layer with a bandgap of 1.6 eV is used (Figure 5.16b). A feature of this cell produced by the United Solar Company is that it is applied to a flexible stainless steel film. It thus represents a substrate cell (*sub*: Latin for "under"). The layer of silver between the film and TCO is for reflecting the light upward. Figure 5.17 shows the external quantum efficiency (see Section 4.3) of the whole cell with the contributions of the respective individual cells. The given short circuit currents show clearly that the current matching has been quite successful. The cell possesses a starting efficiency of 14.6% and a stabilized efficiency of 13%. As the individual pin cells are switched in series, the result is a relatively high open circuit voltage of 2.3 V [64].

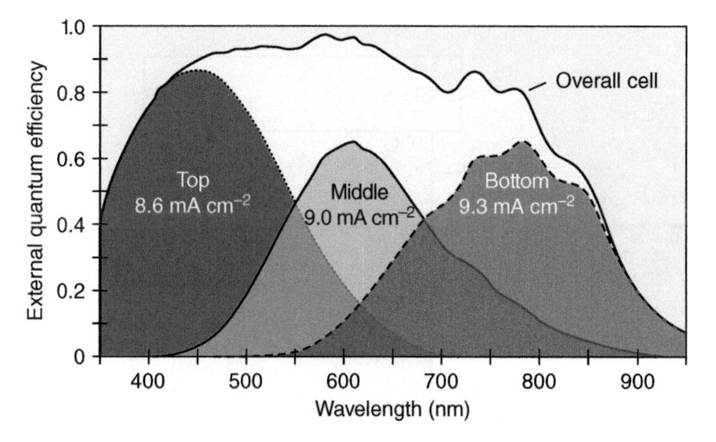

Figure 5.17 External quantum efficiency of the triple cell in Figure 5.16: The individual pin cells are responsible for different wavelength regions [64].

The stabilized efficiency of the record triple cell lies only about 10% below that of the initial efficiency. Why do we have so little degradation in this case?

The example emphasizes another advantage of the stacked cells. The individual pin cells are optimized for the respective spectral regions and can, therefore, be made thinner. However, a thin cell leads to a high electrical field. As seen in Figure 5.14, this also leads to a reduction of the degradation.

At this point, it should be mentioned that the upper limit of the theoretical efficiency η_T discussed in Section 4.6 applies only for simple cells. Stacked cells of different materials can exceed this limit without problems (see also Section 5.4).

5.2.6 Combined Cells of Micromorphous Material

A relatively new development is the combination of amorphous and microcrystalline silicon (μc-Si). This technology was developed in the 1990s at the University of Neuchâtel and has, meanwhile, reached industrial maturity. The term microcrystalline designates a material that contains silicon particles (nanocrystals) in a size significantly less than 1 μm. These nanocrystals occur by PECVD deposition at certain silane concentrations. The result is a conglomerate of nanocrystals that are embedded in an a-Si:H environment (see Figure 5.18).

The material behaves similarly to crystalline silicon with a bandgap of 1.12 eV. Thus, it is very well suited for a combination with a-Si in order to cover a large part of the solar spectrum. Besides this, the material shows practically no degradation. However, like c-Si, it has a low absorption coefficient. In order to still use it in the thin film region, one needs relatively large layer thicknesses and must also significantly lengthen the light

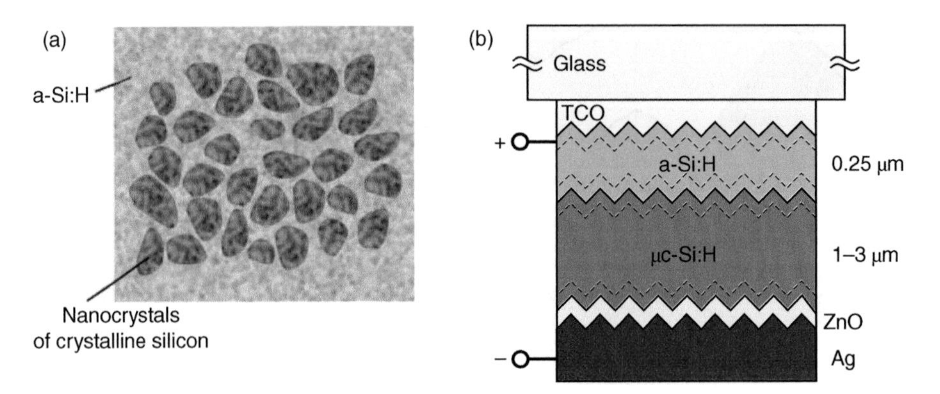

Figure 5.18 Structure of microcrystalline silicon and build-up of the a-Si:H/μ-Si:H tandem cell [32, 58].

path by means of effective light trapping. Figure 5.18b shows the example of a micromorphous tandem cell (*micromorph*: an artificial word made up of microcrystalline and amorph).

Micromorphous laboratory cells have reached stable efficiencies of 12% and large-area modules are available with 10% efficiency. However, a number of companies throughout the world have ceased production of this thin film technology as CdTe and CIS offer much higher efficiencies (see Section 5.3).

5.2.7 Integrated Series Connection

A big advantage of the thin film technology is that the cells can be connected to a whole module during production. Figure 5.19 shows the individual process steps for this. After applying the TCO, it is divided by means of laser into individual partitions on which the pin cells are then deposited. These cells must then again be separated and provided with a rear contact. After this has been structured, the electric connection is complete. Important for a small series resistance is a sufficient width of the pit in the semiconductor structuring, as after filling with the rear contact, the cells are connected in series through this.

The last picture in Figure 5.19 shows the portion of the unusable area caused by this series connection. The depiction is not to scale. Depending on the technology, the unused portion is only approximately 5–10% of the active cell surface. The encapsulation of the module is carried out as for normal c-Si glass–glass modules by means of EVA and a rear pane (see Section 5.1.4). The integrated series connection leads to a very homogeneous looking strip design so that this type of module is appreciated for architecturally demanding solutions such as facades or semitransparent glazing (see Figure 5.20a).

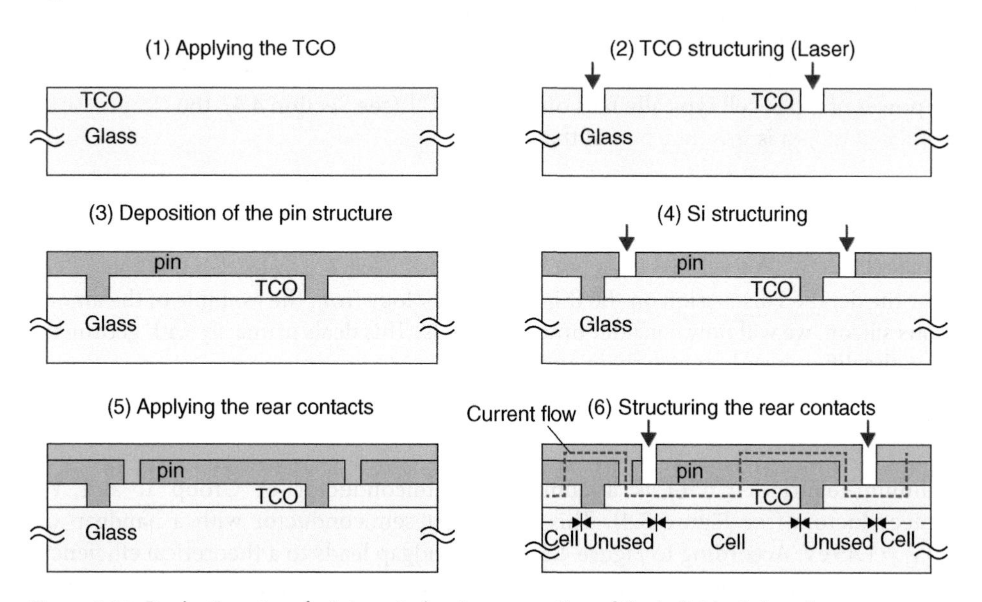

Figure 5.19 Production steps for integrated series connection of the individual pin cells.

(a)

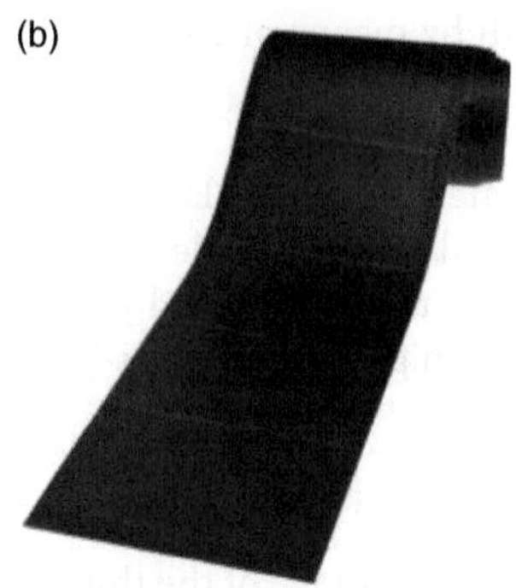
(b)

Figure 5.20 View of a-Si modules: (a) Semitransparent module. Photo – Taiyo Kogyo Corporation; (b) Flexible laminate for roof membranes. Photo – United Solar.

There is, however, also a great disadvantage with this process: The deposition of the silicon layers must be fully homogenous over the whole of the module area (e.g. 2 m^2). If this is not the case, then every inferior cell lowers the current through all the others. For this reason, the efficiencies of thin film modules are often more than 10% below those of individual cells. This is not so in the case of the crystalline wafer technology, as the produced cells are first divided into quality classes and only then are the cells of the same class combined into a module.

The United Solar Company followed a middle path in its a-Si module production. It generated integrated connected solar cells of DIN A4 size on a narrow long roll (see Figure 5.20b). In order to make a module out of this, two strips each with 10 cells were placed next to each other and then electrically connected and encapsulated into a module frame. Unfortunately, the company was not able to compete with the worldwide competitors; therefore, this cell is no longer produced.

Cells of amorphous silicon also have an important advantage compared to their c-Si representatives: The temperature dependency of the efficiency is much less. Whereas the power of a c-Si cell typically falls off by 0.5% K^{-1} (see Section 4.4), the temperature coefficient of a-Si is less than half of that.

5.3 Further Thin Film Cells

After the detailed discussion on the thin film technology from the example of the amorphous silicon, we will now consider other materials. This deals primarily with getting to know the differences between them and a-Si cells.

5.3.1 Cells of Cadmium-Telluride

Cadmium-telluride (CdTe) is a compound semiconductor of Group II and VI semiconductors (see Figure 3.4). This is a direct semiconductor with a bandgap of $\Delta W_G = 1.45$ eV. According to Figure 4.25, this bandgap leads to a theoretical efficiency of 29.7% and is therefore very near to the optimum. A great advantage of this material is that it can be deposited in various ways with good quality as a thin film. The usual method is thermal evaporation over a short distance (CSS – Close-Spaced-Sublimation).

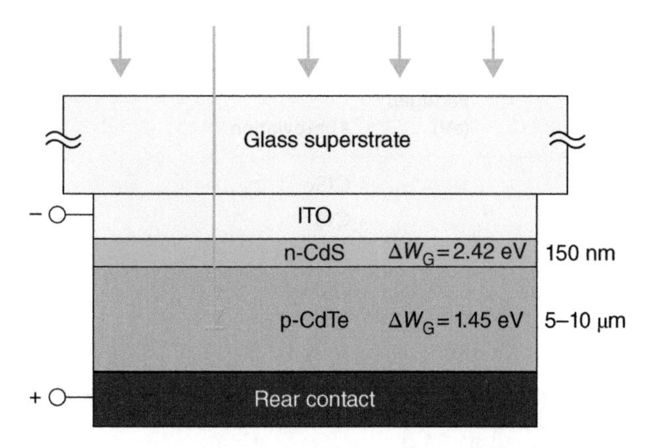

Figure 5.21 Structure of the CdTe superstrate cell: The cadmium-sulfide layer serves at the same time as n-region and as a window layer [65].

In this process, the semiconductor sources are heated to approximately 500 °C. At this temperature, the semiconductors vaporize and deposit on the somewhat lower temperature substrate. Figure 5.21 shows a typical cell structure. As the CdTe can only be badly n-doped, a window layer of n-doped cadmium sulfide (CdS) is grown on after applying the transparent electrode of ITO, then the actual absorber layer of polycrystalline CdTe is followed. The two materials form a so-called hetero junction as the bandgaps of n- and p-doped regions are different (*hetero*: Greek for "other").

The properties of absorber and junction are at first relatively poor but can be much improved with cadmium-chloride treatment. For this purpose, $CdCl_2$ is applied and diffused into the absorber layer by means of tempering. In the first years of CdTe cell development, this treatment seemed to be more or less applied alchemy. Nowadays, however, the effects are fully understood.

In Figure 5.21, it can be seen that the absorber layer is relatively thick with 5–10 μm, although CdTe has a very high absorption coefficient. The reason is the difficulty of generating thin films with high crystal quality on large surfaces. A reduction of the layer thickness is thus still the subject of research [66].

The record efficiency of CdTe cells is 21%, and modules with efficiencies of up to 17% can be found on the market (*www.firstsolar.com*).

We will deal with the question of the environmental aspects of CdTe in Section 5.7.

5.3.2 CIS Cells

The final thin film technology we will deal with consists of materials of the Chalcopyrite group that are generally summarized under the abbreviation of CIS or CIGS. What they have in common is that they have the lattice structures of Chalcopyrite (copper pyrites – $CuFeS_2$). As shown in Table 5.1, there are different ternary (consisting of three elements) compound semiconductors.

Research on CIS cells has been carried out since the 1970s. In the year 1978, ARCO Solar was successful in the production of CIS cells with 14.1% [67]. But disappointment soon followed: The efficiency sank drastically in the transfer to larger areas. Only in 1990,

Table 5.1 Material combination of the CIS family.

Material combination	Name	Bandgap (eV)	Abbreviation
$CuInSe_2$	Copper-indium-diselenide	1	CISe
$CuInS_2$	Copper-indium-disulfide	1.5	CIS
$CuGaSe_2$	Copper-indium-gallium-diselenide	1.7	CIGSe
$CuGaS_2$	Copper-indium-gallium-disulfide	1.55	CIGS

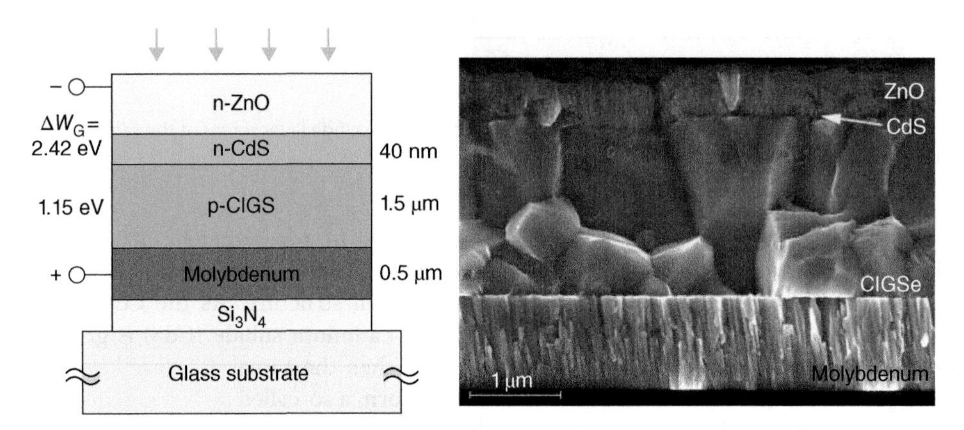

Figure 5.22 Principle of the structure of a CIGS cell and photo of a CIGS cell: The polycrystalline structure of the absorber material can be clearly seen. Photo – Hahn–Meitner Institute; with kind permission of John Wiley & Sons, Ltd. [68].

with better knowledge of the properties of the material was it possible to produce solar modules with an efficiency of 10%.

The most promising material is $CuIn_xGa_{(1-x)}Se_2$. Here x is the portion of indium in the material combination. With $x = 1$, $CuInSe_2$ with a bandgap of 1 eV is obtained, and with $x = 0$ the result is a corresponding $CuGaSe_2$ with 1.7 eV. By changing the indium portion, the bandgap can, therefore, be varied between the two extreme values and thus the efficiency can be optimized.

A typical cell structure is shown in Figure 5.22. This is a substrate configuration. The bottom glass is merely a supporting material; molybdenum acts as the rear electrode. Between the two is a layer of silicon nitride that acts as a barrier for foreign atoms, which could diffuse during the manufacturing process from the glass into the absorber layer. The p–n junction is formed, as with CdTe, from the absorber layer and a thin CdS layer. For a long time, there have been attempts to replace the unpopular cadmium from the CIS cells, but this has only been achieved at the expense of efficiency.

Co-vaporization is mostly used as the method of deposition, where the individual elements are vaporized at temperatures of around 500 °C and deposit themselves on the substrate. Here, too, a bit of "magic" is used: With the addition of sodium and additional tempering, there is an improvement of the crystal structure and the electronic properties of the polycrystalline CIGS.

The record efficiency for laboratory cells is 21%, as in the case of CdTe [69]. Modules are sold with efficiencies up to 16%.

A technology that can dispense with vacuum and higher temperatures would be desirable for economical mass production. An enticing idea is to place the semiconductors as nanocrystals into a watery solution and then to print them out as normal ink. The Nanosolar Company followed such a concept and reportedly achieved efficiencies of 14.5% [70]. However, it only stayed on announcements; commercial modules were never brought on the market. Independently of this, it is to be expected that CIGS technology will make substantial progress in the coming years in economic production of thin film modules.

5.4 Hybrid Wafer Cells

After considering the thin film cells, we will now turn to two technologies that combine different materials on the basis of wafer cells in order to achieve high degrees of efficiency.

5.4.1 Combination of c-Si and a-Si (HIT Cell)

An interesting hybrid form of c-Si and a-Si material has been developed by the Panasonic Corporation. Figure 5.23 shows the structure of these so-called HIT cells (HIT: Heterojunction with Intrinsic Thin-Layer).

On the wafer, which is n-doped on both sides, there are deposited an intrinsic and then a doped layer of a-Si material. A transparent electrode (TCO) is deposited on this. As the TCO is relatively high resistive, additional normal metal contact strips must be used.

What is the advantage of this cell concept? The a-Si layers act as very effective passivating layers for the c-Si wafer cell, resulting in an open-circuit voltage of more than 700 mV. At the same time, the high open-circuit voltage results in an improved temperature dependency of $TC(P) = -0.23\%\ K^{-1}$ compared to $-0.5\%\ K^{-1}$ for a normal c-Si cell. In addition, in the production of the cell, there is the advantage that one can dispense with the energy-intensive diffusion step for emitter production. Temperatures of below 200 °C are sufficient because the a-Si layers are applied in the PECD process. The structure shown in Figure 5.23 also permits light from the bottom of the cell. In

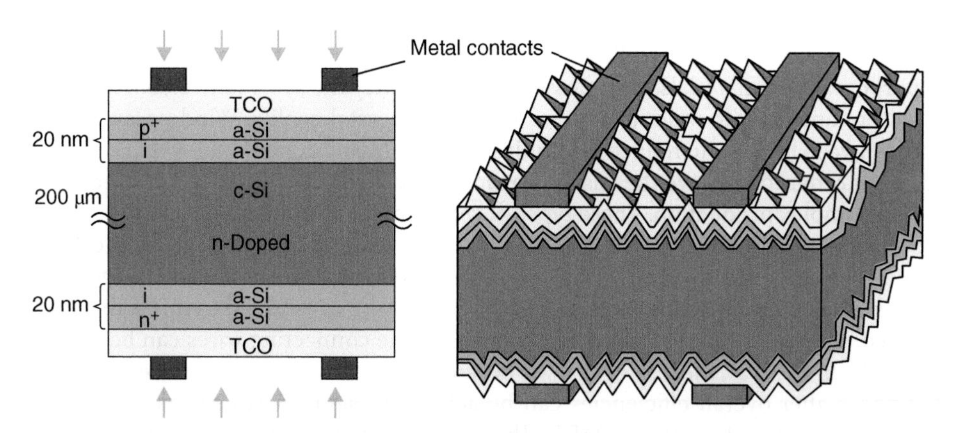

Figure 5.23 Structure and cross section of the HIT cell from Panasonic: The a-Si layers serve primarily for very efficient surface passivation [71].

particular applications, these types of bifacial cells can result in an increased yield of 10% [72]. Meanwhile, Panasonic could present a record cell with an efficiency of 25.6%.

Since important patents for the HIT cell expired in 2010, a number of companies are now adopting the cell concept. Kaneka has shortly presented a heterojunction cell with interdigitated back contacts (IBC) with a world record efficiency of 26.6%, the highest efficiency for a silicon cell so far.

By the way, why actually the good surface passivation of the HIT cell leads to a high open-circuit voltage?

This can be very well explained with the aid of the simplified equivalent circuit model (Figure 4.11). A bad surface passivation leads to recombinations at this surface. The recombined electrons then have to be delivered additionally over the p-n junction. Thus the saturation current I_S increases. According to equation (4.16) this again has a direct influence on the open-circuit voltage: with rising I_S the open-circuit voltage is reduced. In other words: the less recombinations take place (both in the semiconductor and at the surface) the higher is V_{OC}.

The open-circuit voltage is hence a measure of the quality of a solar cell. Extremely good lab cells get up to 740 mV, very good cells in commercial modules reach 710 mV.

5.4.2 Stacked Cells of III/V Semiconductors

For particular applications, for example, space use or concentrator cells (see Section 5.6), maximum efficiency is required, and costs play a subsidiary role. Thus, for instance, use can be made of relatively expensive GaAs wafers as this material with a bandgap of 1.42 eV lies very near to the theoretical optimum (see Figure 4.24). The best lab cells of GaAs thus achieve an efficiency of 31%.

It is even better to combine several materials in a cell as already seen in the triple cell. For instance, the Spectrolab Company has specialized in this technology. The structure of such monolithic stacked cells of III/V semiconductors is shown in Figure 5.24a. Monolithic means that the middle and upper cells are grown onto the bottom cell. For this purpose, these materials must possess approximately the same lattice constants as the germanium wafer in order to achieve a sufficient crystal quality. Window layers are placed between the cells to reduce recombination, and highly doped tunneling layers are inserted in order to improve the charge transport between individual cells. This results in a very complex production. These types of cells achieve efficiencies of up to 38%.

A second method is to stack different cells mechanically on top of each other (so-called mechanically stacked multijunction cells). On the one hand, this has the advantage that the individual materials can possess different lattice constants. Added to this is that the cells need not necessarily be connected in series as the connecting wires can be led to the outside. In this way, the losses fall away because of the current matching (see Section 5.2.5) and higher overall efficiencies can be achieved. Figure 5.24b shows a concept of the Belgian research institute IMEC. The top cell, which again consists of a mono-

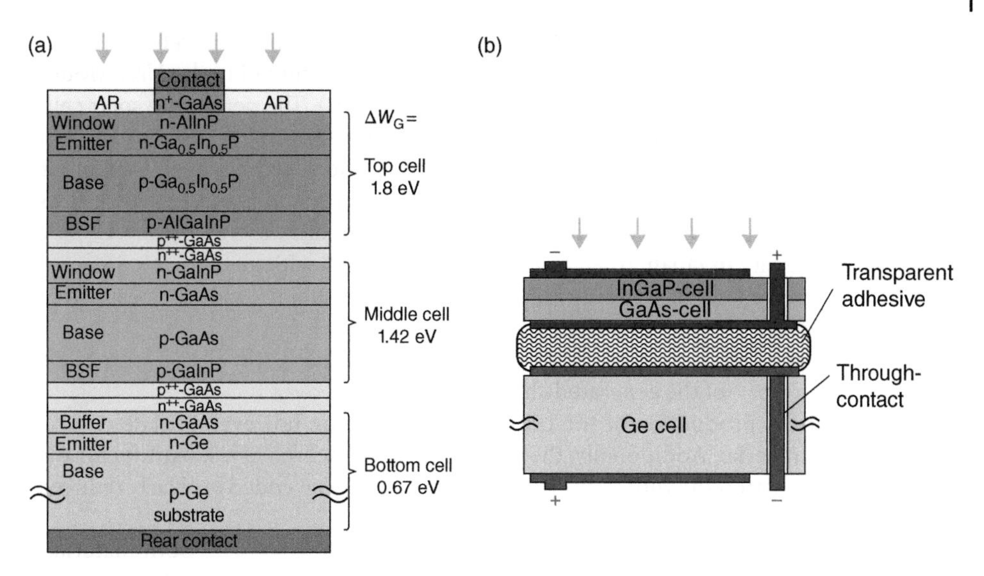

Figure 5.24 Structure of a monolithic GaInP/GaAs/Ge stacked cell of the Spectrolab Company (a) and a depiction of the principle of a mechanically stacked triple cell (b) [73, 74].

lithic tandem cell, is connected to the bottom cell by means of a transparent adhesive. The rear contacts are connected to the surface by means of a through-contact in order to connect the respective cells from outside.

5.5 Other Cell Concepts

Besides the cell technologies already discussed, there is still a series of other materials and cell concepts that can be of importance in the future.

The Dye-Sensitized Cell (DSC) was discovered at the start of the 1990s by Professor Michael Grätzel at the Ecole Polytechnique in Lausanne and is, therefore, called the Grätzel Cell. This cell has long played the role of "Eternal Talent." The specialty of this concept is a relatively simple manufacturing process and the use of cheap materials. Meanwhile, laboratory cells have efficiencies of 12% and minimodules have up to 10% [69]. A hindrance for commercialization has hitherto been the lack of stability. High temperatures lead to the formation of gas and leaks because the cell contains a liquid electrolyte. An exact description for the build-up and method of operation of the DSCs can be found in [75].

A further candidate is the organic solar cell that uses polymers in place of semiconductor materials. As with the DSCs, here too there is hope in the future of producing much cheaper electric energy than with the presently available technologies. The record efficiency for minilaboratory cells is held by Toshiba with 11%; small submodules achieve an efficiency of about 10% [69]. Further success can be expected in the coming years, as there are a number of companies worldwide researching in the field of polymer electronics. Further information on organic solar cells can be found, for instance, in [76].

A very stormy development has taken a new material class of solar cells: The perovskites. The name perovskite originally described a mineral mined in the Ural Mountains, which is generally known as calcium titanate ($CaTiO_3$). The perovskite solar cells have the crystal structure of this mineral; however, they consist of different materials. On the one hand, an organic portion consists of carbon, hydrogen, and nitrogen; on the other hand, an inorganic portion consists of lead, iodine, and chlorine. This organic–inorganic hybrid cell was invented from the team around Tsutomu Miyasaka from Toin University in Japan. In 2009, they presented a cell with an efficiency of 3.8%. Many other research groups worldwide stepped into the subject with the result that efficiencies of about 20% were reached within a few years.

Beneficial is the high absorption coefficient of the material with, at the same time, large diffusion lengths of the generated charge carriers. Moreover, no high temperatures are needed for the production of the cells. A large problem, however, to date is poor stability of the material. Additionally, the developers are in search of a substitute for the heavy metal lead. Attempts with alternative materials, however, ended in clearly reduced efficiencies.

Particularly appealing would be a tandem cell of perovskite and c-Si. Both materials complement each other with respect to their bandgaps: While c-Si covers the longwave part of the spectrum (red and infrared), the perovskite cell placed on top could convert the shortwave part (blue and green) into solar power efficiently.

A good introduction into the subject perovskite solar cells can be found, e.g. in [77].

5.6 Concentrator Systems

The idea of concentrator systems is to concentrate sunlight by means of mirrors or lenses and then to divert it to a solar cell. We will look at this technology in more detail in the following.

5.6.1 Principle of Radiation Bundling

The two most important principles of concentration of light radiation are shown in Figure 5.25. In the case of lens systems, use is made of Fresnel lenses that were discovered by the French Physicist Augustin Jean Fresnel (1788–1827). With these lenses

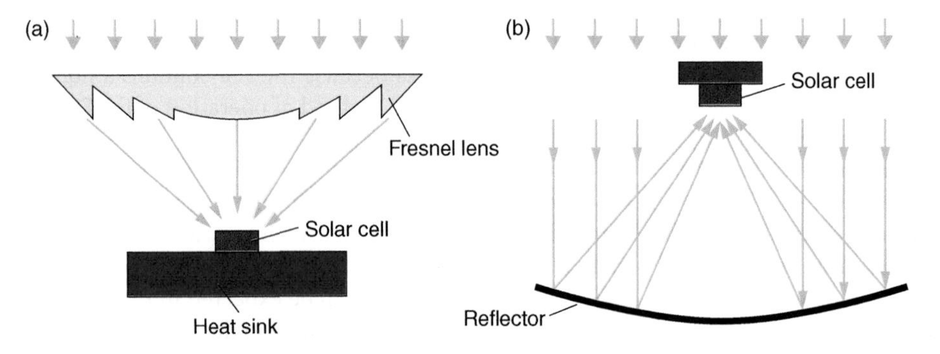

Figure 5.25 Principle of light concentration by means of a Fresnel lens (a) and parabolic mirror (b).

that were originally developed for lighthouses, one can achieve a relatively small focal length without the lens becoming too thick. This occurs with the use of circular steps on the lens surface. The resulting picture errors would be catastrophic for cameras, but for photovoltaics, the only concern is to be able to concentrate radiation on the solar cell.

Another principle is the use of mirror systems. Here, the light is reflected by a curved mirror surface and concentrated on the solar cell. In the optimum case, use is made of parabolic mirrors as they concentrate incident rays on a focal point.

5.6.2 What Is the Advantage of Concentration?

The purpose of light concentration on solar cells is the reduction of production costs. If the lenses or mirrors can be produced cheaply, then an actual cost advantage can accrue due to the drastic reduction of the required solar cell area. But there is still a further advantage: The efficiency of the solar cell increases with the irradiance. How is this to be explained? This can be seen in Figure 5.26. The cell curve moves upward with the increase in radiation. As the short-circuit current behaves proportionally to irradiance, an increase, for instance, by a factor of $X = 2$ effects a doubling of the short-circuit current. However, this does not result in a change in efficiency as double the optical power also hits the cell. But it is known that the open-circuit voltage increases at the same time by the logarithm of the irradiance. With the resulting increase of the MPP voltage, this leads to an overproportional rise in power and thus to an increase in efficiency.

In order to calculate the effects more accurately, we make use of Equation (4.16) for the open-circuit voltage. The increase in the irradiance by the factor X leads to a changed voltage V'_{OC}:

$$V'_{OC} = m \cdot V_T \cdot \ln\left(\frac{I_{SC} \cdot X}{I_S}\right) = m \cdot V_T \cdot \ln\left(\frac{I_{SC}}{I_S}\right) + m \cdot V_T \cdot \ln(X), \quad (5.1)$$

$$V'_{OC} = V_{OC} + m \cdot V_T \cdot \ln(X). \quad (5.2)$$

Thus, for instance, a concentration of the sunlight by the factor $X = 100$ results in an increase in the open-circuit voltage by 120 mV (assumption: $m = 1$). Referenced to a typical c-Si voltage of 600 mV this means an increase of 20%.

The increase in efficiency can be verified for particular cell types. This is shown by the stacked cell presented in Section 5.4.2 of the Spectrolab Company; under standard

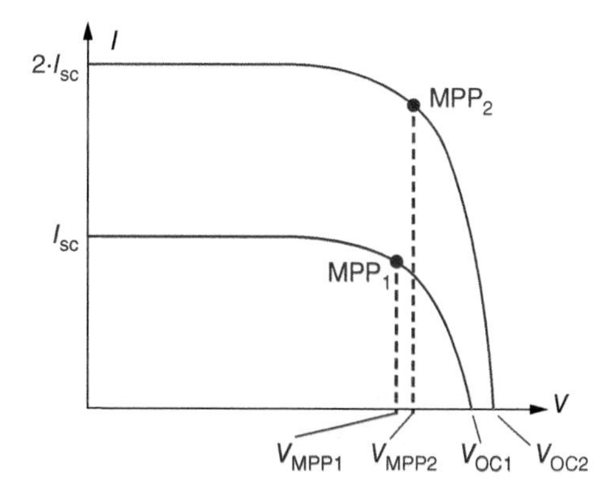

Figure 5.26 Displacement of the solar cell characteristic curve for double irradiance: Besides the short-circuit current, there is also an increase in the open-circuit voltage and thus the MPP voltage.

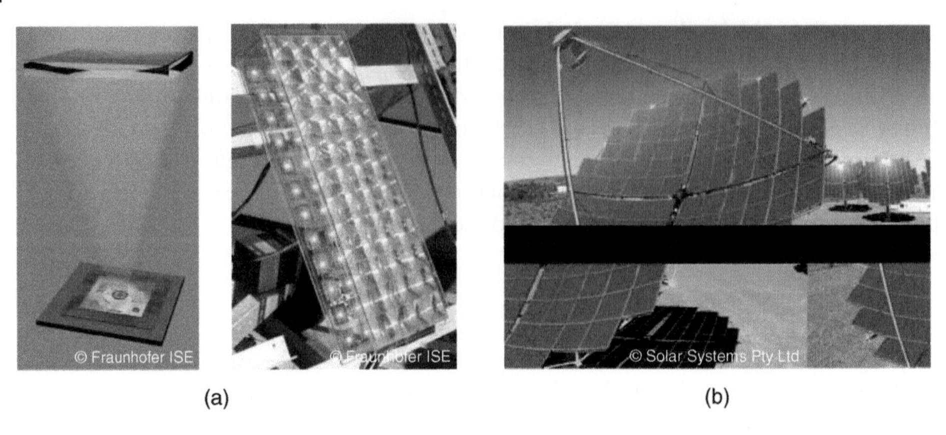

Figure 5.27 Examples of concentrator systems: Path of rays and photo of a Flatcon concentrator module (a) and parabolic mirror system (b).

conditions, an efficiency of 29.1% is achieved, and under 66 times concentrated sunlight, this increases to 35.2%.

The worldwide highest efficiencies show a cell from Soitec: Their four-junction cell delivers an efficiency of 46% under a light concentration of $X = 297$. However, the efficiency does not climb continuously with X. For one thing, the cell heats up with high irradiance and even an active cooling cannot prevent this. On the other hand, the electrical losses of the series resistances rise with the square of the operating current.

5.6.3 Examples of Concentrator Systems

Figure 5.27a shows the application of the lens principle by the Soitec Company. This installs a multitude of extremely small cells (3 mm^2) into a box-shaped module (product name *Flatcon*). The Fresnel lenses imprinted on the upper glass sheet concentrate the light up to 500 times on these cells. The efficiency of the offered modules is about 32%. Using a fourfold stack cell, the best lab module shows a record efficiency of 36.7%.

An example of parabolic mirror collectors is shown in Figure 5.27b. The reflector is made up of 112 individual curved mirrors, which feed the sunlight with 500 times concentration onto the receiver. This, again, consists of an array of III/V stacked cells that are kept to a maximum of 60 °C by means of active cooling. The size of the reflector is gigantic with a diameter of 12 m. In this way, the system achieves an electric power of 35 kWp.

Besides the highly efficient cells of III/V semiconductors, cells of crystalline silicon can also be used. For instance, there is the Point-Contact cell from the SunPower Company (see Figure 4.33). As the contacts are only on the rear side, they can be as thick as desired and can thus conduct extremely high currents without noticeable losses.

5.6.4 Advantages and Disadvantages of Concentrator Systems

Whether concentrator systems are more economical than conventional solar modules depends greatly on the costs of the concentrating elements. They can only be competitive when, despite the high requirements (mechanically and optical stability over more than 20 years) of the used surface, they are significantly cheaper than the standard modules. Added to this is the important disadvantage that concentrator systems can only use the direct portion of the global radiation. As diffuse radiation from many directions arrives at the mirror or reflector, it cannot be bundled on the solar cell. Added to this is that concentrator systems usually require a Sun tracking system, which adds to further cost. It must, therefore, always be weighed up as to whether the utilization of concentrator technology is really worthwhile.

5.7 Ecological Questions on Cell and Module Production

5.7.1 Environmental Effects of Production and Operation

Various materials are used in the production of solar cells and modules, and their environmental friendliness will be discussed.

5.7.1.1 Example of Cadmium-Telluride

The CdTe cells presented in Section 5.3.1 pose a special hazard for the environment. Cadmium is a poisonous heavy metal that is classed as carcinogenic. Thus, it should not be released into the environment. However, cadmium combined with tellurium is a very stable water-insoluble compound that only melts above 1000 °C. In normal operation of a photovoltaic plant, it presents almost no danger for the environment. The quantities used are astonishingly small due to the thinness of the active layer in the cell. Only 7 g are used per square meter, which represents approximately the cadmium content of two Mignon-type NiCd batteries.

However, there could be a hazard in the case of a fire. The temperature reached could be so high that gaseous cadmium would be released into the surroundings. On the other hand, in the case of a house fire, many other poisonous substances (dioxins, etc.) are released so that cadmium would only be one problem among many. At the same time, we note in comparison that, e.g. the German coal-fired power stations emit more than 1.4 t of cadmium into the air every year, and at the same time, they produce approximately 100 t of cadmium in slag form [78].

Special attention is paid to recycling solar modules. This is indispensable in order to justify the use of CdTe modules. The First Solar Company has instituted a free-return system for all sold modules.

According to them, the recycling process developed by the company achieves a recycling rate of 95% of cadmium.

5.7.1.2 Example of Silicon

Silicon looks much better regarding environmental friendliness. It is nonpoisonous and available in unlimited amounts in the form of quartz sand. However, in its production, a series of etching chemicals such as trichlorosilane are used for the silane process (see Section 5.1). Potassium hydroxide and hydrofluoric acid are used in the cleaning of the mono- and multicrystalline wafers. Also phosphoric and boric acids are used for doping the wafer. An increasing number of chemical companies are offering reprocessing plants for these materials in order to increase recycling quotas [79]. In the ideal case, integrated multi-industry companies can make use of the wastes produced in their own

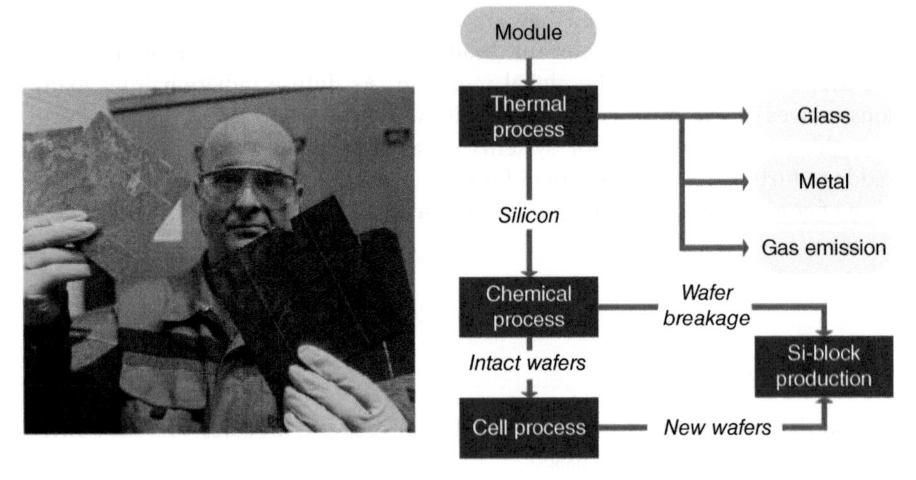

Figure 5.28 Recycling solar modules: Besides the recycling of glass and metal, the solar cells are chemically cleaned and made into "new" wafers again. Photo – Deutsche Solar GmbH; sketch is from [81].

companies. Thus, for instance, at Wacker, the used-tetrachlorosilane is converted into silicon dioxide and is then used for wall paints or even toothpaste [80].

Recycling of c-Si modules is also possible and has been carried out, e.g. by the Solar-World daughter Deutsche Solar GmbH (Figure 5.28). For this, the modules are first heated up to 500 °C so that the EVA laminate is dissolved. The cells are then separated manually and cleaned of the doping substances by means of etching. Originally, the plan was to reuse whole wafers. However, the cells of older modules often have sizes and thicknesses that are no longer in use today. Besides, the thin wafers used today break easily when being separated. For this reason, typically, wafer pieces are melted into ingots again and new wafers are produced in the block-casting process. The remaining materials such as glass, aluminum, copper, and silver can also be reused. In total, a recycling quota of 90% was achieved [82]. However, the Deutsche Solar GmbH ceased recycling solar modules after the prices for solar silicon fell sharply.

Meanwhile, there exists the European Directive 2012/19/EU on waste electrical and electronic equipment (WEEE). It states that every producer has to register his solar modules and to ensure their recycling after end of life.

5.7.2 Availability of Materials

Besides the environmental relevance of the substances used, their availability plays an important role. If photovoltaics are to be a support in the worldwide supply of energy, then the materials necessary for the production of solar modules must be available in sufficient quantities.

5.7.2.1 Silicon

In the case of silicon solar cells, the situation is very relaxed. Silicon is the second most common element on the Earth's surface and can be produced fairly easily from quartz sand (see Section 5.1). In recent years, however, there has been talk of a "scarcity of silicon." This, though, always referred to the already produced and highly refined polysilicon. The producers had underestimated the demand so that the price rose to more than 200€ kg^{-1}. Since massive capacity was added, there is now sufficient polysilicon available, which has reduced prices to below 20€ kg^{-1}.

5.7.2.2 Cadmium-Telluride

Regarding the availability of CdTe, tellurium is a critical material, as on Earth it is almost as scarce as gold. The estimate of the total available quantities is 21 000 t worldwide. Annually about 130 t are extracted mainly as a by-product in the extraction of copper and nickel, which is used primarily in steel production. Some studies assume that the extraction of tellurium could be increased to 600 t a^{-1} [83]. Let us assume that at best 500 t a^{-1} would be available for photovoltaics. How much photovoltaic power could be generated with this?

One needs approximately 7 g of tellurium for a square meter of module area. If in the future, a module efficiency of 15% is achieved, then the result is a requirement of 50 g tellurium per kilowattpeak. With the assumed extraction of 500 t a^{-1}, we would have a possible PV production of

$$(500\ \text{t a}^{-1})/(50\ \text{g kWp}^{-1}) = 10\ \text{GWp a}^{-1}.$$

Is that a lot or little? The quantity corresponds approximately to one-fourth of the solar module production of the year 2013. But what would the situation be in 2030? If we assume an annual growth of the PV market of 25%, then the yearly production would have increased to approximately 1650 GWp. CdTe modules would contribute to this world market only a maximum of 0.6% (see Figure 5.29).

With cadmium, the situation regarding availability is unproblematic. The annual production is 20 000 t a^{-1}; moreover, the use of cadmium is generally being reduced.

5.7.2.3 Cadmium Indium Selenide

With the CIS modules, it is mainly the indium that has limited availability as it occurs almost as rarely as silver [84]. The estimate of the total extraction is 6000–11 000 t, but actually, about 950 t are extracted, of which 800 t a^{-1} are required by other industries. The market for flat screens especially takes up a large part as indium is used there in indium oxide as a transparent electrode.

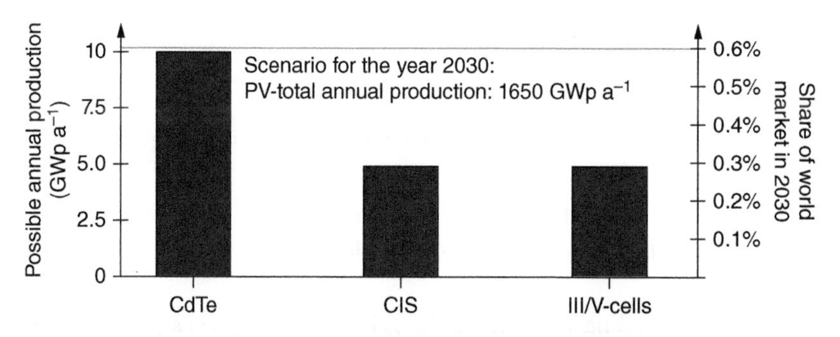

Figure 5.29 Maximum possible annual production of various PV technologies: They will only be able to make up a small portion for the assumed world market in 2030.

Since the triumphal development of the flat screen, the price of indium has risen 10-fold to 1000€ kg^{-1}. If one uses the remaining 150 t a^{-1} for CIS modules, then for a demand of 30 g kWp^{-1}, one arrives at a installable power of 5 GWp a^{-1} [83].

The selenium required, like tellurium, is also a by-product of copper and nickel production. The current annual production of 1500 t a^{-1} can easily be increased so that here there are no bottlenecks for PV production.

5.7.2.4 III/V Semiconductors

With the stacked cells discussed in Section 5.4.2, there is a problem with the availability of germanium. It occurs in relatively small concentrations in the rare earths and can only be extracted with much effort. The extracted quantities are about 90 t a^{-1}. Prices have risen in recent years as germanium is used in modern optical components. Assume that, despite this, half of the annual extraction could be used for photovoltaics. With the utilization of cells in concentrator systems, we can assume a future efficiency of 45%. For a concentration factor of 500, the demand for germanium is 9 g kWp^{-1} [83]. The possible annual production is then:

$$(45\ \text{t a}^{-1})/(9\ \text{g kWp}^{-1}) = 5\ \text{GWp a}^{-1}.$$

In summary, one can say that no availability problems are foreseen up to 2020. After that, one should expect raw material bottlenecks for CdTe, CIS, and III/V semiconductors.

5.7.3 Energy Amortization Time and Yield Factor

There is a persistent rumor that more energy is required for the production of photovoltaic plants than the energy generated by the plant in the course of its life. If this were actually the case, then one could hardly call photovoltaics an option for the solution of energy problems. It would then only be suitable for providing power for areas far from an electric grid (space, rural areas in the developing world).

In order to discover the amount of energy required for the production of a photovoltaic plant, let us assume a rooftop installation in Germany. This consists of multicrystalline solar modules, support structure, cables, and inverters. In a study by Erik Alsema in 2006, it was shown that the primary energy demand w_{Prod} for producing the installation was 7830 kW h kWp^{-1}. The cells under consideration had an efficiency of 13.2% and a wafer thickness of 285 μm, which corresponded to the state of production of 2004. Figure 5.30 shows that approximately three-quarters of the energy was required for the production of the cells. Of this, again, the largest part was for production of polysilicon.

The energy amortization time is more informative than the primary energy demand.

The energy amortization time, T_A, is the time that a solar power plant must work until it has generated as much energy as was required for its production.

If a photovoltaic plant is feeding into the public grid, then it replaces power from conventional power stations. Thus, every kilowatt hours that is fed-in is included in the number of kilowatt hours of primary energy it replaces. The primary energy factor, F_{PE},

Figure 5.30 A portion of the primary energy demand for the production of a PV plant with multicrystalline modules in year 2004.

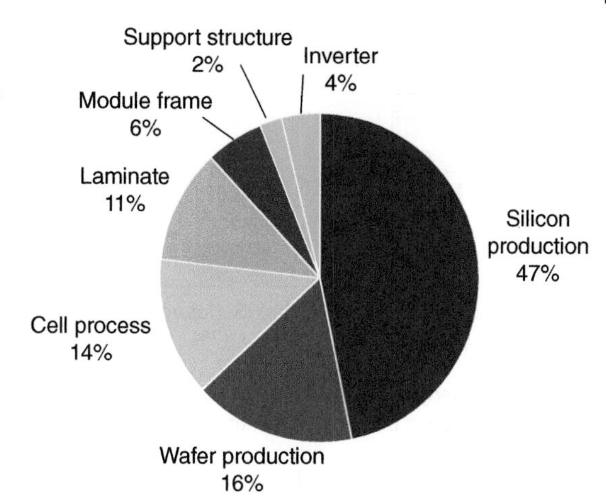

describes this relationship. For the typical Middle-European electric grid, we assume a value of $F_{PE} \approx 3$ (see also Chapter 1).

$$T_A = \frac{w_{Prod}}{w_{Year} \cdot F_{PE}} \tag{5.3}$$

with w_{Prod}: Specific production effort (primary energy); w_{Year}: Specific annual yield per PV plant.

In Germany, the specific annual yield of a plant is $w_{Year} = 900\,\text{kW h(kWp a)}^{-1}$. For the example of the rooftop installation with multicrystalline cells, the result is then

$$T_A = \frac{w_{Prod}}{w_{Year} \cdot F_{PE}} = \frac{7830\ \text{kWh kWp}^{-1}}{900\ \text{kWp (kWp a)}^{-1} \cdot 3} = 2.9\ \text{a}. \tag{5.4}$$

In Germany, the installation must be in use for almost 3 years in order to produce as much energy as was required for its production. This is a very good result in view of the plant lifetime of about 25 years.

This consideration immediately leads to the definition of the Energy Returned on Energy Invested (ERoEI).

The ERoEI is the amount of energy that a solar power plant generates in its lifetime, T_L, in comparison to the required production energy (always referenced to primary energy).

$$\text{ERoEI} = \frac{\text{Total generated energy}}{w_{Prod}} = \frac{T_L \cdot w_{Year} \cdot F_{PE}}{T_A \cdot w_{Year} \cdot F_{PE}} = \frac{T_L}{T_A}. \tag{5.5}$$

In our example, this results in

$$\text{ERoEI} = \frac{T_L}{T_A} = \frac{25\ \text{a}}{2.9\ \text{s}} = 8.6. \tag{5.6}$$

Thus, in the course of its period of operation, the plant generates more than eight times the energy required for its production.

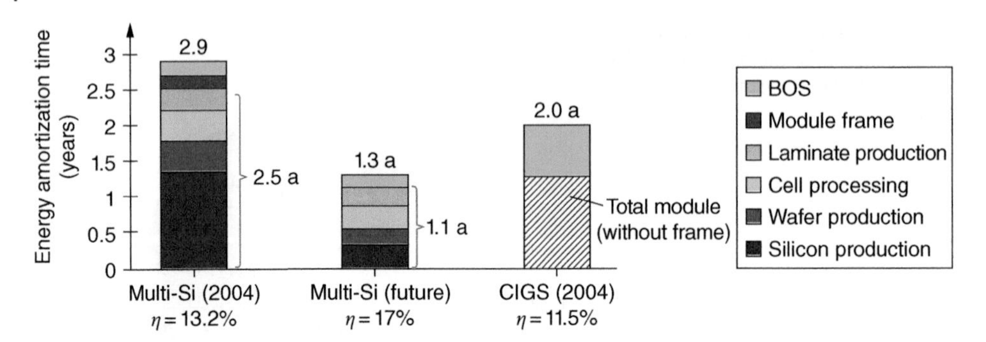

Figure 5.31 Energy amortization times of various types of plants: Modern production techniques substantially reduce the energy demand for plant production [49].

However, development does not stand still. Figure 5.31 shows the energy amortization times of three different technologies. The left bar shows the situation in the already considered multi-Si plant of 2004. Besides this is an analysis of a modern plant. Compared to the old plant, use was made of thinner wafers (150 μm) and higher efficiencies (17%). In addition to this, the silicon production of solar-grade silicon was carried out with the aid of a fluidized bed reactor instead of the older Siemens reactor. These framework conditions are largely normal practice in modern plants.

The result is a reduction of the energy amortization time from 2.9 to 1.3 years, which corresponds to an *ERoEI* of 19.

Figure 5.31 shows the results of a third plant with CIGS thin film modules from year 2004. Here an amortization time of 2 years has already been achieved. Because of the relatively low efficiency, the area-dependent material portions (glass of the modules, support structure, etc.) have a strong influence. In the region of thin film technology, future improvements in energy amortization time will be achieved mainly by means of higher efficiency.

Remarks: The term *Balance of System* (BOS) combines the components of the system technology (support structure, cables, inverter, etc.).

Possibly, the results of Figure 5.31 are too optimistic. A current study from 2013 states that the energy amortization time of 1.3 years will be only attained in 2020 [85]. The authors, among other things, assume that the cell efficiencies then have raised to 23% as a consequence of the optimization measures known from Chapters 4 and 5 (HIT concept, IBC structure, etc.). The wafer thickness is expected to reduce to 120 μm. This scenario results in an energy amortization time in Germany for scarcely one year for the module. For a complete PV plant, it is differentiated between a small rooftop plant of 2.5 kWp and a 4.6 MWp open air plant. The resulting amortization times are 1.4 and 1.3 years, respectively.

Up to now, we have always considered Germany as the site of photovoltaic plants. Operation of a plant in southern Europe (e.g. Spain) would reduce the energy amortization time by a factor of 1.7 again.

Would the energy amortization time be reduced further if recycled material is used in production?

At present, module recycling is not a part of the calculation of the energy amortization time. In fact, it would effect a further improvement of the environmental balance of photovoltaic plants in the future. The materials silicon, glass, and aluminum can be reused very efficiently. A saving of more than 95% can be achieved especially with aluminum.

What is the actual energy amortization of other power-generating plants such as wind turbines or coal-fired power stations?

Depending on their location, wind turbines have an energy amortization of only 3–5 months for a typical life of 15 years [86]. Conventional power plants never produce more energy than is used for their production and operation as ever more primary energy in the form of coal or gas or uranium must be provided. Their energy amortization is, therefore, infinite to a certain extent.

5.8 Summary

After having considered the individual cell technologies in detail, it is now time for a summary. Table 5.2 shows the efficiencies of the types of cells together with their most important advantages and disadvantages. In the second column, the respective peak efficiencies of laboratory cells are shown, and the third column shows the largest efficiency of the modules delivered on the market.

Figure 5.32 shows the development of the worldwide photovoltaic market with reference to the various cell technologies. The two crystalline technologies for years have dominated photovoltaic production with a market share between 80% and 90%. Interim, thin film technology could make up ground again especially driven with the CdTe production of First Solar. However, in the last years, the portion of c-Si cell installations went up again. It can be expected that in the thin film market, only CdTe and CIS will make their way. At the same time, the trend to high efficiencies will possibly raise again the portion of monocrystalline silicon.

Finally, let us look at Figure 5.33. This presents the development of the best cell efficiencies over the past 40 years. It shows the great advances that have been made since then.

In the upper part, we naturally find the stacked cells of III/V semiconductors under concentrated sunlight. The winner is the already known cell from Soitec described in Section 5.2 with an efficiency of 46%.

With the crystalline cells, we find the PERL cell with an efficiency of 25% developed by Martin Green at the UNSW in Australia (see Section 4.7). This one was shortly topped with a 23.3% cell from Fraunhofer ISE.

Table 5.2 Comparison of the various cell technologies.

Cell technology	η_{Cell_Lab} (%)	η_{Module} (%)	Important advantages and disadvantages
Mono c-Si	25	21.5	+ Very high efficiencies + Unlimited availability – Presently high-energy amortization time
Multi c-Si	21.9	17	+ High efficiencies + Unlimited availability + Acceptable energy amortization time
a-Si (single)	10.2	7	+ Low temperature coefficient – Efficiency too low
a-Si (triple)	13	8.2	
a-Si/μc-Si	11.8	10	+ Potential for improvements – Low efficiencies
CdTe	22.1	17	+ Medium efficiencies – Availability problem + Potential for improvements – Image problem + Low energy amortization time
CIS	22.6	16	+ Acceptable efficiencies + Potential for improvements + Low energy amortization time – Availability problem
Mono c-Si/a-Si (Hetero junction)	26.6	19.4	+ Very high efficiencies + Great potential for improvements
III/V semiconductors	39	n.a.	+ Extremely high efficiencies (with concentration over 40%) – Possible availability problem – Only sensible in concentrator systems

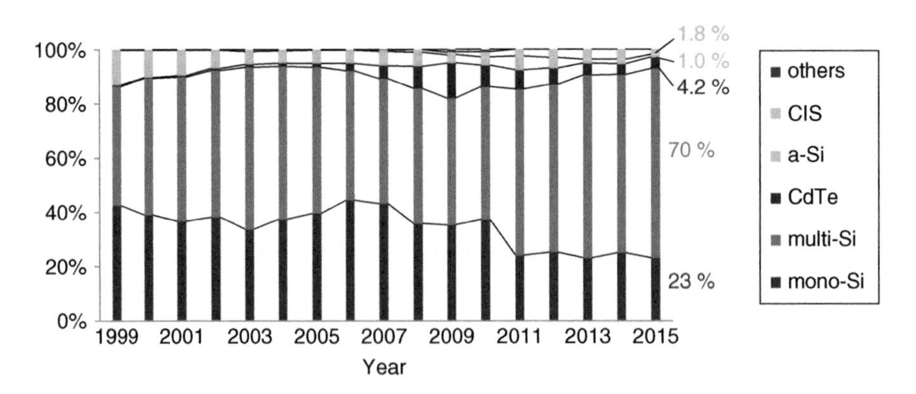

Figure 5.32 A portion of various cell technologies in percentage over the years. Crystalline silicon continues to dominate the world market with a share of more than 90% [87].

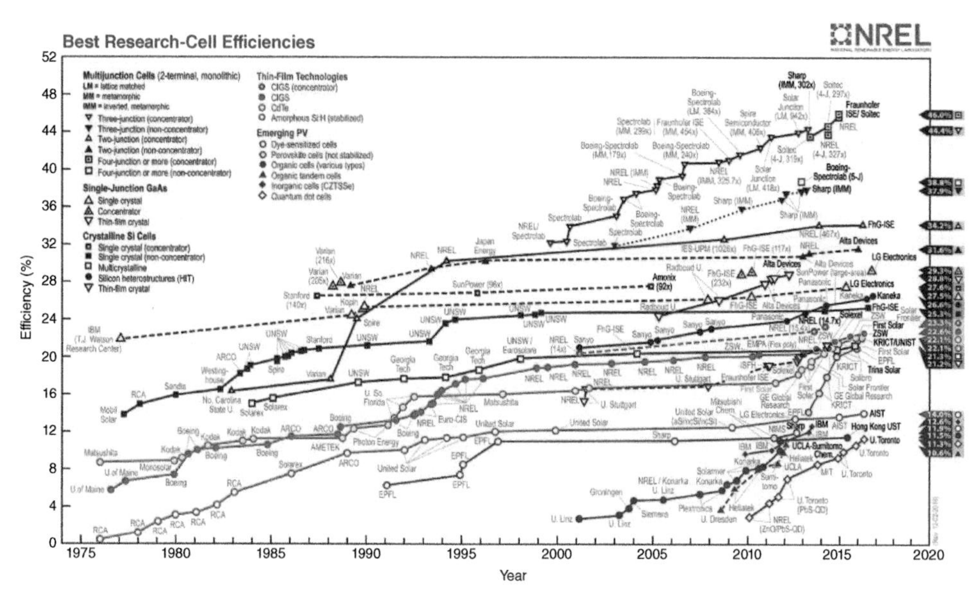

Figure 5.33 Development of record cell efficiencies in the past 40 years: Practically, all technologies continue to show increasing efficiencies. Source: NREL.

Also noticeable are the rapid advances in organic cells: Within only 15 years, the efficiencies have increased almost fourfold from 3% to nearly 12% today.

The steepest rise of efficiencies, of course, shows the perovskites with efficiencies from 0% to 22% (not stabilized) in 4 years.

As we are now very well acquainted with the principles and production of solar cells, we will now turn our attention in the next chapter to the application of solar generators.

6

Solar Modules and Solar Generators

In this chapter, we will become acquainted with the construction of solar generators, namely the interconnection of solar modules in series and parallel, connected into a direct current source. First, we will deal with the features of solar modules and the problems that can arise in the connection of modules. Then we will consider the special components of direct current technology and, subsequently, look at different structural variations of photovoltaic systems.

6.1 Properties of Solar Modules

The properties of solar modules (temperature coefficient, efficiency, etc.) are mainly determined by the solar cells used in those modules and, in addition, the type of interconnection in the module. Here parallel and series connections have different effects, especially in the case of partial shading.

6.1.1 Solar Cell Characteristic Curve in All Four Quadrants

If several cells are interconnected, then reverse voltages or reverse currents can easily occur in individual cells. The result is that these cells operate not only in the first but also in the second or fourth quadrant. As a reminder, Figure 6.1 shows the characteristic curves of a solar cell in all quadrants. Here, the generator reference-arrow system has been used again.

The first quadrant is called the active region, as this is where normal operation occurs and in which power is generated. The reverse voltage region is situated in the second quadrant: With rising reverse voltages, the starting avalanche breakthrough of the p–n junction (see also Section 3.5.4) can clearly be seen. The pass region is situated in the fourth quadrant. It is sometimes called the reverse current region as the current flows in the reverse direction to the normally flowing photocurrent.

6.1.2 Parallel Connection of Cells

We will first consider the parallel connection of cells in a solar module. Figure 6.2 shows a mini-module that will consist of three parallel-connected solar cells. The parallel connection forces all the cells to have the same voltage. At the same time, the individual

Photovoltaics – Fundamentals, Technology, and Practice, Second Edition. Konrad Mertens.

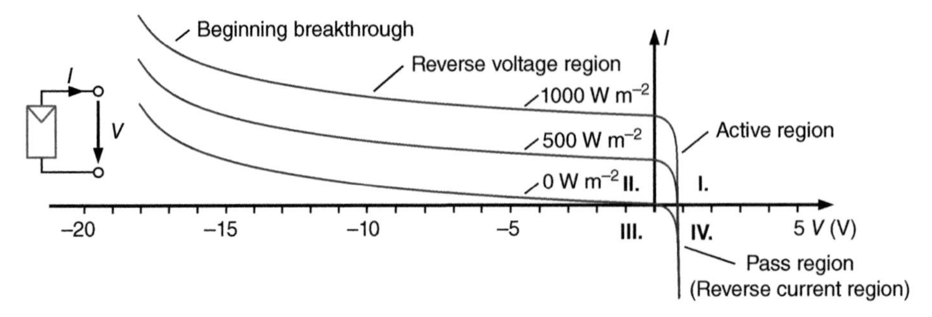

Figure 6.1 Solar cell characteristic curves in all quadrants in the generator reference-arrow system.

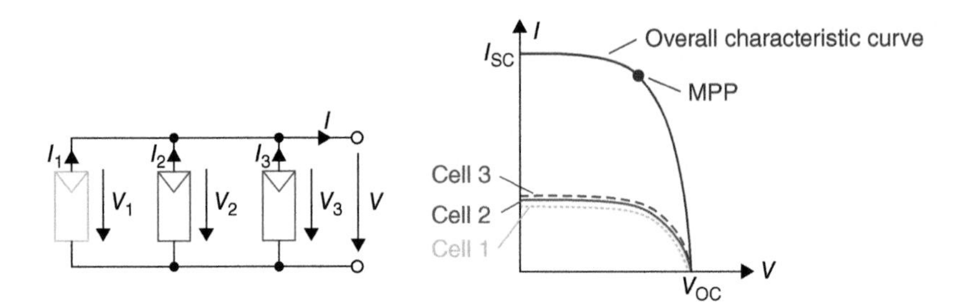

Figure 6.2 Parallel connection of solar cells: The voltage is the same in all cells, whereas the currents add up.

currents are added up:

$$V = V_1 = V_2 = V_3, \tag{6.1}$$

$$I = I_1 + I_2 + I_3. \tag{6.2}$$

The simplest method for drawing the overall characteristic curve (module characteristic curve) is to prescribe voltage values and then add the individual currents.

What happens when one of the cells is partly shaded? In the following, we will call this cell "Shady" (see Figure 6.3). We will assume that Shady is three-quarters shaded. We know from Chapter 4 that the open-circuit voltage of Shady changes only slightly, but the short-circuit current will decay by about three-quarters.

Figure 6.3 shows the effect on the overall characteristic curve of the solar module: It also decays by approximately the amount of the current loss of Shady, whereas the

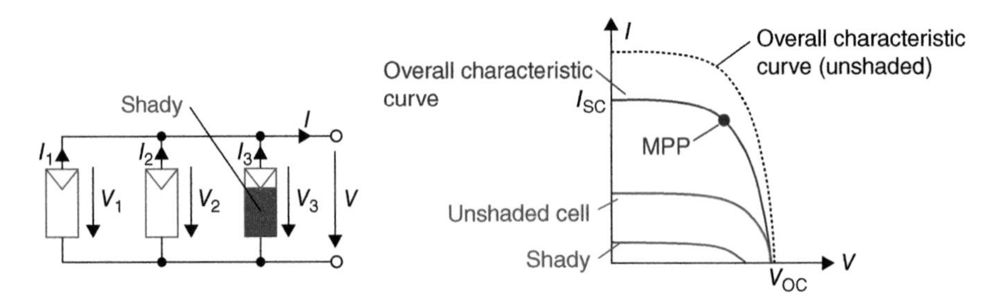

Figure 6.3 Shading one of the three cells: The current of the overall characteristic curve decays by the same amount as the current of Shady.

open-circuit voltage hardly changes. The power loss of the solar module is thus approximately one-quarter and corresponds to the proportion of the area of the module that was shaded. Therefore, the parallel connection reacts in a relatively benign way to the partial shading. When considering the series connection, we will see that this behaves in a very much worse fashion.

Why don't we connect all cells in parallel?

If all cells were connected in parallel, then the module would have an open-circuit voltage of only 0.6 V and a short-circuit current of, for instance, 100 A. To transport this current, we would need an extremely thick cable. In addition, typical photovoltaic plants (especially solar plants connected to the grid) would need much higher voltages that could only be generated from the 0.5 V with great effort.

6.1.3 Series Connection of Cells

As already described, one connects many cells in a module in series in order to achieve "decent" voltages. Figure 6.4 shows the effect of the series connection on an example of a three-cell mini-module: The current in all cells is the same, and the overall voltage is made up of the sum of the individual voltages:

$$I = I_1 = I_2 = I_3, \tag{6.3}$$

$$V = V_1 + V_2 + V_3. \tag{6.4}$$

The overall characteristic curve of a series connection can be determined graphically by adding the individual voltages for fixed current values.

What happens when one of the cells is partly shaded? For this, we will assume that we connect three equal cells in series, of which again one is shaded by three-quarters (see Figure 6.5).

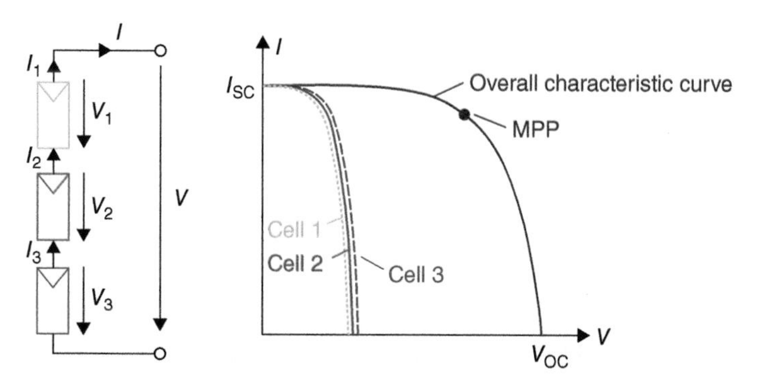

Figure 6.4 Series connection of solar cells: The voltages of individual cells are added together.

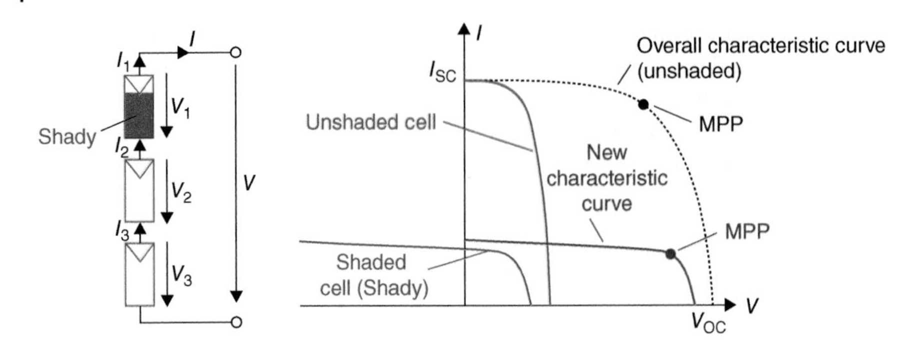

Figure 6.5 Partial shading of a cell with series connection: As Shady acts as the bottleneck, the overall current decreases strongly.

The two fully irradiated cells attempt to press their current through Shady. This causes a negative voltage at the shaded cell, and it is, therefore, operated partly in the second quadrant.

If we again add all the cell voltages at prescribed currents, then we obtain the new overall characteristic curve shown in Figure 6.5. The current is almost completely determined by Shady. The maximum power point (MPP) power of the module has been reduced by approximately three-quarters compared with the unshaded case, although just a single cell was shaded by three-quarters.

6.1.4 Use of Bypass Diodes

6.1.4.1 Reducing Shading Losses

The cells in modern modules are usually connected in series; this normally involves cell numbers of 36, 48, 60, or 72, leading to MPP voltages of between 18 and 36 V. Figure 6.6 shows a typical solar module with 36 cells and the resulting characteristic curve. Here, too, one shaded cell leads to a drastic power reduction from MPP_1 to MPP_2. Such a

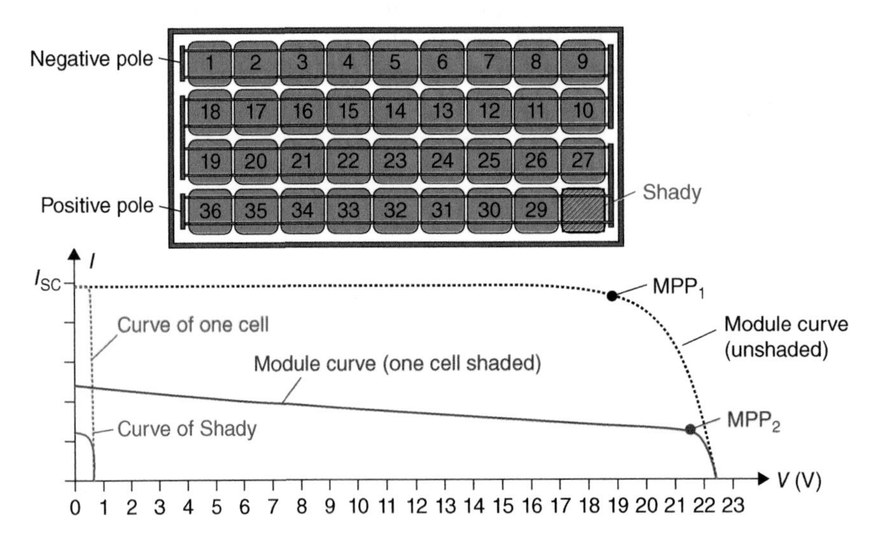

Figure 6.6 Solar module with 36 cells: The module power decreases drastically in the case of shading of a single cell.

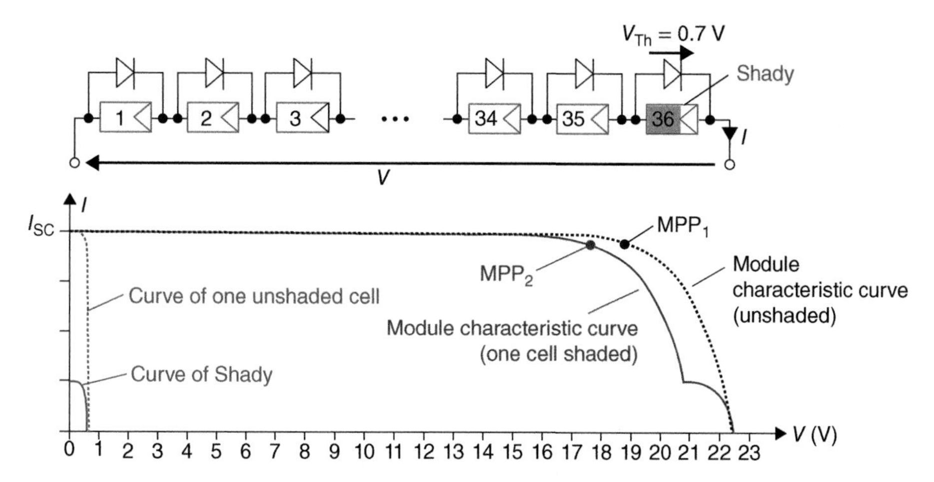

Figure 6.7 Solar module with 36 cells and bypass diode over each cell: The power loss is a minimum in the case of shading of any cell.

loss is unacceptable, which is the reason why so-called bypass diodes are included as additional components.

This is shown in Figure 6.7 for the example of Figure 6.6: A bypass diode is connected antiparallel to each solar cell. Provided that there is no shading, all cells have a positive voltage. This voltage acts as a reverse voltage for the diodes; they conduct no current and create no disturbance. If Shady is now again three-quarters covered by shade, then this cell has a negative voltage. This means that the diode conducts and Shady are bridged. The remaining 35 cells can, therefore, conduct their full current. However, the bypass diode has a threshold voltage, V_{Th}, of approximately 0.7 V, which is approximately the open-circuit voltage of a solar cell. Only when the current drawn off outside the module is as small as the current still deliverable by Shady will Shady's voltage become positive again. As a result, the bypass diode blocks, and Shady can still deliver a portion of the voltage (see the characteristic curve at the bottom right of Figure 6.7).

The adjusting MPP_2 in the case of shading is approximately two cell voltages lower than MPP_1. The power loss due to shading is

$$\frac{P_{MPP2} - P_{MPP1}}{P_{MPP1}} \approx \frac{I_{MPP} \cdot 34 \cdot V_{Cell} - I_{MPP} \cdot 36 \cdot V_{Cell}}{I_{MPP} \cdot 36 \cdot V_{Cell}} = \frac{-2}{36} = \frac{-1}{18} = -5.6\%. \quad (6.5)$$

The losses have been drastically reduced due to the bypass diodes.

However, only a few bypass diodes are used in real solar modules. If one wished to equip all cells with their own diodes, they would have to be accommodated in the very thin ethylene-vinyl acetate (EVA) encapsulation. It would hardly be possible to dissipate there the heat created in the diodes in the case of shading. Added to this is the fact that they could not be replaced in the case of a defect.

For this reason, the bypass diodes are situated in the module connection box. Typically, only one bypass diode is provided for 12, 18, or 24 cells.

The disadvantage of this solution is that the shading of a cell has a much stronger effect than is the case shown in, for instance, Figure 6.7. This is seen in Figure 6.8. Depending on the number of bypass diodes, a more or less large cell string falls away with shading. In the case of only two diodes per module, the decrease in module power with the shading of only one cell is approximately half.

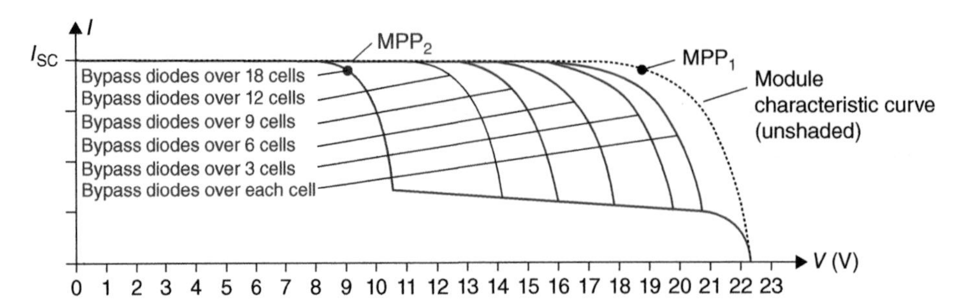

Figure 6.8 Characteristic curve of the solar module of Figure 6.7 with a different number of bypass diodes: With only a few diodes, the shading has a particularly strong effect.

6.1.4.2 Prevention of Hotspots

The bypass diodes are used for another reason besides the reduction of shading losses: To prevent the existence of hotspots. This is understood to be the massive heating of a shaded cell caused by the other series-connected cells. An explanation is given in Figure 6.9, which again shows a solar module with 36 cells of which one is shaded.

Let us assume that the module is being operated in the short-circuit mode. In this case, the 35 cells attempt to push their power through Shady. The current in Shady is still positive, but the voltage is negative. In order to find the operating point, we mirror the original characteristic curve of Shady at the axis of the current. This results in an operating point with a transferred power, which is a multiple of the normal MPP power of a cell. The result is massive heating of Shady. The temperatures thus obtained can cause damage to the EVA encapsulation or even lead to the destruction of the cell.

If a maximum of 24 cells is used per diode, then experience has shown that in this case, the heat power generated will not damage the module. Figure 6.10 again shows the module of Figure 6.9, which is now equipped with a bypass diode over each of the 18 cells. Because of the high internal resistance of Shady, the current of the lower 18 cells is pressed through the upper bypass diode. A loop in the upper mesh provides the existing voltage V_{Shady} at Shady:

$$V_{\text{Shady}} = (z - 1) \cdot V_{\text{Cell}} + V_{\text{Th}}, \tag{6.6}$$

where z is the number of cells under a bypass diode.

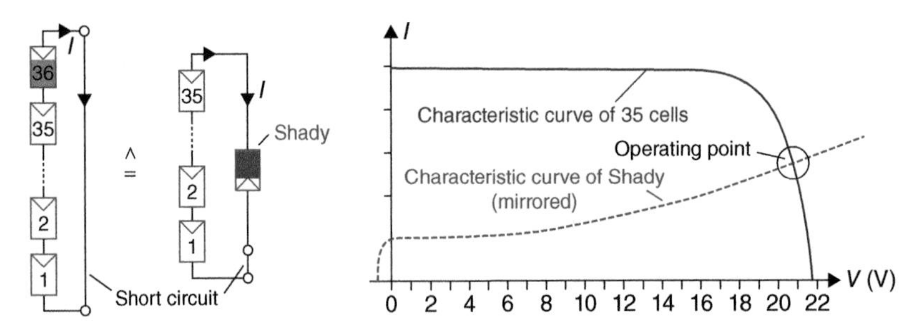

Figure 6.9 Solar module with 36 cells without bypass diodes: Shady acts as a load that is massively heated by the remaining 35 cells.

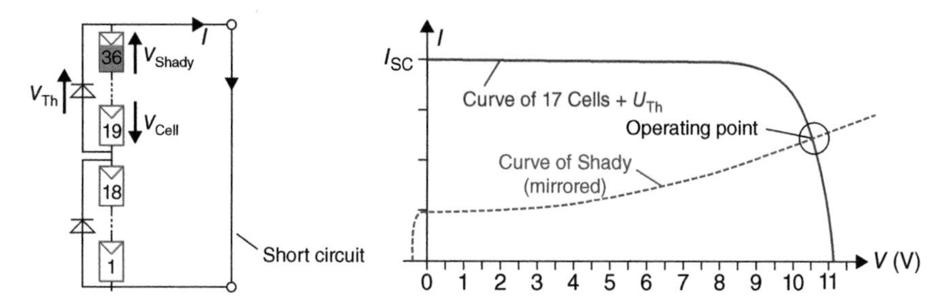

Figure 6.10 Solar module with 36 cells and two bypass diodes: The voltage at Shady clearly sinks compared with Figure 6.9 so that the heating is reduced.

Hence, Shady has a voltage of 17 cells and additionally, the threshold voltage, V_{Th}, of the bypass diode.

The new operating point is approximately at half the voltage compared with Figure 6.9; hence, the transmitted heat power is halved.

[?] The short-circuit case is always used in the last two figures, but a short circuit occurs only in the case of an error. Does the assumed heating also occur without a short circuit?

[!] Normally, the modules are connected in series with other modules and connected to an inverter. This operates the modules in the MPP. The MPP current of a module is only slightly below the short-circuit current. Thus, the actual heating of Shady is actually almost as great as in the short-circuit case.

[?] In which case does Shady heat up most: With greater or lesser shading?

[!] In general, this is hard to say. For this, we must check where the operating point sets with the respective shading. Figure 6.11 shows the characteristic curve for Shady in various degrees of shading. It can be clearly seen that the maximum power occurs for degrees of shading between one-quarter and half. However, different cell types have very different reverse characteristics that also depend on the temperature. One can generally only say that maximum heating occurs with medium shading.

[?] Is there also a hotspot by Shady in the partial shading of parallel-connected cells (Figure 6.3)?

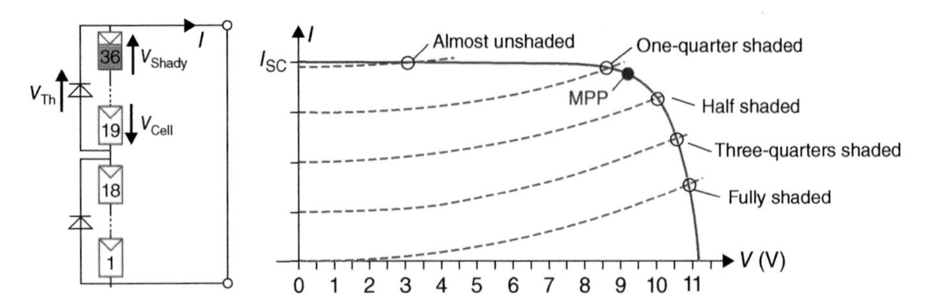

Figure 6.11 View of various degrees of shading: The transferred heat power reaches a maximum for degrees of shading between one-quarter and half of the cell.

We will have to look at this in greater detail. The worst case for Shady is certainly the open-circuit point of the module. In this case, the two unshaded cells will try to press their current through Shady (Figure 6.12). The direction of the current through Shady is therefore negative, whereas the voltage remains positive, and this corresponds to the load reference-arrow system. We will, therefore, find the self-adjusting operating point when we mirror the characteristic curve of Shady about the *x*-axis. The power imparted at this operating point to Shady is within the range of the MPP power of an unshaded cell. Therefore, no hotspot will arise at Shady.

This may not be critical for two parallel connected cells. But what happens when many cells are connected in parallel?

In this case, too, we can make an estimate based on Figure 6.12. Further, parallel cells lead us to add many cell characteristic curves to the shown characteristic curve of the two cells. It becomes ever steeper at the right edge and the operating point wanders upward. In the extreme case of an infinite number of parallel cells, there will arise an intersection point of Shady with a vertical line through V_{OC}. Even this operating point is not critical as regards the power.

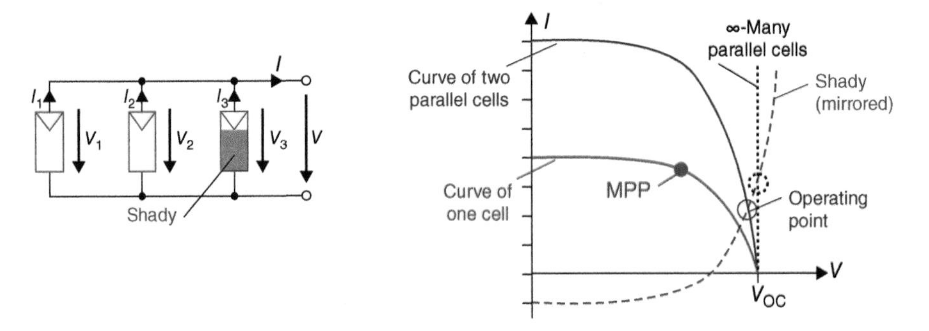

Figure 6.12 Characteristic curve of Shady and the two unshaded cells connected in parallel: In the case of an open circuit in the module, the resulting operating point shown receives little heat from Shady.

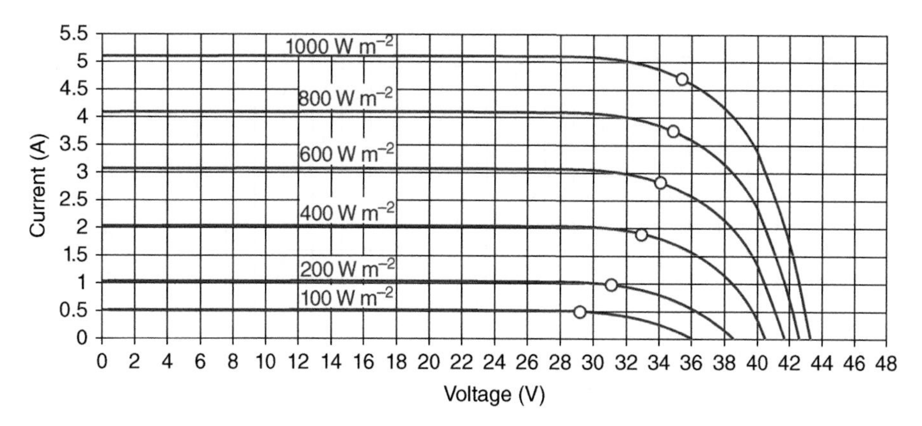

Figure 6.13 Characteristic curve of the SW-165 module at various irradiances and a constant module temperature of 25 °C (spectrum: AM 1.5).

6.1.5 Typical Characteristic Curves of Solar Modules

6.1.5.1 Variation of the Irradiance

Figure 6.13 shows the typical characteristic curve of the SolarWorld SW-165 solar module at a cell temperature of $\vartheta = 25\,°C$ and various irradiances. As already mentioned for the characteristic curve in Chapter 4, the short-circuit current increases linearly with the irradiance, whereas the open-circuit voltage alters little.

The purchaser of a solar module is not interested only in the nominal power of the module under standard test conditions (STC). Because of the high contribution of the diffuse radiation to the annual yield, the weak-light behavior of the solar module should also be taken into account. In the example of the SW-165, the MPP power for an irradiance of 200 W m^{-2} is only around 31 W. The efficiency has been reduced by

$$\frac{\eta_{200} - \eta_{1000}}{\eta_{1000}} = \frac{31\ \mathrm{W}/(200\ \mathrm{W\ m^{-2}}) - 165\ \mathrm{W}/(1000\ \mathrm{W\ m^{-2}})}{165\ \mathrm{W}/(1000\ \mathrm{W\ m^{-2}})} = -6\%. \tag{6.7}$$

This relatively small degradation of the efficiency is also shown in many module data sheets. The cause of this degradation is the dependence of the open-circuit voltage on the incident light described in Sections 4.4.2 and 5.6.2. For very small irradiances, the shunt resistance, R_{Sh}, of the standard equivalent circuit is also noticeable (Section 4.5). For a small photocurrent, there is only a small voltage, V_D, at the diode of the equivalent circuit so that it hardly conducts. Instead of this, the photocurrent is partly converted in the shunt resistance, R_{Sh}, into heat.

6.1.5.2 Temperature Behavior

In addition to the irradiance, the temperature behavior of solar modules is also of interest. As an example, Figure 6.14 shows the characteristic curve of the SW-165 at constant irradiance and different temperatures. The characteristic values from the data sheets are used here: TC(V_{OC}) = −0.39% K^{-1} and TC(I_{SC}) = 0.04% K^{-1}. Starting from 25 °C, the open-circuit voltage decreases from 43.2 to only 35.6 V at 75 °C. At the same time, the MPP power is reduced from 175 to 131 W.

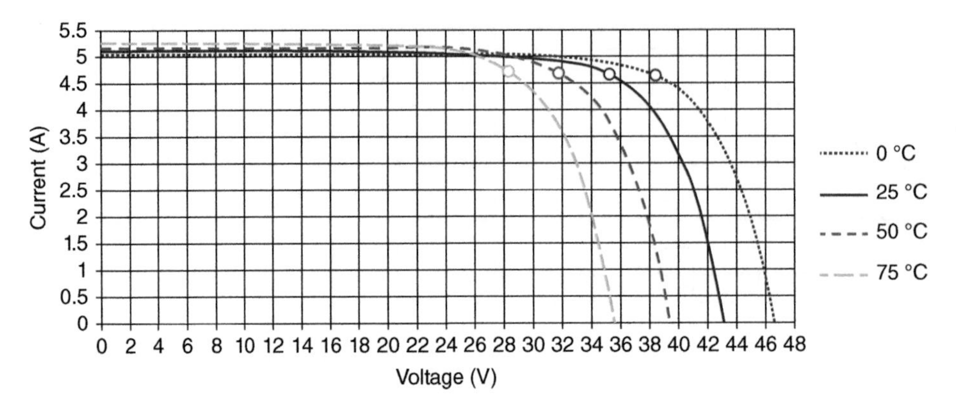

Figure 6.14 Characteristic curve of the SolarWorld SW-165 module at various module temperatures (irradiance, 1000 W m^{-2}; spectrum, AM 1.5).

In the data sheets of solar modules, the temperature coefficients TC(V_{OC}), TC(I_{SC}), and TC(P_{MPP}) are often designated by α, β, and γ.

In Figure 6.13, the irradiance was varied and the temperature was kept constant at the same time. These types of characteristic curves can be obtained in the laboratory with a module flasher that irradiates the module for only a few milliseconds with an air mass 1.5 (AM 1.5) spectrum and measures the characteristic curve at the same time (see Chapter 8). The module can barely heat up in this short period of time. However, things look quite different in the operation of a solar module in a photovoltaics plant. The solar module heats up differently with various irradiances. Added to this are influences such as types of construction and assembly of the modules, ambient temperature, wind velocities, and so on.

The data sheet gives the nominal operating cell temperature (NOCT) for estimating the self-heating of a particular module. NOCT is defined as the temperature that is arrived at for the following conditions:

- Irradiance $E = E_{NOCT} = 800$ W m^{-2}
- Ambient temperature $\vartheta_A = 20\,°C$
- Wind velocity $v = 1$ m s^{-1}

The typical NOCT temperature of c-Si modules is in the region of 45–50 °C.

If the NOCT temperature is known, then one can calculate approximately the expected cell temperature ϑ_{Cell} for a given irradiance and ambient temperature ϑ_A:

$$\vartheta_{Cell} = \vartheta_A + (\text{NOCT} - 20\,°\text{C}) \cdot \frac{E}{E_{NOCT}}. \tag{6.8}$$

Here, for the sake of simplicity, we assume that the temperature increase against the ambient temperature is proportional to the irradiance.

Example 6.1 ***Actual module power on a summer's day***
The 200 W Bosch c-Si M48-200 solar module has a NOCT temperature of 48.6 °C. What module power can be expected on a nice summer's day ($E = 1000$ W m^{-2}, $\vartheta_A = 30\,°C$)? ■

The actual cell temperature is

$$\vartheta_{\text{Cell}} = 30\,^{\circ}\text{C} + \left(48.6\,^{\circ}\text{C} - 20\,^{\circ}\text{C} \cdot \frac{1000\ \text{W m}^{-2}}{800\ \text{W m}^{-2}}\right) = 30\,^{\circ}\text{C} + 35.75\ \text{K} = 65.75\,^{\circ}\text{C}.$$

With the temperature coefficient $\text{TC}(P_{\text{MPP}})$, we obtain as actual power

$$\begin{aligned} P &= P_{\text{STC}} \cdot [1 + \text{TC}(P_{\text{MPP}}) \cdot (\vartheta_{\text{Cell}} - 25\,^{\circ}\text{C})] \\ &= 200\ \text{W} \cdot (1 - 0.47\%\ \text{K}^{-1} \cdot 40.75\ \text{K}) = 161.7\ \text{W}. \end{aligned}$$

The 200 W module thus generates a power of only 161.7 W.

6.1.6 Special Case Thin-film Modules

As already comprehensively described in Chapter 5, modules of thin-film materials often possess properties that deviate strongly from those of c-Si modules. Of special interest to us are the characteristic curve and the shading tolerance. As an example, Figure 6.15 shows the characteristic curve of the First Solar FS-275 CdTe module. Noticeable in comparison with c-Si modules is the low fill factor of 68%. This results mainly from the relatively high series resistances that are a consequence of the integrated series connection of cells. The power reduction at 50 °C is only 6.25% owing to the low power temperature coefficient of −0.25% K^{-1}.

Thin-film modules are also clearly differentiated from c-Si modules regarding the shading tolerance. As the individual cells are long and narrow, they are often only partly shaded.

Figure 6.16 shows the case where one module is shaded in the longitudinal direction (crosswise to the cell strips) by 25%. The current through all the cells is reduced by 25% so that also approximately only 25% of the module power is lost. This is substantially different from the case with lateral shading (parallel to the cell strips). The left cells are fully shaded so that their current collapses. As no bypass diodes are connected, even the remaining cells are hardly effective. The module power in the example decreases by approximately 90%.

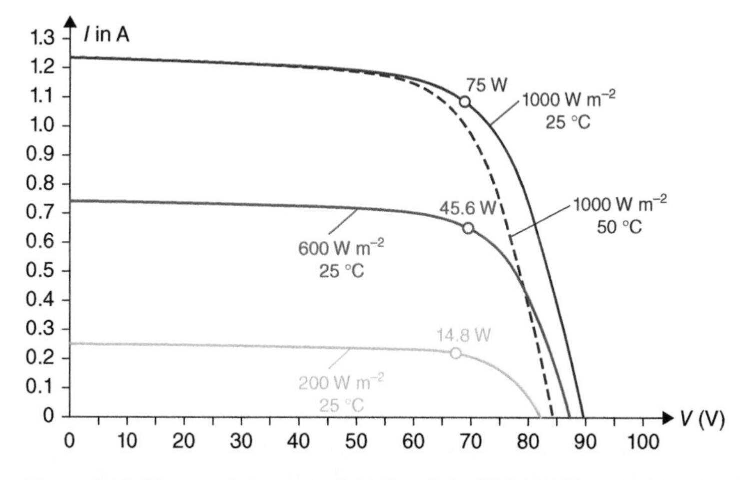

Figure 6.15 Characteristic curve of the First Solar FS-275 CdTe module: A disadvantage is the low fill factor of 68%, but the power decay is relatively small at high temperatures.

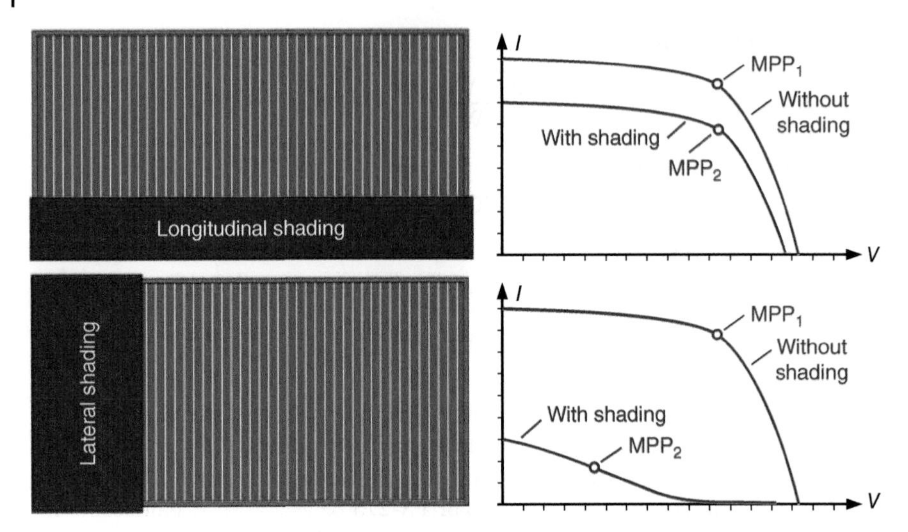

Figure 6.16 Comparison of longitudinal and lateral shading: In the case of lateral shading, the module power declines drastically as complete cells fail. *Source*: Own measurements [22].

Are bypass diodes used in thin-film modules? In addition, could hotspots occur in addition to shading losses?

Only a single bypass diode is installed in most thin-film modules. This is meant to prevent the shading of a module from reducing the power of the whole string. Hotspots do not occur as the thin-film cells have a relatively low breakdown voltage of 3–12 V. If a negative voltage occurs in a cell due to shading, then its reverse current increases until the breakdown voltage is reached. Thus, even without bypass diodes, no voltages higher than 12 V occur.

6.1.7 Examples of Data Sheet Information

Table 6.1 lists the most important technical data for some solar modules.

6.2 Connecting Solar Modules

6.2.1 Parallel Connection of Strings

To construct a solar generator, first a row of modules are connected in a string in series. This string can again be connected in parallel to further strings. Figure 6.17 shows a typical structure for this.

The string diodes shown are meant to prevent the situation that with the appearance of a short circuit or an earth fault in a string, all the other parallel-connected strings drive a reverse current through the defective string. A disadvantage of the string diodes, however, is the voltage drop associated with them: This causes a continuous power loss also in normal operation of the plant. For this reason, string diodes are seldom used nowadays and have been replaced by string fuses.

Table 6.1 Technical data for some solar modules.

Designation	SW-280 mono	Q.PRO- G3 270	SPR-X21345-COM	HIT N245	US-64	U-EA120	FS-102A	MS-175 GG-04
Manufacturer	SolarWorld	Hanwa *Q* Cells	SunPower	Panasonic	United Solar	Kaneka	First Solar	Miasolé
Type of cell	mono-Si	multi-Si	mono-Si	HIT	a-Si	μc-Si/a-Si	CdTe	CIGS
Nominal power P_{STC} (Wp)	280	270	345	245	64	120	102.5	175
Nominal current I_N (A)	9.07	8.82	6.02	5.54	3.88	2.18	1.47	8.33
Nominal voltage V_N (V)	31.2	30.8	57.3	44.3	16.5	55	70	21
Short-circuit current I_{SC} (A)	9.71	9.47	6.39	5.86	4.8	2.6	1.57	9.15
Open-circuit voltage V_{OC} (V)	39.5	38.9	68.2	53.0	23.8	71	88	27.8
Temperature coefficient $TC(I_{SC})$ (% K^{-1})	0.04	0.04	0.033	0.03	0.1	0.056	0.04	−0.03
Temperature coefficient $TC(V_{OC})$ (% K^{-1})	−0.3	−0.33	−0.25	−0.25	−0.31	−0.39	−0.28	−0.35
Temperature coefficient $TC(P_{MPP})$ (% K^{-1})	−0.45	−0.42	−0.3	−0.29	−0.21	−0.35	−0.29	−0.4
NOCT (°C)	46	45	41.5	44	46	45	45	45
Module efficiency η_M (%)	16.7	16.2	21.2	19.4	7.5	9.8	14.2	16.3
Number of cells	60	60	96	72	11	106	216	88
Number of bypass diodes	3	3	3	3	11	1	0	44
Length L (mm)	1675	1670	1559	1580	1366	1210	1200	1611
Width W (mm)	1001	1000	1046	798	741	1008	600	665

Source: www.enfsolar.com/pv/panel.

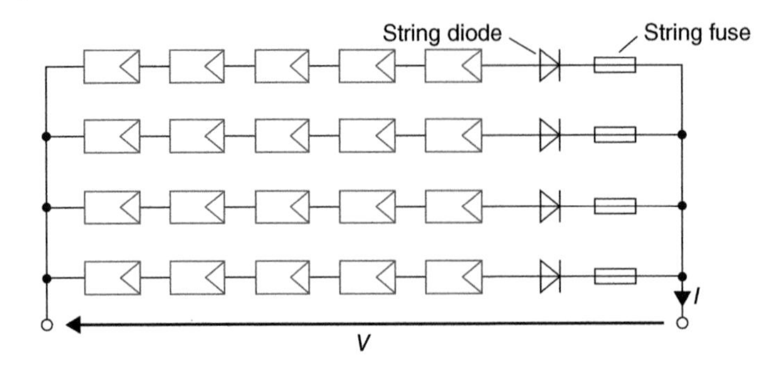

Figure 6.17 Structure of a solar generator with several strings.

String fuses are special sand-filled DC fuses that safely extinguish the electric arc when fuses burn through. They are typically dimensioned with double the nominal string current as solar modules can easily accommodate reverse currents of double or triple the nominal current. However, this means that generators with up to three parallel strings do not need string fuses, as in this case a maximum of two strings would feed their current through the defective string. At the same time, it must be ensured that the cables are designed to handle the increased strength of current.

6.2.2 What Happens in Case of Cabling Errors?

Assume that during the cabling of a solar generator an unequal number of modules per string are interconnected. For an example of this, see the generator in Figure 6.18: It contains two strings each with four modules and one string with only two modules.

What current will flow in the right-hand string? The worst case is certainly the open-circuit case in which both the left-hand strings press their current through the right-hand string. Thus, in a manner similar to Figure 6.12, we could depict the right-hand string in the load reference-arrow system. There arises an operating point at which a reverse current in the right-hand string is set to almost double the module short-circuit current. This operating condition is not critical for both modules.

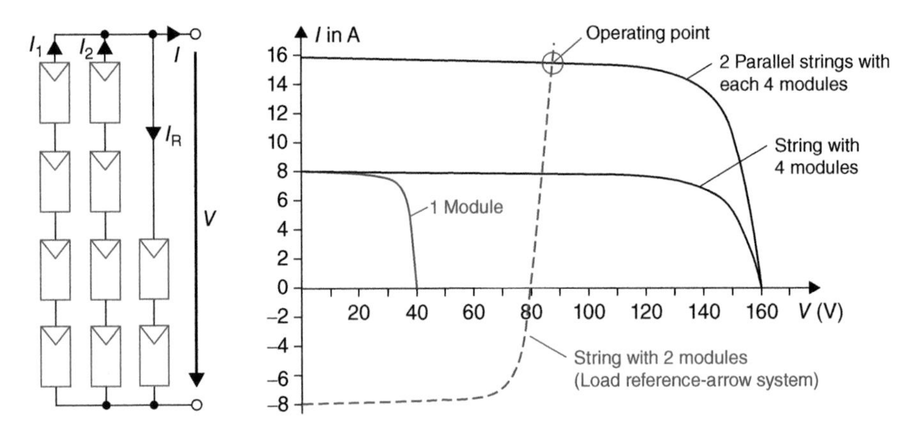

Figure 6.18 Structure of a solar generator with cabling error: The right-hand string has two modules too few so that a reverse current of almost double the module short-circuit current is set.

At the generator connections, in this example, in the open-circuit case, one would measure a voltage of approximately 90 V instead of the expected 160 V.

This type of cabling error would be discovered by a simple measurement with a multimeter.

6.2.3 Losses Due to Mismatching

If one purchases several modules of the same type, then they are not identical in their voltages and currents because of manufacturing tolerances. For this reason, when connecting the solar modules to a generator, it can occur that the overall power of the modules is not the same as the sum of the individual powers. The losses occurring in this way are called mismatch losses.

In order to explain this phenomenon, we will assume a string of four 100 W modules (Figure 6.19). One of the modules will have a nominal power of only 90 W owing to manufacturing tolerances. In principle, we can see this as though a normal 100 W module is only irradiated with 900 W m^{-2}. The I/V characteristic curve shows a new MPP_2 that lies at 384 W, which is about 6 W below the sum of the module power of 390 W.

The loss occurs because the three good modules can no longer contribute their full current to the MPP_2.

Thus, one can state for the installation of a photovoltaic plant:

> If the purchased modules show clear manufacturing tolerances, then they should be sorted in such a way that always modules with the same short-circuit current are combined into a string.

6.2.4 Smart Installation in Case of Shading

In Section 6.1, we discussed the shading losses in solar modules in detail. When connecting many solar modules, thought must also be given to how the losses of shading can be kept as small as possible.

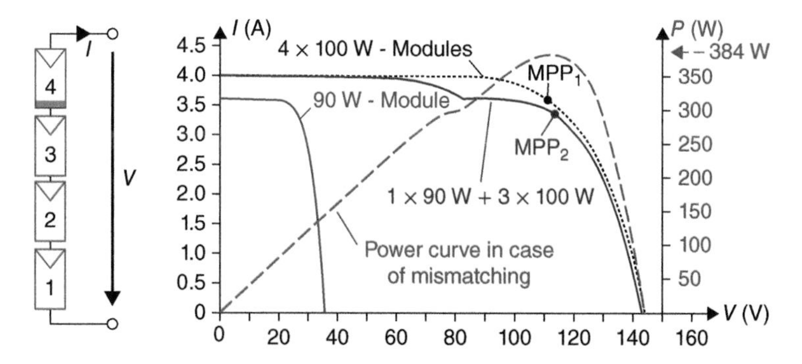

Figure 6.19 Effect of mismatching losses: The series connection of three good and one bad module leads to an overall power of only 384 W instead of the expected 390 W.

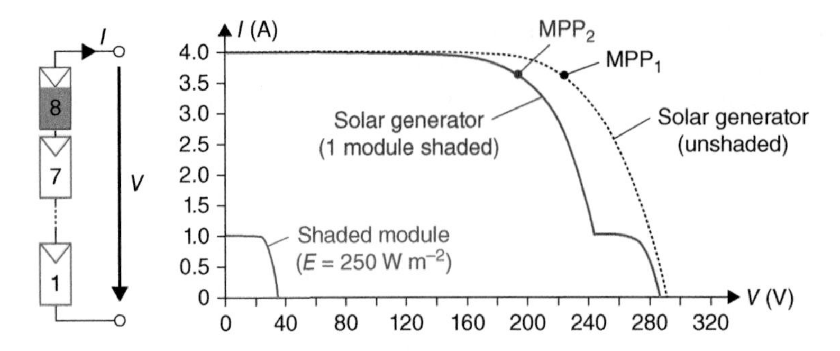

Figure 6.20 Series connection of all eight modules into a string: The shading of a module has only a small effect on the overall power.

Let us assume that we wish to build up a solar generator of eight modules, each with 100 W. The first possibility is series connection of all modules into a single string (Figure 6.20). The unshaded modules will produce approximately 800 W in the MPP (MPP_1 in Figure 6.20). If module 8 is shaded by three-quarters, then the current also drops by three-quarters. As the module is equipped with bypass diodes, the remaining modules feed their power past module 8. Thus, the MPP moves by approximately one module voltage to the left so that the result is a new MPP power:

$$P_{\mathrm{MPP2}} = 196\ \mathrm{V} \cdot 3.57\ \mathrm{A} = 700\ \mathrm{W}.$$

This means that the shaded module makes practically no contribution to the overall power.

What happens now if we build up the solar generator from two strings each with four modules (Figure 6.21)? In the unshaded condition, the result is again the maximum power of 800 W. If module 8 is shaded again as before, then here too the MPP will move by almost one module voltage to the left. The result is the new MPP_2 from Figure 6.21:

$$P_{\mathrm{MPP2}} = 89\ \mathrm{V} \cdot 7.1\ \mathrm{A} = 632\ \mathrm{W}.$$

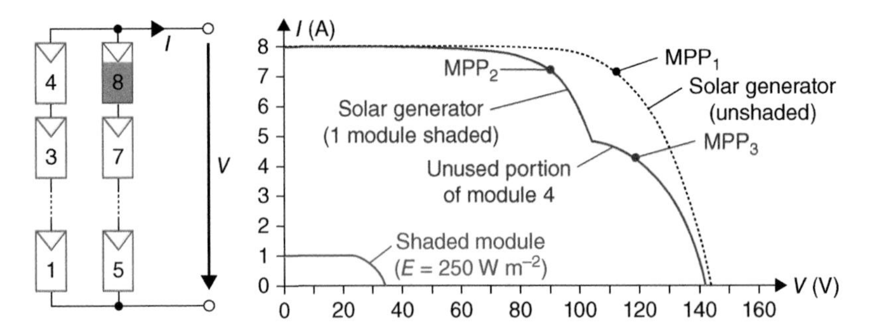

Figure 6.21 Connection of eight modules into two strings each with four modules: The losses for shading of a module doubles compared with Figure 6.20.

Compared with the first case, we have lost almost 70 W.

Why do we have greater power losses now? As before, only one of eight modules is shaded.

Due to the division of the solar generator into two strings, the shading of module 8 also affects the left string. Module 4 can make almost no contribution to MPP_2 power as a large part of its power is unused (see Figure 6.21). The shading of module 8 thus acts as if module 4 is shaded as well as module 8.

In general, one can derive a rule of thumb from this:

If there is a danger of shading in a solar plant, then all possible modules should be combined into a single string. This string should possess its own MPP tracker (see Chapter 7).

The effects of mismatching or of partial shading can be visualized with the aid of the software tool *PV-Teach* (see Figure 6.22). It offers the possibility of connecting different modules in series and again connecting the resulting string in parallel to further

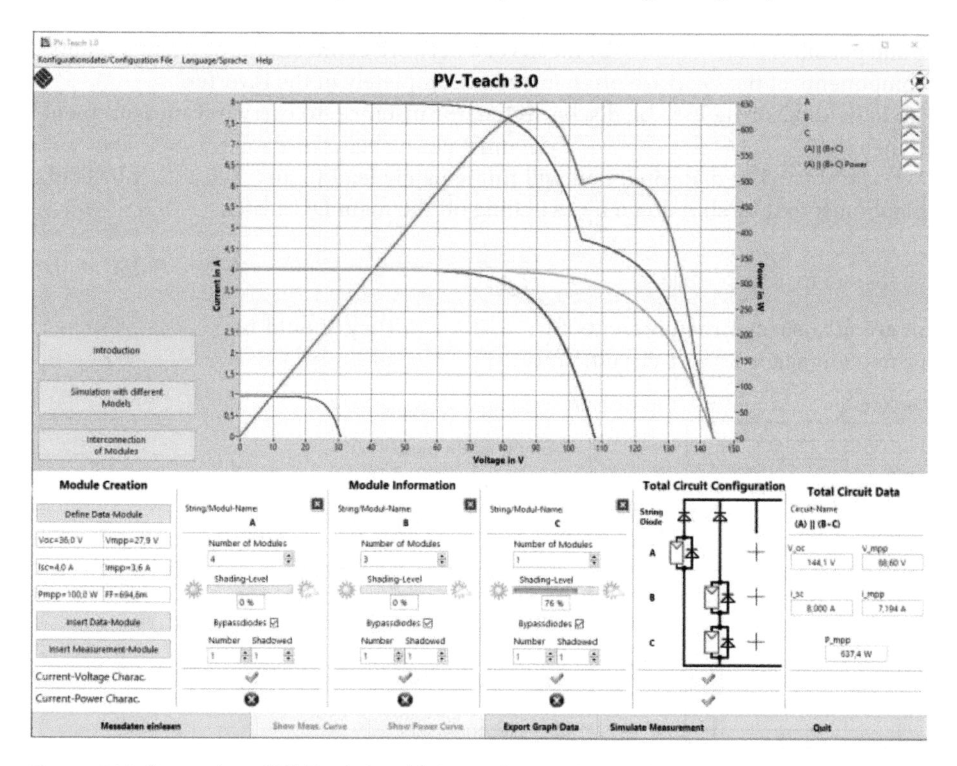

Figure 6.22 Screenshot of PV-Teach: In addition to the simulation of mismatching and partial shading, the influence of bypass diodes and string diodes can also be examined.

strings. Moreover, the influence of bypass diodes and string diodes can be examined. In Figure 6.22, the circuitry of Figure 6.21 is shown as an example. *PV-Teach* can be downloaded for free from website *www.textbook-pv.org*.

6.3 Direct Current Components

6.3.1 Principle of Plant Construction

The typical overall structure of a photovoltaic plant connected to an electric grid is shown in Figure 6.23. The current of the series-connected solar modules is fed via the string line to the generator connection box (GCB). In addition to the already discussed string fuses, this possibly contains a DC disconnector with which each individual string can be disconnected. The two varistors ensure protection against thunderstorm-caused voltage peaks.

From the GCB, the main DC line leads to the inverter. The safety standards for the installations of photovoltaic plants typically claim a main switch in order to disconnect the solar generator from the inverter safely. This main switch must be specially designed for direct current because with direct current the resulting electric arc is not self-extinguishing when disconnected.

Figure 6.24 shows the interior of a GCB. In addition to string fuses and overload voltage protection, the main DC switch was also directly built in here. In addition, the model contains the measurement instrumentation that monitors the string currents and string fuses and permits remote control by means of a network connection. For smaller plants, the components of the GCB are often included completely in the inverter.

If an individual string is to be disconnected, for instance, to carry out measurement on it, then the DC disconnector terminals must not be pulled under any circumstances while under load: The disconnection will immediately lead to an electric arc. Instead, the plant must first be shut down by switching off the main DC switch.

6.3.2 Direct Current Cabling

There are also special requirements for DC cables. They should be double insulated for protection against short circuits; mostly positive and negative lines are carried in

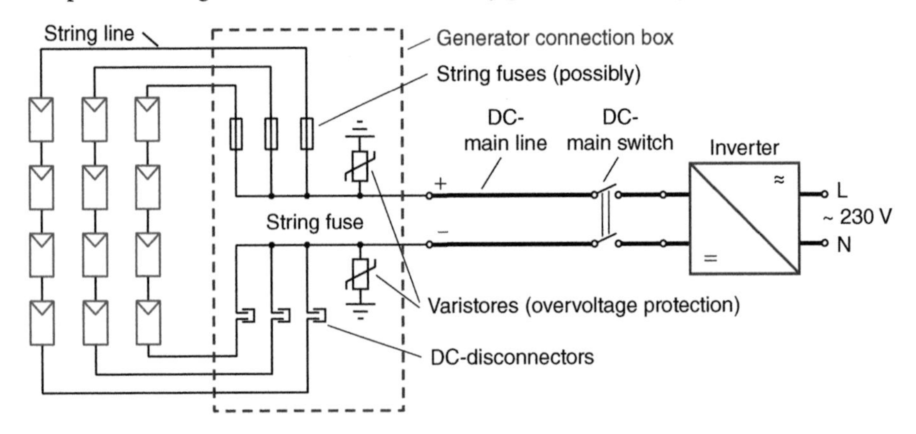

Figure 6.23 Structure of a typical photovoltaic plant connected to the electric grid: The individual strings are combined in the generator connection box and further connected via the main line to the inverter.

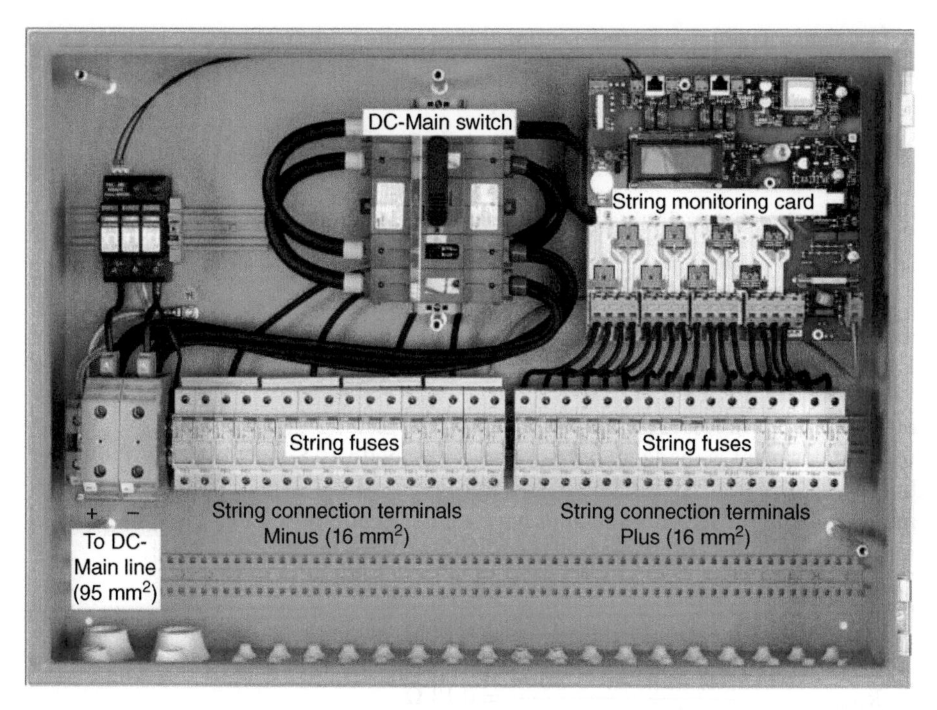

Figure 6.24 Interior of a modern generator connection box: In addition to the typical components of DC main switch, string fuses, and overload voltage protection, it also contains electronic instrumentation for remote monitoring of strings. *Source*: Sputnik Engineering AG.

different cables. As the string cables are subject to the weather, solar radiation, and high temperatures, they must be UV resistant, flame retardant, and designed for high operating temperatures. Solar connectors have been developed for connecting the modules and these provide simple and safe connections. They are designed such that no accidental touching of the contacts can occur. Figure 6.25 shows an example of the connectors from Multicontact that has become a quasi-standard. The type MC-3 has mostly been replaced by the newer type MC-4 as this possesses a lock against accidental disconnection.

In order that the expensive electrical energy is not immediately turned into heat, care must be taken to ensure a sufficient cross-sectional area of the cables. In order to determine the losses, resistance, R, of the cables is first calculated:

$$R = \frac{\rho \cdot l}{A}, \tag{6.9}$$

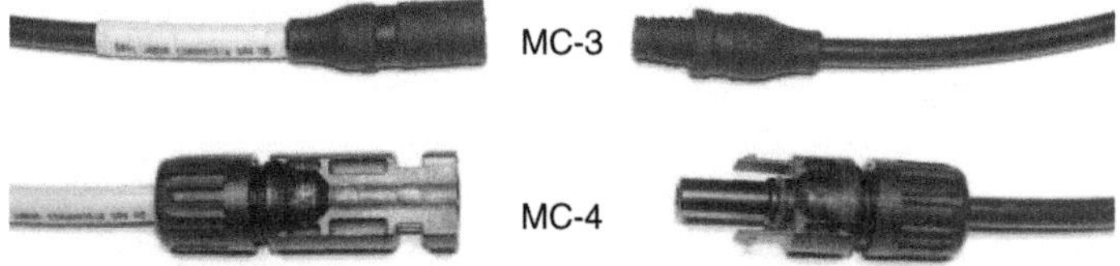

Figure 6.25 Solar plug assortment from Multicontact: The MC-4 has a mechanical lock to prevent accidental disconnection.

where l = length of cable; A = cross-sectional area of the cable; ρ = specific resistance of copper: $\rho_{Cu} = 0.0175\,\Omega\,mm^2\,m^{-1}$.

The length l must be the overall length of positive and negative lines. The cable losses can then be determined with

$$P_{Loss} = I^2 \cdot R. \tag{6.10}$$

There is a simple dimensioning rule for the maximum cable losses:

The electrical cable losses on the DC side should be a maximum of 1% of the nominal power of the plant.

Example 6.2 ***Cable losses on the DC side***

A photovoltaics plant is made up of two strings each with 10 SolarWorld SW-280 solar modules (see Table 6.1). Each string has a cable length of 2 × 10 m and the length of the DC main line is 15 m. The module lines have a cross-sectional area of 2.5 mm^2 and the DC main line 4 mm^2. Are the losses tolerable?

We first calculate the resistance of the lines:

$$R_{String} = \frac{0.0175\ \Omega\ mm^2\ m^{-1} \cdot 20\ m}{2.5\ mm^2} = 0.14\ \Omega,$$

$$R_{Main} = \frac{0.0175\ \Omega\ mm^2\ m^{-1} \cdot 30\ m}{4\ mm^2} = 0.13\ \Omega.$$

The losses are as follows:

$$\begin{aligned} P_{Loss} &= 2 \cdot I_{String}^2 \cdot R_{String} + I_{Main}^2 \cdot R_{Main} \\ &= 2 \cdot (9.07\ A)^2 \cdot 0.14\ \Omega + (18.14\ A)^2 \cdot 0.13\ \Omega, \\ &= 23.0\ W + 42.8\ W = 65.8\ W. \end{aligned}$$

■

Compared with the plant nominal power of 5.6 kWp the loss is 1.2% and is thus slightly too high. Here a DC main line with a cross-sectional area of 6 mm^2 would be advisable.

The loss of 1.2% calculated in the example does not mean that the operator of the plant loses 1.2% of the generated power every year. Seen over the year, the plant mostly works in part-load operation. As the losses increase with the square of the current, then, for instance, with a half-power operation only one-quarter of the calculated value of 1.2% is lost. In considering the typical occurrence of the various power levels, a power loss coefficient, k_{PL}, of approximately 0.5 [40] is defined. With the calculated power loss of 1.2%, the energy loss would be approximately 1.2% × 0.5% = 0.6%.

6.4 Types of Plants

Because of the modularity of the photovoltaic technology, solar power plants can be erected in fully different sizes and structural environments. We will look at the most important variants in the following.

6.4.1 Ground-mounted Plants

Ground-mounted plants are mostly built as large plants such as solar parks in the megawatt range. As an example, Figure 6.26 shows the citizen's solar power plant at

(a) (b)

Figure 6.26 View of two solar power plants: (a) Citizen's solar power plant at Hofbieber-Traisbach with 1.3 MWp of multicrystalline modules. *Source*: IBC SOLAR AG. (b) 3.5 MWp solar power plant at Mehring on the Moselle with CdTe modules. *Source*: juwi Solar GmbH.

Hofbieber-Traisbach in Bavaria, Germany, which has a total power of 1.3 MW with 5586 multicrystalline modules. The second example is a 3.5 MW solar power plant near Mehring on the Moselle consisting of 45 560 CdTe modules installed by First Solar.

The types of construction of ground-mounted plants range across a wide spectrum. The foundation of the overall structure is usually a ram foundation in which a long steel rod or profile is driven into the ground. The material for the module support structure is mostly aluminum or galvanized steel (Figure 6.27). The modules are fixed to these rails by means of module clamps.

A variant of the ram foundation is the screw-in foundation, in which a spiral-shaped tube is screwed into the ground to ensure a secure support (Figure 6.28). If the ground is rocky or if construction must progress especially quickly, then use can be made of concrete foundations, which will support the overall construction with their own weight.

Some ground-mounted plants are also built with tracking systems. Here one differentiates between dual- and single-tracker plants. The dual-axis plant provides a much greater yield that is approximately 30% in Germany (see Chapter 2). However, the mechanically complex design means a substantial price rise. In addition, the individual systems must be erected at relatively greater distances apart in order to prevent

Figure 6.27 Typical construction of a ground-mounted plant: Galvanized steel posts are rammed into the ground and serve as the base for the module panel support. *Source*: Schletter GmbH.

Figure 6.28 Alternative ground fixings: Screwed foundation and concrete heavy load foundation. *Source*: Krinner Schraubfundamente GmbH, Schletter GmbH.

mutual shading with a low-lying Sun. Most the plants make use of astronomic trackers in which the module always faces the Sun even when it is not visible. The brightness tracker, on the other hand, is based on a light sensor, in order to fix on a point in the sky on cloudy days that is brighter than in the direction of the Sun. Which system is actually better suited will probably continue to be a matter of faith.

Figure 6.29 shows Solon-Movers on the left, each with 8 kWp power. The solar power plant Gut Erlasee in Germany was equipped with 1500 Solon-Movers from Solon that follow the Sun in a dual-axis manner. However, the relatively expensive dual-axis trackers are hardly used today owing to large price decreases for solar modules in the past few years.

An alternative is single-axis tracker plants. Figure 6.29 shows on the right, a system with a horizontal axis in which the south-facing modules always only follow the height of Sun (elevation). Here a hydraulically operated pusher rod moves up to 12 module rows simultaneously. These systems are relatively cheap to produce and hardly cause mutual

Figure 6.29 Tracked ground-mounted plants: Dual-axis Solon-Mover and a single-axis system with horizontal axis. *Source*: © SOLON Energy GmbH, Urnato/© SOLON.

shading. The yield increase, according to the information from the producer, is between 12% and 18% in Germany, whereas in southern Europe a yield increase between 21% and 27% is achieved.

6.4.2 Flat-roof Plants

In the case of flat roofs, aluminum supports originally dominated the market. Figure 6.30 shows as an example the photovoltaic teaching plant erected in 1994 on the roof of Münster University of Applied Sciences in Steinfurt. A heavy load foundation was achieved with the use of flagstones so that no roof penetrations were necessary.

Today, more and more use is being made of systems with plastic tubs. Figure 6.31 shows a 25 kWp photovoltaic plant from 2008 at Münster University of Applied Sciences that was erected by the *fair-Pla.net* nonprofit cooperative. The polyethylene tubs are weighed down with paving stones. Horizontal metal tubes serve to carry the string cables, which are thus protected from UV radiation and movement.

In the case of roofs that can carry only small area loads, the classic heavy load foundation is only to be recommended after investigation by a statics expert. As an alternative, there are an increasing number of systems that manage with little weight. An example

Figure 6.30 Photovoltaic plant at Münster University of Applied Sciences from 1994: The aluminum support structure is held down by means of flagstones. *Source*: W. Göbel.

Figure 6.31 Flat-roof system on the roof of Münster University of Applied Sciences in 2008: The modules are placed in plastic tubs that are weighed down by paving stones.

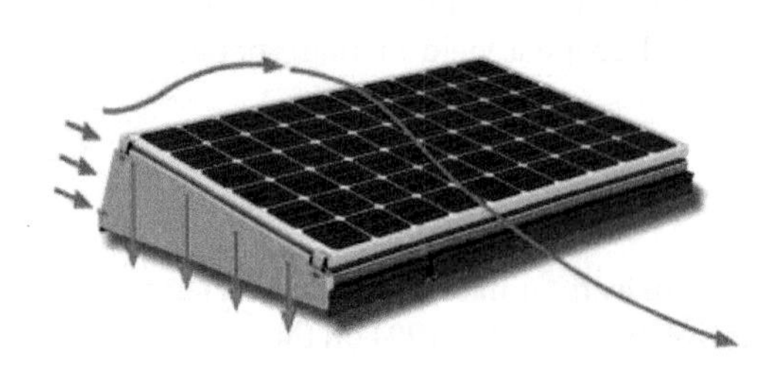

Figure 6.32 Solution for roofs with low bearing capacity: Because of the low closed form, the system offers hardly any resistance to the wind; instead, the elements are pressed down by the wind. *Source*: LORENZ-Montagesysteme GmbH.

is shown in Figure 6.32 with the LORENZaero10 model: The low, enclosed shape provides the wind with hardly any surface of attack. Instead, the elements are pressed down on the ground by the wind from the front (southerly wind). If, on the other hand, the wind blows from the back on to the system, then openings on the top side cause suction that keeps it on the ground. As all elements are interconnected, there is a high degree of stability for the overall system.

6.4.3 Pitched-roof Systems

Photovoltaic installations on pitched roofs are the most common form of photovoltaic systems and are found equally in urban and rural areas. They are relatively cheap to install as the roof is already available as a base and they fit harmoniously into the environment. An example of an on-roof system is shown in Figure 6.33: The modules are above and beside the dormer, and there is sufficient spacing to minimize shading. At the side,

Figure 6.33 On-roof system on a private house: There should be at least one row of tiles spacing at the side in order to reduce the possibility of attack by the wind.

there is a distance of approximately one row of tiles to the edge of the roof in order to limit the possibilities of wind attack as far as possible. The panel at the bottom right provides the power for a solar-driven watercourse pump.

The fixing of the module supports is carried out by means of stainless-steel roof hooks that are screwed to the rafters of the roof (Figure 6.34). A sufficient distance of the hook from the tiles is important, as the hook could bend with the weight of snow and damage the roof tiles below it. The horizontally arranged rails are made of special aluminum profiles that are usually also suitable for holding the module connection cables.

An alternative to erecting a photovoltaic system on the roof is to integrate it into the roof. Such an in-roof system (or roof-integrated system) consists of solar modules that form the actual roofing (Figure 6.35). With the InDaX system of Schott Solar, the individual solar modules are provided with special frames and then placed over each other like roof shingles. The model of Schüco, in contrast, uses frame strips that are used for sealing between the individual modules.

A feature of this system is the combination with skylights and solar thermal collectors (to be seen at the left and right of the skylights). This provides a very harmonious overall view.

Figure 6.34 Fixing the on-roof system: The roof hooks are screwed to the rafters and thus form the base for holding the module supports. A sufficient distance between roof hooks and roof tiles is important in order to avoid damaging the tiles. *Source*: Ch. Niemann, A. Schroer.

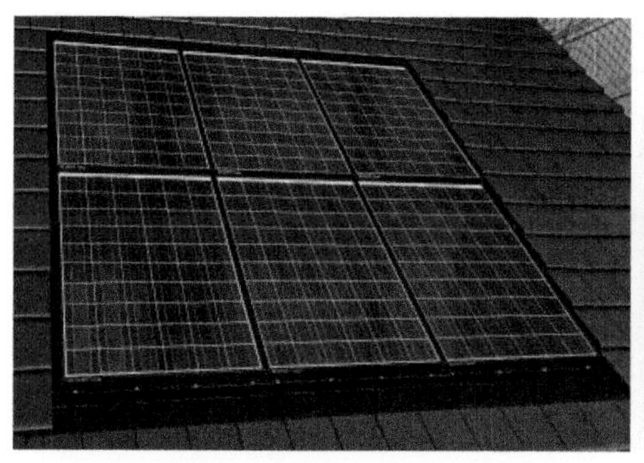

InDax system (source: Schott solar AG)

In-roof system from Schüco (photo: Mertens)

Figure 6.35 Examples on in-roof systems: The solar modules replace the actual roofing and are harmoniously integrated into the overall architecture.

(a) (b)

Figure 6.36 View of two facade installations: (a) A 50 kWp system of polycrystalline modules in Freiburg, Germany. *Source*: Solarfabrik AG. (b) Cold façade Prosol TF+ of a-Si modules. *Source*: Schüco International KG.

When using the in-roof system, special attention must be paid to sufficient back ventilation of the plant, otherwise the module temperature rises and the annual yield is less than that of a similarly arranged on-roof plant. A further problem is the need for participation of various tradespeople during the installation of the plant. This requires their cooperation and clarification of who is responsible for the warranty in case of a leaking roof.

6.4.4 Facade Systems

Facade systems are mostly installed on industrial or office buildings. Figure 6.36a shows a solar facade erected in 2000 in the course of a renovation of a block of apartments in Freiburg, Germany. With almost 200 multicrystalline modules, the system delivers 50 kWp of power and is designed as a back-ventilated hanging facade.

An alternative with thin-film modules, the cold facade ProSol TF+ by Schüco can be seen in Figure 6.36b. It consists of large (approximately 6 m^2) thin-film modules of amorphous silicon.

As can be seen from Table 2.4, the south-oriented facade provides only about 70% of an optimally arranged solar system. However, there are further limiting effects on the annual yield: Shading due to trees, fire escapes, other buildings, and so on.

7

System Technology of Grid-connected Plants

We are now well acquainted with the power generation chain of photovoltaics:

Solar radiation → cell → module → string → PV generator

Now, we will deal with how the electric power provided by the generator can best be utilized completely for feeding into the electric grid or other loads. For this purpose, we first consider the possibilities of adapting the photovoltaic generator to the available electric load. After that, we discuss the different plant concepts and describe the construction and the functional principles of inverters. The chapter is completed with considerations of the requirements of the grid operators and safety aspects.

Sometimes the expression balance of systems (BoS) is used to summarize all the components that are needed additionally to the solar modules to construct a complete photovoltaic system.

7.1 Solar Generator and Load

The power produced by the solar generator can be used by various consumers of electricity. Typical examples are for charging a battery, a solar-driven water pump or a public supply grid. These different types of loads always have their own requirements for the provided voltages and currents. In most cases, therefore, a component must be interposed that makes the necessary adaptation possible.

7.1.1 Resistive Load

The easiest load to study is an ohmic resistance. In the I/V characteristic curve it is described by a linear equation:

$$I = \frac{1}{R} \cdot V. \tag{7.1}$$

Figure 7.1 shows the case of a direct connection of an ohmic load (e.g., a light bulb) to a solar module. If the solar generator is operated at 1000 W m^{-2}, then in this example, the operating point 1 is near the MPP$_1$ of the module. If the irradiance falls to half, then a new operating point 2 is adjusted that is, however, far away from the actual optimum MPP$_2$. In this case, the solar module can contribute only a portion of the actually

Photovoltaics – Fundamentals, Technology, and Practice, Second Edition. Konrad Mertens.

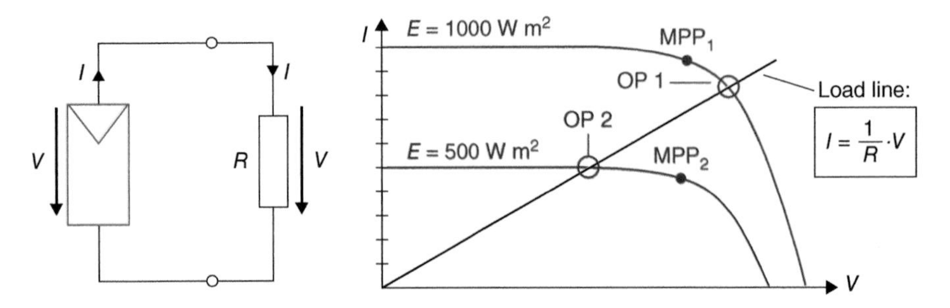

Figure 7.1 Operation of an ohmic load at a solar module: In the case of half the Sun's irradiance ($E = 500\ \text{W m}^{-2}$), the operating point (OP 2) is far away from MPP2.

available power to the load. It is, therefore, desirable to decouple the voltage at the solar generator from the voltage at the load. For this purpose, we use an electronic adaptation circuit, the DC/DC converter (direct current converter).

7.1.2 DC/DC Converter

7.1.2.1 Idea

A DC/DC converter converts an input voltage V_1 into an output voltage V_2. The result is that the voltage at the solar module can be selected almost independently of the voltage at the load. Thus, for instance, in Figure 7.2 the voltage V_1 at the solar module can be kept constant. The self-adjusting operating point in both cases is very near to the respective MPP.

A direct current converter will never work without loss. However, good converters achieve efficiencies of more than 95%, the rest being converted into heat. In the case of an ideal converter with an efficiency of 100%, the input and output powers are the same:

$$P_1 = V_1 \cdot I_1 = V_2 \cdot I_2 = P_2. \tag{7.2}$$

If we choose V_1 differently to V_2, then the currents I_1 and I_2 must also be different. Thus the voltage converter works at the same time as an impedance transformer.

7.1.2.2 Buck Converter

Modules are often connected in series in order to obtain a high voltage. If this voltage is to be reduced, then a buck converter (step-down converter) is used. The principle of the

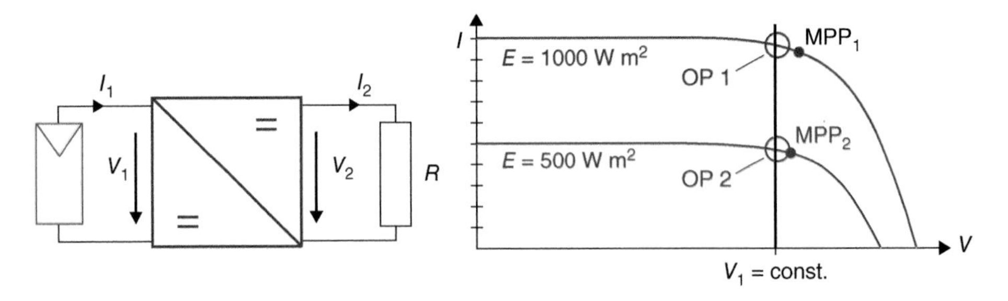

Figure 7.2 Application of a DC/DC converter: The voltage at the solar generator can be selected independently of that at the load; for example, it can be left constant.

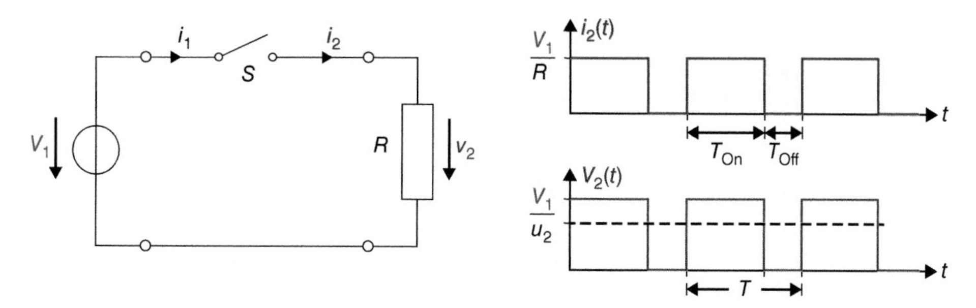

Figure 7.3 Simple model of a buck converter: At the output, the result due to the pulse width modulation in the time average is a reduced voltage $\bar{v}_2$ compared with V_1.

buck converter is to switch through the input voltage V_1 only for a certain time period T_{On} to the output voltage (so-called pulse width modulation (PWM); see Figure 7.3).

The result is a pulsed voltage $v_2(t)$ at the output that has a mean value of

$$\bar{v}_2 = \frac{T_{\mathrm{On}}}{T} \cdot a \cdot V_1, \tag{7.3}$$

where

T_{On} = on-time;
T = time *period;*

and the quantity a is the duty cycle or duty factor:

$$a = \frac{T_{\mathrm{On}}}{T} \tag{7.4}$$

In practical applications, a pulsed output component cannot be accepted. For this reason, smoothing elements for current and voltage must be added. Figure 7.4 shows the complete circuitry.

The choke coil L is used to ensure a continuous current i_{L} and the capacitor C_2 is used for smoothing the output voltage. The capacity should be large enough that we can assume a direct voltage V_2 at the output.

The function of the switch from Figure 7.3 is carried out in practice by a semiconductor switch, for example, a power MOSFET (metal oxide semiconductor field effect transistor). It can be switched like a normal switch via a positive potential at its gate terminal. The capacitor C_1 serves to prevent the source of the voltage from being loaded at the input by pulsating currents.

In detail, how does this function? First, we will look at the case of the MOSFET being switched on. For the voltage v_{D} at the diode there applies

$$v_{\mathrm{D}} = V_1. \tag{7.5}$$

The voltage at the choke coil then follows:

$$v_{\mathrm{L}} = v_{\mathrm{D}} - V_2 = V_1 - V_2. \tag{7.6}$$

Thus, there is a constant voltage at the coil so that according to the induction law:

$$v_{\mathrm{L}}(t) = L \cdot \frac{di_{\mathrm{L}}}{dt}. \tag{7.7}$$

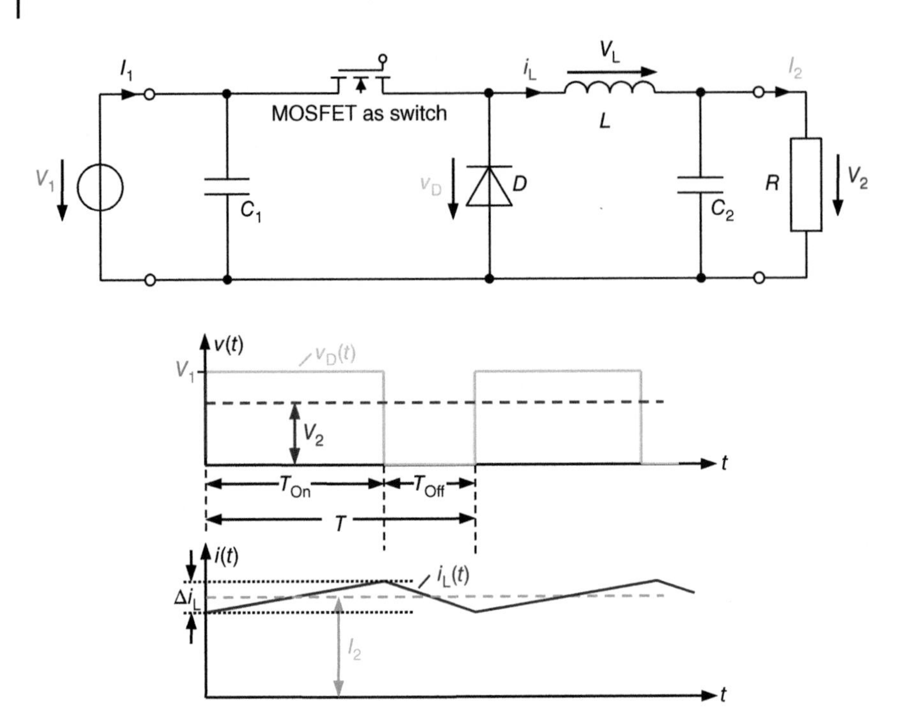

Figure 7.4 Circuit and current and voltage curves of the buck converter [88].

We can say that for the time T_{On}, the current will vary linearly with time. The current rises with the constant slew rate (see Figure 7.4):

$$\frac{di_L}{dt} = \frac{v_L}{L} = \frac{V_1 - V_2}{L}. \tag{7.8}$$

If the transistor is switched off at time $t = T_{On}$, then the coil attempts to maintain the current i_L. It drives the current further via the flyback diode D. If we assume that this is an ideal diode ($v_D = 0$), then in an analogous manner to that mentioned previously, we can derive the slew rate of the current:

$$\frac{di_L}{dt} = \frac{v_L}{L} = -\frac{V_2}{L}. \tag{7.9}$$

Thus, the current decays linearly with time. The value of the starting point is obtained from the principle that there can be no decrease in direct voltage at an ideal coil. The constant voltage V_2 is derived from the mean average of the voltage v_D over time:

$$V_2 = a \cdot V_1 \tag{7.10}$$

where a is the duty factor.

Conclusion: By means of the variation of the duty factor, the output voltage of the buck converter can be set almost arbitrarily between 0 and V_1.

These results do not apply when the current i_L decays to zero during the time period T_{Off}. In this discontinuous mode, the output voltage would no longer depend just on the duty factor but also on the load current I_2. In order to prevent the undesired discontinuous mode, one selects the switching frequency of the transistor to be as large as possible (e.g. 20 kHz). In this way, a good quality of the output voltage can be achieved even for small values of L, C_1, and C_2. However, the cutoff frequency of the transistor presents an upper limit for the switching frequency. In addition, the switching losses increase with the increase in the switching frequency. Further details can be found in, for instance, [88].

7.1.2.3 Boost Converter

It is often necessary to convert a small solar generator voltage into a higher voltage, for example, in order to feed into the public grid. In this case, use is made of a boost converter (step-up converter). The circuit is shown in Figure 7.5. First, the transistor is switched on. With $v_S = 0$, it follows that $v_L = V_1$ so that the slew rate of the current can be immediately given as

$$\frac{di_L}{dt} = \frac{v_L}{L} = \frac{V_1}{L}. \tag{7.11}$$

The transistor is switched off after the time period T_{On}. The choke attempts to maintain the current. We will assume that V_2 is greater than V_1. In this case, the choke drives the current slowly down via the diode D at the output. The voltage at the choke is then

$$v_L = V_1 - V_2 \tag{7.12}$$

and the current slew rate is

$$\frac{di_L}{dt} = \frac{v_L}{L} = \frac{V_1 - V_2}{L}. \tag{7.13}$$

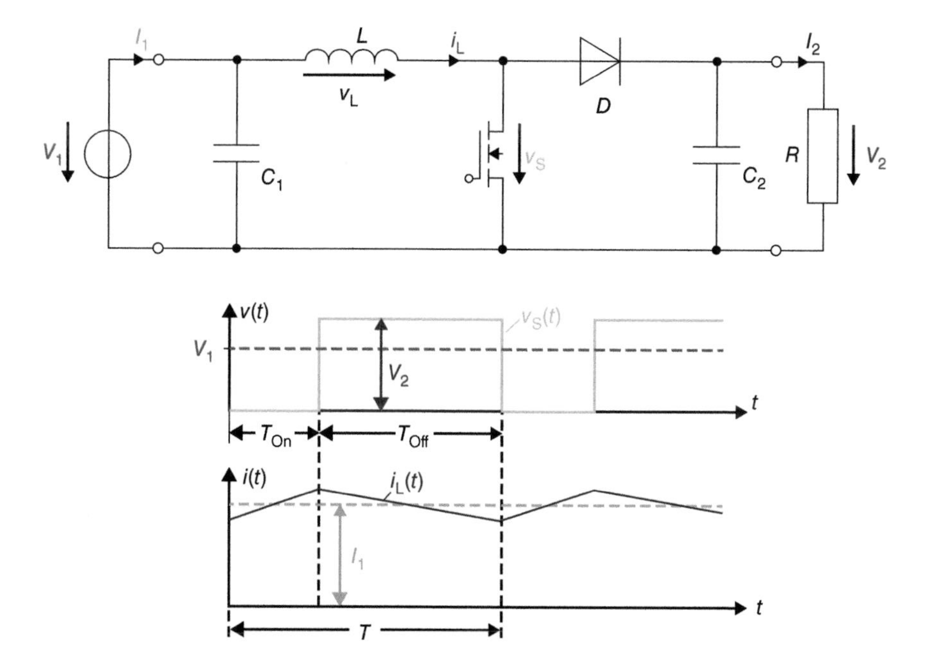

Figure 7.5 Connecting a boost converter with current and voltage progressions [88].

As i_L decreases, the voltage v_L will be negative according to the induction law Equation (7.7). If one solves Equation (7.12) for V_2, then it will be clear that the output voltage must be greater than the input voltage, which will confirm our assumption:

$$V_2 = V_1 - v_L. \tag{7.14}$$

Here, too, the following applies: the choke coil cannot accept direct current so that the temporal mean value of $v_S(t)$ must be equal to V_1. Therefore, we obtain from the voltage progression in Figure 7.5:

$$V_2 \cdot T_{Off} = V_1 \cdot T. \tag{7.15}$$

For the output voltage, there is thus

$$V_2 = \frac{T}{T_{Off}} \cdot V_1 = \frac{T}{T - T_{On}} \cdot V_1 = \frac{1}{1 - T_{On}/T} \cdot V_1 = \frac{1}{1-a} \cdot V_1. \tag{7.16}$$

Example 7.1 ***Various duty cycles with the boost converter***
Apply an input voltage of 10 V to a boost converter and adjust the duty cycles one after another to 0.1, 0.5, and 0.9. The output voltage results are 11.1, 20 and 90 V. ■

Conclusion: By varying the duty cycles, one can set the output voltage of the boost converter by a multiple of V_1.

[?] Actually, it is strange that with a longer switch-on phase of the transistor, there is a greater output voltage of the boost converter. Should it not be the other way around?

[!] During the switch-on phase of the transistor, the current at the choke coil rises. The longer the switch-on phase lasts, the more power is stored in the choke, which it can then pass on in the switch-off phase to the output. To a certain extent, the coil takes more impetus.

[?] What is the purpose of the diode *D* in Figure 7.5? In contrast to the buck converter, we do not need a flyback diode here.

[!] Without the diode, the output would be connected directly in parallel to the transistor. As soon as the transistor is switched on, the capacitor would immediately be unloaded and the output voltage would be reduced to zero.

7.1.3 MPP Tracker

Now we have become acquainted with the DC/DC converter, we can use it for MPP tracking (MPP control). Figure 7.6 shows the basic principle: at the output (or input) of the DC/DC converter, the actual power is determined by means of measurement of current and voltage. The operating point can be varied by varying the duty factor a.

There are various methods of finding the MPP, of which the *perturb and observe method* has become the most popular [89]. The flow diagram of the algorithm can be seen in Figure 7.7. Most of the MPP trackers start at the open-circuit point of the I/V curve. First, the actual power is determined, then the duty factor is increased. If the new power is greater than the old value, then the tracking was correct, and the duty factor is raised further. If the MPP is exceeded, then the measured power is decreased

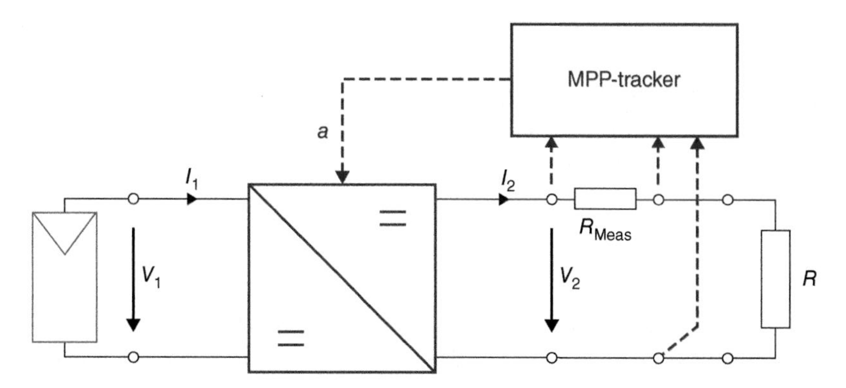

Figure 7.6 Principle of MPP tracking: The output power is maximized by measuring the current and voltage with simultaneous variation of the duty factor.

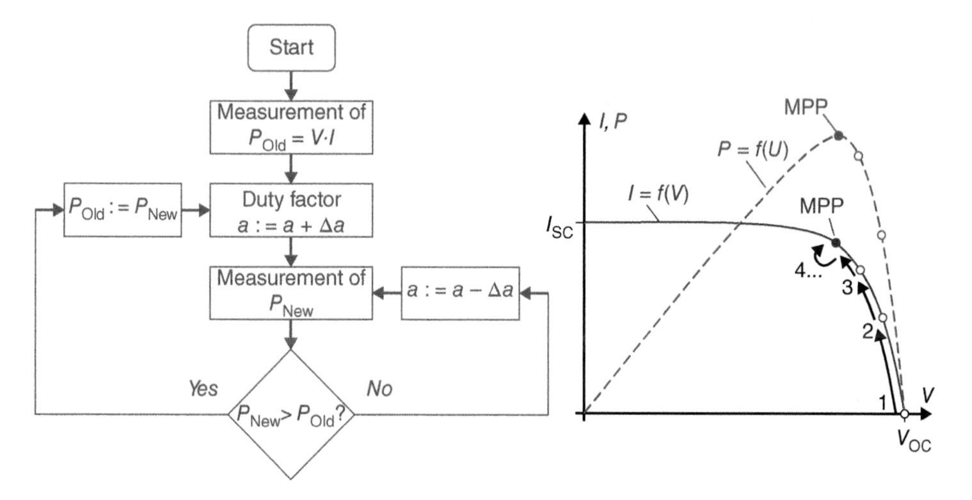

Figure 7.7 Algorithm of the *perturb and observe method*: Starting from the open-circuit point, the duty cycle is changed, the new power is determined, and the duty cycle is further optimized depending on the result until finally the MPP is reached.

and the duty factor is reduced again. Thus the actual operating point varies slightly about the MPP.

7.2 Construction of Grid-connected Systems

Although in the beginning the main uses of photovoltaics were applications separated from the electric grid (see Chapter 1), today systems connected to the grid play a dominant role. In this, the electric power supply grid is used to a certain extent as storage that takes up the power that is fed to it.

The technical conditions of the worldwide energy supply grids differ from country to country. For example, the grid voltage of most countries in Europe is 230 V with a frequency of 50 Hz.

In contrast, the electric grids in the United States, Central America, and a few other countries have a voltage of 110 V and a frequency of 60 Hz. This book deals with the European electric grid and the grid connection conditions refer to a large extent to Germany.

7.2.1 Feed-in Variations

The classic arrangement of a plant connected to the grid is shown in Figure 7.8. As most of the plants are refinanced by feed-in tariffs, the inverter feeds the generated power completely into the grid via a feed-in meter. Separated from this, the operator of the plant measures their normal domestic energy demand via a consumption meter.

However, at present, the feed-in tariffs in Germany are below the power consumption price for normal power tariff customers. For this reason, it pays to use as much generated energy as possible for one's own use (so-called self-consumption). Hence, today, a bidirectional meter is normally installed that measures both the fed-in energy and the energy taken from the grid separately (Figure 7.9).

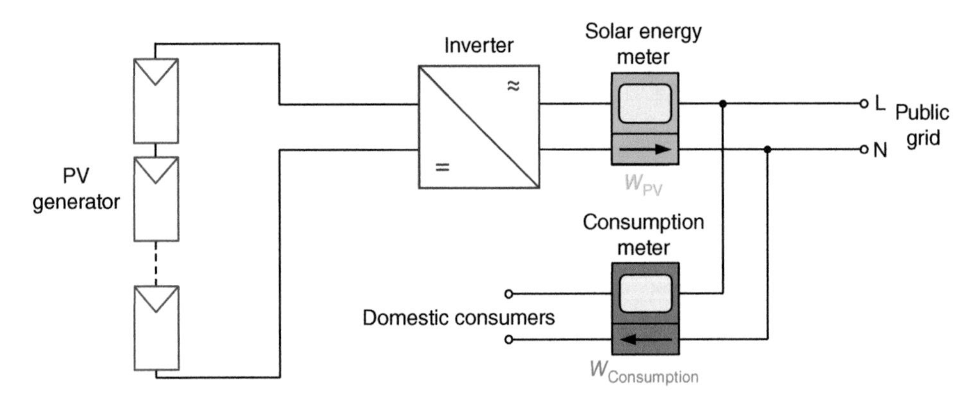

Figure 7.8 Classic connection of a photovoltaic installation to the public grid. The whole solar electricity is fed into the public grid via a feed-in meter while the domestic consumption is measured separately by means of a consumption meter ("full feed-in").

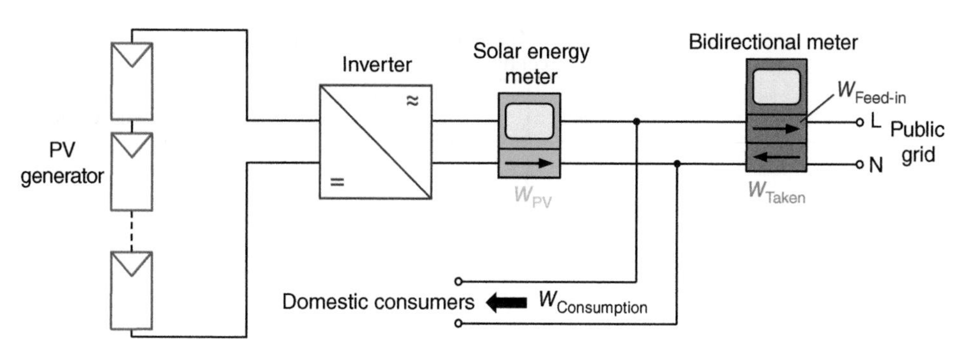

Figure 7.9 Use of bidirectional meter for separate acquisition of the fed-in energy and the energy taken from the grid. The solar energy meter is installed additionally if the overall generated solar energy also is to be measured ("feed-in with self-consumption" or "excess feed-in").

If one still wished to know how much solar energy has been generated, then a separate solar energy meter is installed between the photovoltaic installation and the bidirectional meter as shown in Figure 7.9. This offers the possibility of measuring the respective self-consumption rate at the end of the year. This self-consumption rate a_{Self} is the relation between the self-consumed solar energy $W_{Consumption\_PV}$ and the total generated solar energy W_{PV} (see Figure 7.9):

$$a_{Self\text{-}consumption} = \frac{W_{Consumption\_PV}}{W_{PV}} = \frac{W_{PV} - W_{Feed\text{-}in}}{W_{PV}}. \tag{7.17}$$

Often, additionally, the degree of self-sufficiency $a_{Self\text{-}sufficiency}$ is of interest. This indicates how much of the energy consumed in a household is self-generated (see Figure 7.9):

$$a_{Self\text{-}sufficiency} = \frac{W_{Consumption\_PV}}{W_{Consumption}} \tag{7.18}$$

In Chapter 8, we will go into details about how the degree of self-consumption and the degree of self-sufficiency can be made as large as possible.

7.2.2 Plant Concepts

The various concepts for constructing a solar power system suitable for connecting to the public grid are shown in Figure 7.10. The first type of installation with a central inverter is already familiar to us from Chapter 6. The individual strings are connected in parallel in the generator connection box, and the generated power is fed into the grid via a central inverter.

The advantage of this concept is that only a single inverter is needed. However, we also know the disadvantages: If the individual strings are differently shaded, then the parallel connection leads to mismatching losses. Added to this is the high effort (and possible losses) in direct current cabling.

The second concept is much more elegant as it is based on string-inverters (Figure 7.10b). In the consistent application of this concept, only one string is connected to each inverter.

A generator connection box is not necessary as each string is individually MPP controlled and, in any case, is easy to monitor. Compared with the central inverter variant,

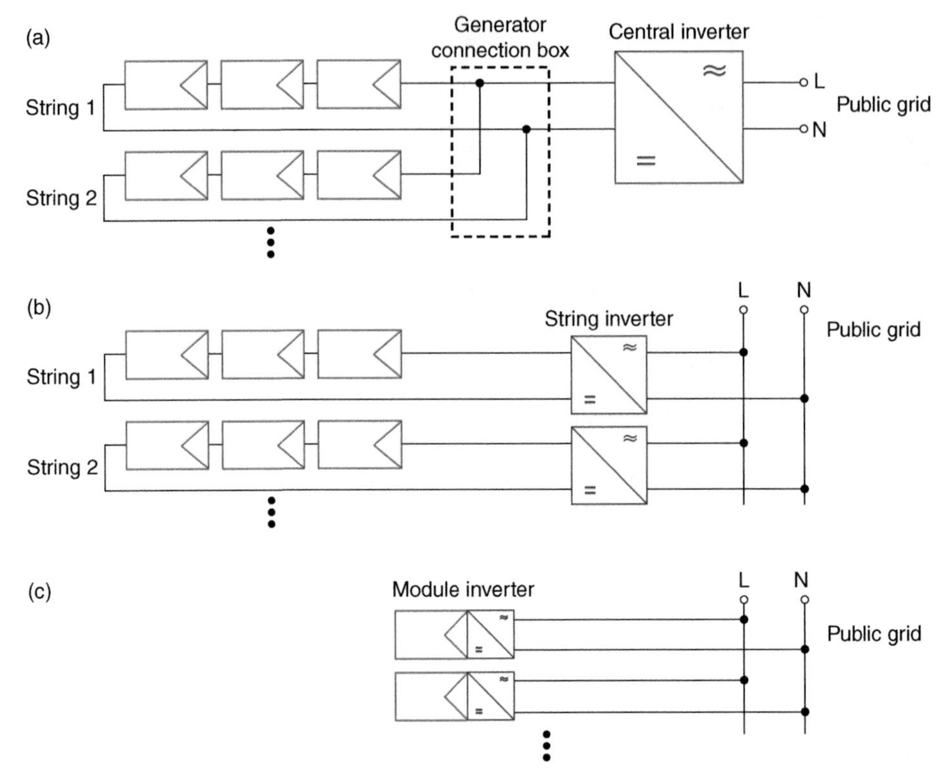

Figure 7.10 Various arrangements of photovoltaic systems connected to the grid. (a) Central inverter; (b) string inverter; and (c) module-integrated inverter.

the cabling effort on the direct current side is much simpler. In practice, two parallel strings are often connected to one string inverter when it is certain that both strings have the same structure and will not be shaded.

One can dispense with direct current cabling altogether in the concept of the module inverter (Figure 7.10c). Here, each module has its own inverter that is attached directly to the rear of the module. This means that each module is individually monitored and can be kept in the MPP. However, this concept also has clear disadvantages. An important problem is that the inverters are installed with the modules on the roof; hence, they are subject to wind, weather, and temperature fluctuations, which may adversely affect the life of the electronic components. Added to this is that much effort is required to replace a failed inverter. These factors have led to module inverters being used almost only in demonstration projects.

7.3 Construction of Inverters

As already mentioned, in addition to the solar modules, the inverter is the heart of a photovoltaic system connected to the grid. For this reason, we will look at it in greater detail in the following.

7.3.1 Tasks of the Inverter

We will now list the most important tasks of the inverter of a photovoltaic plant connected to the grid. The individual subjects will be dealt with in more detail next.

- Converting direct current into a possibly sinusoidal-form alternating current
- Achieving a high degree of efficiency (>95%) in both partial and peak loads
- Feeding the current synchronously with the grid frequency
- MPP tracking
- Monitoring the grid for voltage, frequency, and grid impedance in order to prevent an inadvertent stand-alone operation
- Measures for personnel protection:
  - Inverter with transformer: Insulation monitoring of the solar generator
  - Inverter without transformer: Residual current monitoring of the solar generator
- Preparation of actual condition data of the plant (power, current, voltage, and error codes) via an external data interface.

7.3.2 Line-commutated and Self-commutated Inverter

The classic inverter makes use of thyristors as switching elements. These have the disadvantage that they cannot be turned off by means of the control electrodes. In order to block them, one must wait for the next zero pass of the mains voltage.

For this reason, this type of inverter is called a line-commutated inverter. The thyristors can only be switched on and off once per period, leading to a rectangular form of current flow. In order to fulfill the requirements of electromagnetic compatibility (EMC), the current must be smoothed by means of additional filters.

We obtain far fewer harmonics with self-commutated inverters. This switching principle is now standard for devices up to 100 kW as a series of suitable components that can be switched off are available: GTOs (gate turn-off thyristors), IGBTs (insulated gate bipolar transistors) and power MOSFETs. These permit quick on and off switching (e.g. 20 kHz) and thus a piece-by-piece copy of a sinusoidal current flow (see Figure 7.12). Therefore, we will consider only the self-commutated inverters.

7.3.3 Inverters Without Transformers

We will use the example of an inverter without transformer to consider the complete arrangement of a modern string inverter. Figure 7.11 shows the principle of the circuit as used nowadays in many inverters. The boost converter raises the input voltage according to the requirement of the MPP tracker to a higher DC voltage level. This DC voltage is converted by the PWM bridge, including the two choke coils, into a 50 Hz sinusoidal voltage and fed into the grid.

As there is no galvanic separation between the mains and the PV plant in the case of an inverter without transformer, an all-current sensitive (reacting to errors on the DC and AC sides) residual current protective device (RCD) is prescribed for reasons of personnel safety. This must be specially designed so that it reacts to sudden current changes from 30 mA. A normal RCD is not sufficient as, in the case of larger photovoltaic generators in normal operation, high capacitive leakage currents against ground that can exceed 30 mA are possible.

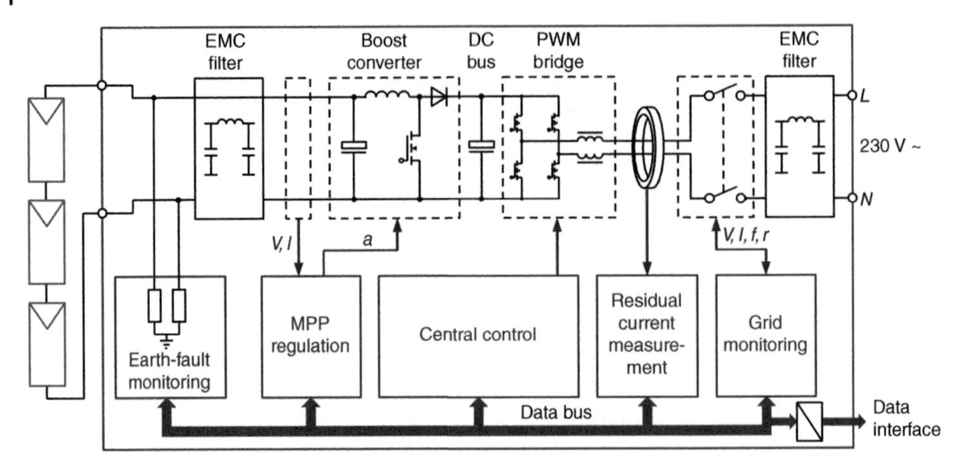

Figure 7.11 Overall arrangement of an inverter without transformer: In addition, the actual grid feed-in it must fulfill a number of other functions such as MPP tracking, residual current measurement, and grid monitoring.

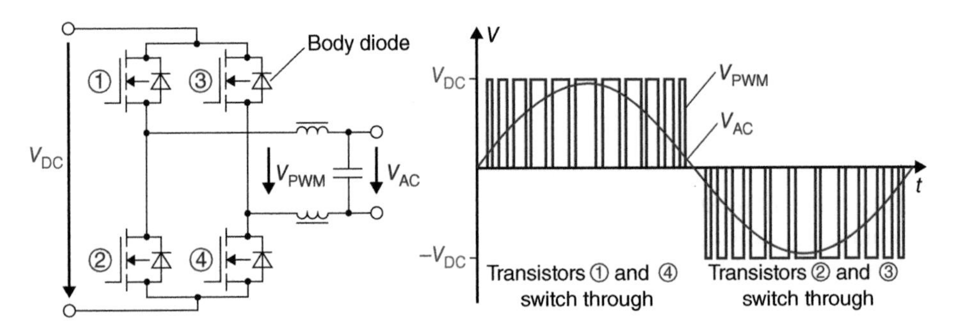

Figure 7.12 Principle of the self-commutated inverter: The direct current is chopped into pulses of varying widths and then filtered by means of a low-pass filter so that a 50 Hz sinusoidal voltage is achieved.

Finally, the grid monitoring must ensure that voltage and frequency are within the permissible ranges so that feeding in takes place only when a proper public grid is available (see Section 7.6.1).

The principle of PWM is explained in more detail in Figure 7.12. The direct voltage V_{DC} is chopped by the MOSFET bridge into pulses of various widths. In the first half-period of the grid AC voltage, the transistors ① and ④ are switched through and in the second half, the transistors ② and ③. The downstream low-pass filter ensures that only the moving average of this component arrives at the output; this is the desired 50 Hz signal. This has almost an ideal sinusoidal form, but because of the small impulses, high-frequency signal parts are generated that could disturb other devices connected to the grid (e.g. radios). For this reason, an EMC filter is installed before the feed-in into the grid.

The transistors in Figure 7.12 all possess a so-called body diode. This ensures that after switching off a transistor, the current is not suddenly zero in the series inductance (induced voltage peak). Instead, the diode of transistor ② takes over the current after switching off transistor ① and ensures a continuous flow of current.

The problem of potential induced degradation (PID) of thin-film modules in connection with inverters without transformers has been known for some years. The background is the fact that the solar generator can have a high potential of approximately −500 V against earth due to the lack of a galvanic isolation. With some thin film modules, this leads to diffusion of positively charged sodium ions from the cover glass to the TCO layer. When, at the same time, water vapor enters the cell, an electrochemical reaction occurs, causing the TCO layer to corrode. The result is permanent damage to the module with substantial power losses.

This problem only occurs with superstrate cells (e.g. CdTe and a-Si; see Chapter 5) in which the TCO layer is applied directly to the cover glass [90]. Meanwhile, there are circuit concepts for inverters without transformers that permit a one-sided earthing of the solar generator [91]. In this case, the sodium ions are moved away from the cell so that electrical corrosion no longer occurs. In the case of thin-film modules, the planner should always first check the plant for whether the selected inverter has been released by the producer for the respective type of module.

c-Si modules can also be affected by a degradation effect because of high existing voltages compared with the earth potential. Here, in a moist environment, there are possible leakage currents from the cell to the frame due to the high existing potential. The results are increased recombinations of the charge carriers generated by the light. The effect, also called PID, is not necessarily connected with damage to the module, as is the case with thin-film modules, but in some types of modules, it can lead to a reduction in power of, for example, 50% (see Chapter 9).

In the following, we will look more closely at the alternatives to inverters without transformers and discuss their advantages and disadvantages.

7.3.4 Inverters with Mains Transformer

Without exception, the first string inverters were equipped with mains transformers. The main reason for this was that solar generators were designed only for safety with extra-low voltage (<120 V), and low generator voltage was easily transformed to the desired level by means of a transformer. In addition, the galvanic isolation between the solar generator and the mains was specifically desired for personnel protection reasons. Unfortunately, apart from these advantages, 50 Hz transformers only have disadvantages: They are large, heavy, expensive, and cause relatively high electric losses. For this reason, one tries to do without them as much as possible. Nowadays, practically all solar modules are built according to protection class II and are, therefore, checked for insulation voltages of 1000 V. Hence, string voltages of, for instance, 400 V are possible.

Transformer inverters are still required for smaller plants with voltages below 200 V. In addition, the galvanic isolation ensures that the DC cabling presents no voltage fluctuations to earth. Therefore, in principle, no electromagnetic radiation can arise such as may occur with inverters without transformers.

Figure 7.13 shows the basic structure of an inverter with mains transformer. In the example, a voltage of 100 V is delivered that is chopped in the PWM bridge into a peak–peak voltage of 200 V. The effective voltage of approximately 70 V caused by this is finally converted by the mains transformer into the desired 230 V.

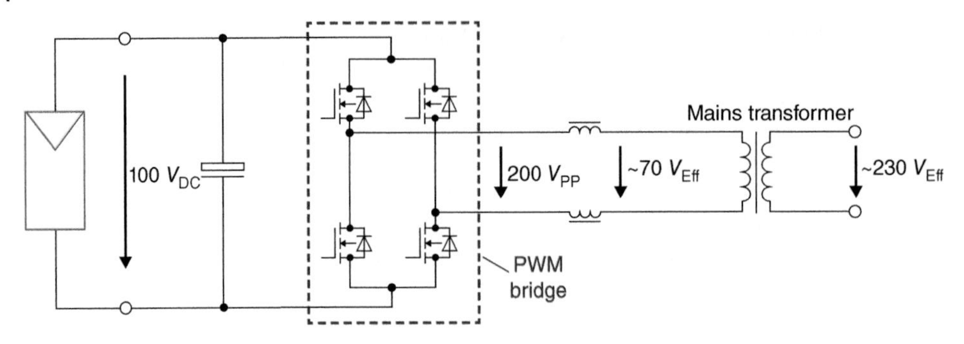

Figure 7.13 Principle of the inverter with mains transformer: The voltage signal received from the PWM bridge is converted to the desired mains voltage by the transformer.

7.3.5 Inverters with HF Transformer

There is a further type of inverter that permits a galvanic isolation of the DC and AC sides and yet prevents the disadvantages of the mains transformer. This is the inverter with a high-frequency transformer (HF transformer). Figure 7.14 shows the arrangement with the occurring voltage progression: The direct voltage is converted by means of a fast PWM bridge into a high-frequency alternating voltage. At this high frequency, the required transformer inductance is several-fold smaller than for the 50 Hz transformer. For this reason, one can use a small, light, low-loss, and cheap high-frequency transformer in order to achieve the galvanic isolation. The high-frequency alternating voltage is then rectified and filtered so that a pulsating half-wave voltage emerges.

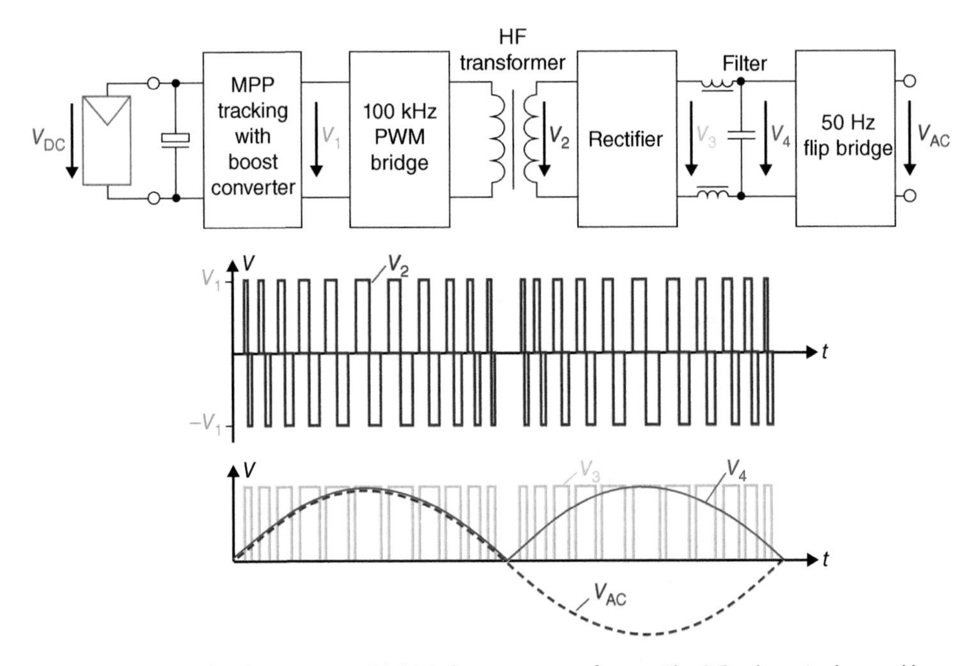

Figure 7.14 Principle of an inverter with high-frequency transformer: The DC voltage is chopped by means of an HF bridge and can thus be galvanically isolated by a low-loss transformer of low inductance. After rectifying and flip bridge, the desired 50 Hz signal is finally available [40].

Table 7.1 Advantages and disadvantages of various types of inverters.

Features	Inverter with grid transformer	Inverter with HF transformer	Inverter without transformer
Galvanic isolation	Yes	Yes	No
Residual current monitoring necessary?	No	No	Yes
EMC radiation of the solar generator	Little	Little	Possibly high
Usage with $V_{DC} < 150$ V?	Quite possible	Possible	Hardly possible
Usage with thin-film modules?	Yes	Yes	Possibly
Size and weight	Large	Medium	Low
Efficiency	Poor	Medium	High

This must then be converted into the desired mains alternating voltage by means of a 50 Hz flip bridge (a bridge that changes the polarity every 10 ms).

Since the HF transformer inverters went into decline at the end of the 1990s, there has been a renaissance owing to the potential problems with thin-film modules described earlier.

Table 7.1 summarizes the advantages and disadvantages of the various types of inverters again.

7.3.6 Three-phase Feed-in

In past years, the trend for on-roof installations has been for larger plants (e.g. for farms or industrial buildings). For this reason, there are increasing numbers of inverters with powers above 5 kW on the market that feed three-phase power into the grid.

Figure 7.15 shows the principle of one such inverter. A PWM bridge of six semiconductor switches is constructed and generates three voltages, V_1, V_2, and V_3, each displaced about 120° between the lines.

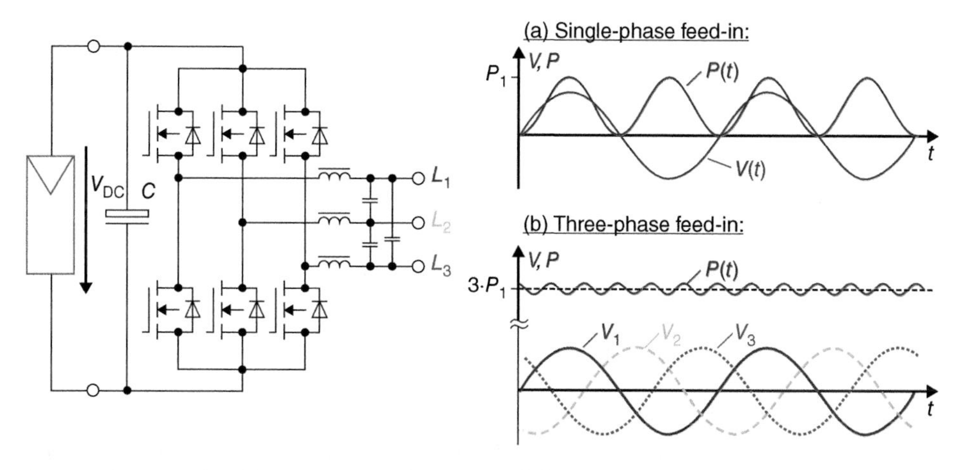

Figure 7.15 Principle of three-phase feed-in inverter: With six instead of four MOSFETs, it is possible to feed in three times the power compared with the single-phase case.

In order to generate a sufficient voltage for the 400 V three-phase grid, there must be at least a direct voltage of

$$V_{\mathrm{DC\_Min}} = 400\ \mathrm{V} \cdot \sqrt{2} = 567\ \mathrm{V} \tag{7.19}$$

at the input of the inverter. Three-phase inverters without boost converters typically have an input voltage between 600 and 800 V. With this input voltage, the overall power fed into the three-phase grid is three times that of a single-phase feed-in. This is where one of the advantages of the three-phase inverter comes to the fore: With two additional power transistors (50% more in comparison with the original four transistors), it is possible to obtain 200% more power. However, the transistors must be designed for the increased voltage region.

A further advantage of the three-phase inverter is that it feeds in equally on all three phases and, therefore, the grid is symmetrically supplied.

Added to this is the timely even feed-in: As the power progressions in Figure 7.15 show, the momentary value of the power fed into the grid in the case of the single-phase feed-in pulses from zero to maximum. However, the solar generator continually supplies its direct current power to the inverter. For this reason, the capacitor C must possess a very high capacity in order to provide intermediate storage of the energy of the solar generator. Relatively expensive and large electrolytic capacitors are necessary for this. In the case of three-phase feed-in, the momentary value of the fed-in power is nearly constant so that only a small storage capacitor is required.

7.3.7 Further Clever Concepts

In addition to the inverter concepts already discussed, there is now a multitude of mixed and special forms. A clever idea for reducing the mismatching losses is the multistring inverter (Figure 7.16). This contains two or three inputs, each possessing a separate MPP tracker. The device is especially suited for plants with partial shading in which the string with shaded modules is to be individually MPP controlled (see Chapter 6). A further application case would be, for instance, plants with modules on a south- and also on a west-facing roof. Here, too, there is the possibility of controlling both solar generators

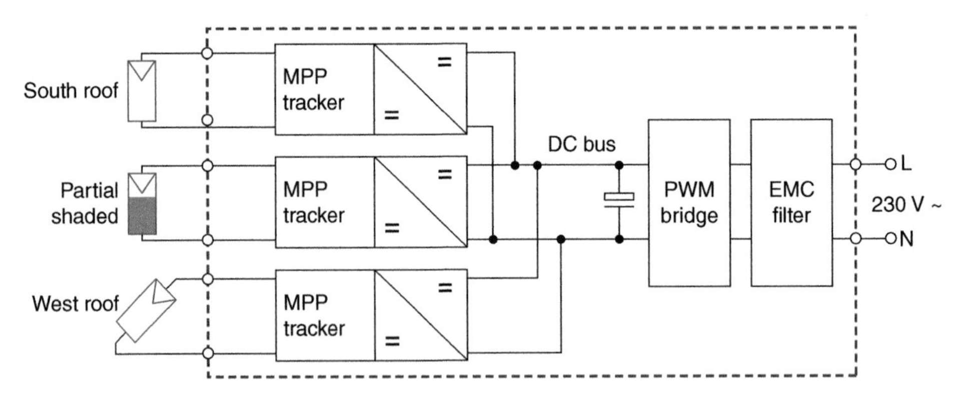

Figure 7.16 Arrangement of a multistring inverter for connecting different part-generators: The three inputs are separately MPP controlled and then fed into a common direct voltage bus.

separately in order to prevent mismatching losses. The alternative would be the application of two string inverters with two housings, double control electronics, and so on, which would result in high costs and low efficiencies. Table 7.2 lists, among others, the data for two multistring inverters.

A second idea for increasing the yield of photovoltaic plants is the Master–Slave concept (Figure 7.17). This is mainly used for large plants (>30 kW). In place of a central inverter, use is made of several individual devices that are possibly positioned in a common housing. At times of low radiation (e.g. mornings, evenings, or cloudy days), the whole of the photovoltaic power is switched to the Master inverter. This then achieves a high loading with a corresponding high efficiency (see Section 7.4). Slave 1 is only added when the Master can no longer take up the solar power by itself. With rising photovoltaic power, Slave 2 is then brought into the game. The control of the overall sequence is carried out by the Master.

It is particularly clever to allocate the role of the Master to a different inverter every day. In this case, all inverters end up with the same mean operating hours. Master–Slave-type concepts are also available for plants up to 30 kW, and these are marketed, for instance, under the names Team or Mix Concept.

7.4 Efficiency of Inverters

Ideally, the complete power provided by the solar generator should be fed into the electric grid, which is not achievable in practice. Each inverter requires a central control with a microcontroller that takes over the entire operation. More definite, however, are the losses in the power part of the device. Each real component possesses nonideal properties. Thus, in addition to the inductance, the choke also has an ohmic resistance that leads to heat losses. Further, there are switching losses in the semiconductor switches, especially in the switch-off process.

7.4.1 Conversion Efficiency

In order to compare various inverters, the conversion efficiency η_{Con} is defined and shows what proportion of the direct current power delivered by the solar generator is fed into the alternating current grid:

$$\eta_{Con} = \frac{P_{AC}}{P_{DC}} \tag{7.20}$$

where

P_{AC} = alternating current power at output of the inverter;
P_{DC} = direct current power at input of the inverter.

Figure 7.18 shows the efficiency curves for various types of inverters. The x-axis gives the actual input power P_{DC} with reference to the input nominal power P_{DC-N} of the inverter. For lower powers, the efficiency curve declines for all types, as is to be expected; in medium and upper ranges, it remains relatively stable. The lowest curve belongs to the SMA Sunny Boy 1100, which came onto the market in 1999 and is thus a real veteran.

Table 7.2 Data for various types of inverters.

Designation	Sunny Boy 1100	SMC 7000 HV	SMC 8000TL	SolarMax 6000S	IG Plus 100	STP 5000TL	STP 20000TL-HE
Manufacturer	SMA Solar Techn. AG	SMA Solar Techn. AG	SMA Solar Techn. AG	Sputnik Engin. AG	Fronius Int. GmbH	SMA Solar Techn. AG	SMA Solar Techn. AG
DC nom. Power P_{DC_N} (kW)	1.1	7.35	8.25	5.3	8.42	5.1	20.5
AC nom. Power V_{AC_N} (kW)	1	6.65	8	4.6	8	5	20
DC nom. Voltage V_{DC_N} (V)	180	340	350	400	390	580	580
MPP range (V)	139–320	335–560	333–500	100–550	230–500	245–800	580–800
DC nom. Current I_{DC_N} (A)	6.1	21.5	25	12	21.05	11/10	36
Max. efficiency η_{Max} (%)	93	96.2	98	97	96	98.0	99
Europ. efficiency η_{EU} (%)	91.6	95.5	97.7	96.2	95.5	97.2	98.7
Number of inputs	2	4	4	3	6	2	6
Number of MPP trackers	1	1	1	1	1	2	1
Feed-in	Single-phase	Single-phase	Single-phase	Single-phase	Single-phase	Three-phase	Three-phase
Remarks	Mains transformer	Mains transformer	Without transformer	Without transformer	HF transformer	Without transformer Multistring	Without transformer Multistring

Source: www.photon.info.

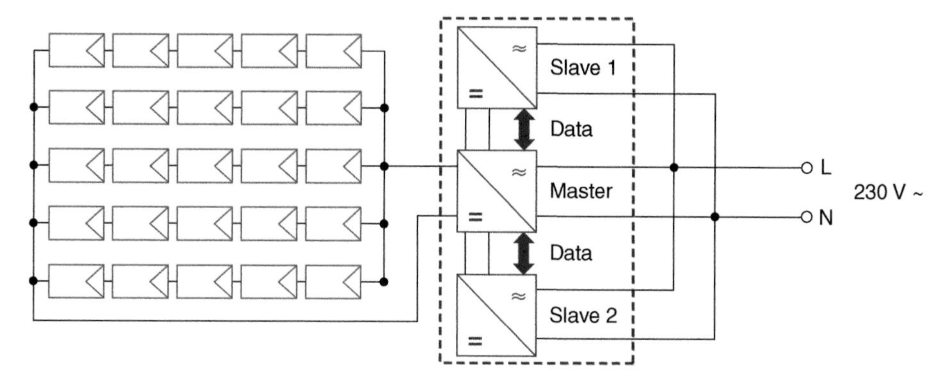

Figure 7.17 Master–Slave concept: With low radiation, the whole photovoltaic power is switched to the Master, which then works with a high part-load. If solar power rises then the Slaves successively take on part of the work.

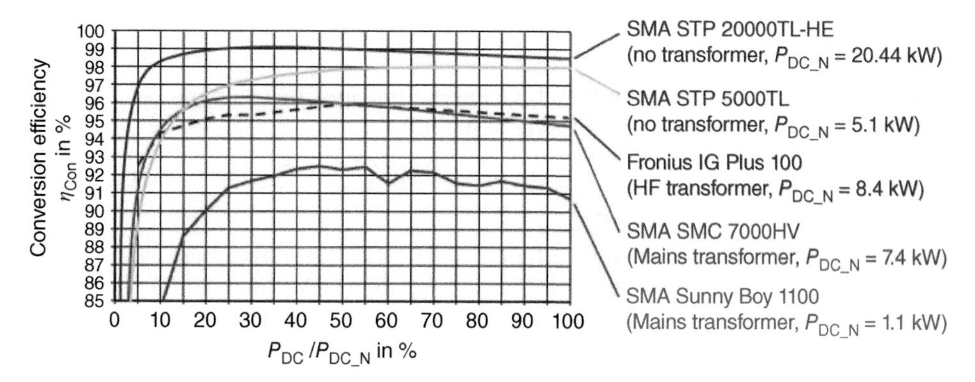

Figure 7.18 Conversion efficiency η_{Con} of different types of inverters: It can be clearly seen that the types without transformers show the highest efficiencies (data sheets, photon inverter tests).

Because of its low power of 1.1 kW, its transformer, and the old-fashioned design, it achieves a maximum efficiency of only 92.4%. A more up-to-date device with mains transformer is the SMA SMC 7000HV, which achieves a maximum efficiency of 95.6%. A slightly better efficiency of 96.2% is reached by the Fronius IG Plus 100. It is equipped with a high-frequency transformer for potential isolation. The somewhat irregular progression of the efficiency curve is due to a switching peculiarity: Depending on the existing input voltage, there is a change-over between several primary windings of the transformer. The next curve belongs to the Sunny Tripower STP-5000TL, a three-phase inverter without transformer that achieves a peak efficiency of 98%.

The absolute top device in Figure 7.18 is the STP-20000TL-HE. The abbreviation "HE" stands for "high efficiency," and this description is appropriate. The inverter shows a peak efficiency of 99%.

The curves shown in Figure 7.18 apply only for a particular existing DC voltage. The efficiency actually varies also for different PV generator voltages. Figure 7.19 shows as an example the efficiency curves of an inverter without transformer (Sputnik SolarMax 6000S). The inverter only reaches its maximum efficiency of 97% for an input MPP

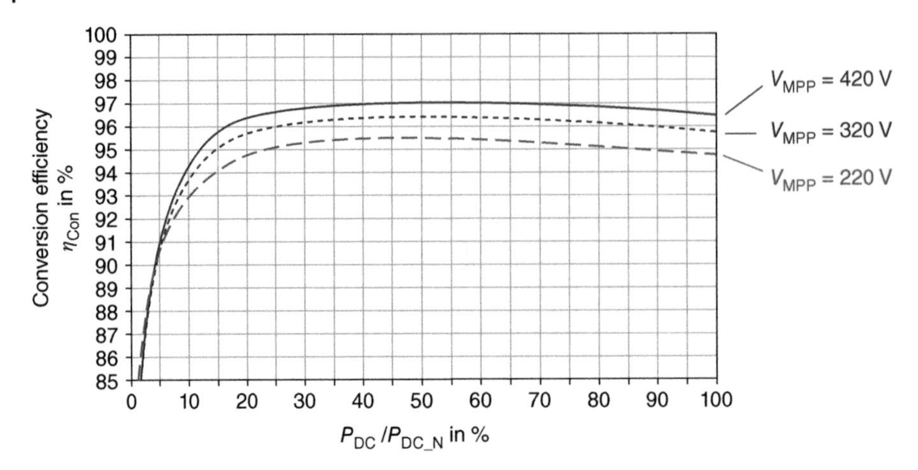

Figure 7.19 Conversion efficiency η_{Con} of the Sputnik SolarMax 6000S: The maximum efficiency depends greatly on the applied input voltage.

voltage of 420 V; at 220 V, it is already 1.5% lower. This must be taken into account by the planner in the selection of the number of modules.

7.4.2 European Efficiency

Actually, the peak efficiency of an inverter is not that important from the point of view of the operator of a photovoltaic plant. Decisive for the yield of the plant is the mean efficiency over the whole year. For the sake of clarity, Figure 7.20 shows an example from 2000 of the measured frequency with which various classes of radiation occurred at a site in Freiburg, Germany (solid line). The vertical bars represent the energy portions that the respective radiation classes contribute to the overall solar annual energy. The first bar, for instance, shows that the irradiance between 0 and 50 W m^{-2} contributes almost 1.5% to the annual energy. The irradiance up to 500 W m^{-2} over the year amounts to approximately 30% of the overall energy.

For the inverter, this means that it is often working in the lower part-load region. For this reason, the DIN EN 50524 standard specifies the European efficiency η_{Eu}, which weights the individual part-load efficiencies according to how often they occur in Central Europe:

$$\eta_{Eu} = 0.03 \cdot \eta_{5\%} + 0.06 \cdot \eta_{10\%} + 0.13 \cdot \eta_{20\%} + 0,1 \cdot \eta_{30\%} + 0.48 \cdot \eta_{50\%} + 0.2 \cdot \eta_{100\%}, \tag{7.21}$$

where $\eta_{x\%}$ is the conversion efficiency at a part-load of x%.

From the equation, it can be seen that the inverter in this model is operated at 20% of its operating time with the nominal power ($P_{DC} = P_{DC-N}$). The efficiency of the inverter at half of its nominal power ($\eta_{50\%}$) is weighted at 48%, and so on. If we apply this equation to the inverters in Figure 7.18 and Figure 7.19, then the results are the values shown in Table 7.2 for the European efficiency. The modern devices all have efficiencies above 95% and for some even show values of >97%.

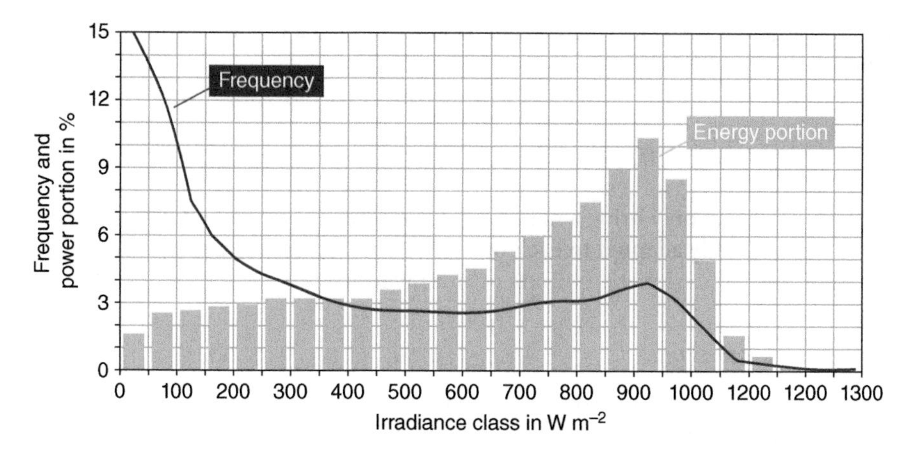

Figure 7.20 Relative frequencies of the radiation in 2000 and annual energy portions of the individual radiation classes in Freiburg: The low radiation classes seen over the year provide relatively high power portions [92]. The radiation measurements were taken at close intervals as "momentary values" (10 s averaged).

Notable again is the STP 20000TL-HE: The efficiency η_{Eu} here is only 0.3% below η_{Max}, which shows the very good part-load behavior of this inverter. This high efficiency is partly achieved because the inverter uses no step-up converter. Therefore, the minimal MPP voltage lies at a relative high value of 580 V.

The efficiencies of photovoltaic inverters have improved continuously over recent years and further improvements can be expected. The availability of new power components made with silicon carbide (SiC) and gallium nitride (GaN) shows especially that the losses can be further reduced.

SiC has a very large bandgap of 3.2 eV that results in a small intrinsic carrier concentration even for high temperatures. Whereas silicon transistors can be used up to approximately 150 °C, the critical temperature for SiC is about twice that. Hence, much smaller heatsinks can be used. Further advantages of the material are high possible reverse voltages, low forward resistances, and reduced switching losses. Further, SiC permits high switching frequencies, which allow smaller choke coils.

The use of flyback diodes of SiC is widespread today. There is a whole palette of SiC transistor types in the region of semiconductor switches. Although they are much more expensive than Si transistors, they can compete owing to their higher inverter efficiencies. Also, the 20 kW top device described above uses SiC power transistors to attain the record efficiency of 99%.

7.4.3 Clever MPP Tracking

In the case of shading, there is the possibility that more than one power maximum in the $P = f(V)$ curve of the solar generator exists. For this purpose, we will look at the case presented in Figure 6.20 (one module shaded). An inverter starting at the open-circuit point would immediately seize up in the local MPP at 270 V and almost 1 A. The actual point of maximum power (MPP_2) would not even be found. Help is provided here by a better algorithm that every now and then searches the whole characteristic curve in order to find the global MPP and then tracks around it. Such a control, for instance, is offered commercially under the name OptiTrac Global Peak.

7.5 Dimensioning of Inverters

A photovoltaic plant can only bring a maximum yield when the photovoltaic generator and inverter are optimally adapted to each other. For this reason, we will now consider the most important dimensional requirements for inverters.

7.5.1 Power Dimensioning

The inverters of the 1990s had relatively poor efficiencies in the lower part-load region. For this reason, they were often underdimensioned by 20%, meaning that for a photovoltaic plant with 2 kWp, an inverter would be used whose maximum input power was 1.6 kW. This corresponds to a solar generator overdimensioning factor k_{Over} of

$$k_{\text{Over}} = \frac{P_{\text{STC}}}{P_{\text{DC\_N}}} = \frac{2\text{ kWp}}{1.6\text{ kW}} = 1.25, \tag{7.22}$$

where

P_{STC} = nominal power of the photovoltaic generator (at STC);
$P_{\text{DC–N}}$ = DC nominal power at the inverter input.

Thus, these inverters achieved average part-load regions even for low radiation and thus higher efficiencies. The disadvantage, naturally, was that for nominal powers of the photovoltaic generator (e.g. on a sunny, cold day in May), there was a limitation of the inverter and much energy was given away. Today's inverters show much better part-load behavior so that underdimensioning of the inverter hardly makes sense any more.

Meanwhile, increasing use is being made of the so-called design factor SR_{AC} (sizing ratio). This refers to the output power $P_{\text{AC–N}}$ of the inverter:

$$SR_{\text{AC}} = \frac{P_{\text{STC}}}{P_{\text{AC\_N}}} \tag{7.23}$$

The reason for this new reference value is that some inverter producers declare input powers that are too high so that the devices often work in overload operation.

What could possibly be the correct design factor? To find the answer, use is made of an investigation carried out by the Fraunhofer Institute for Solar Energy Systems (Fraunhofer ISE). There the radiation measurements of Figure 7.20 were used in order to determine a realistic annual average efficiency of an inverter without transformer. The result is shown in Figure 7.21: Taking into account the hourly mean values, one can afford to use a design factor of 1.1 without any energy losses. However, things are different when using the momentary values (10 s mean values): Here it is seen that SR_{AC} should be a maximum of 1.0 in order not to reduce the yields.

Why does the timely resolution of the weather data have an influence on the results of dimensioning the inverter?

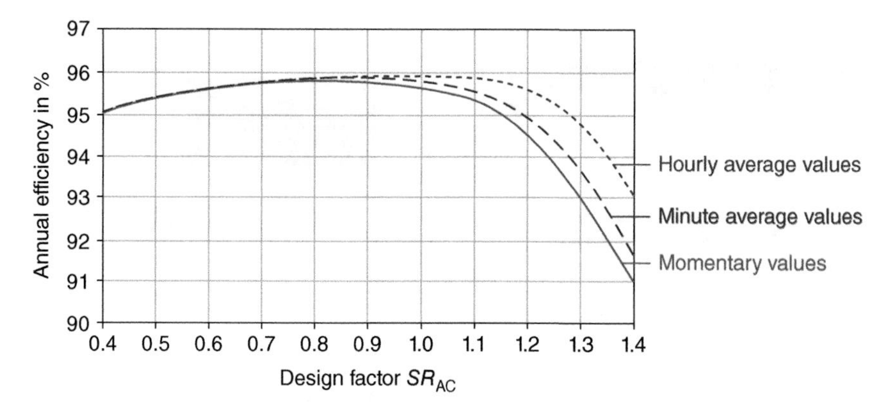

Figure 7.21 Annual efficiency of an inverter without transformer as a function of the design factor: When viewing the radiation momentary values, it is seen that the design factor should be less than 1.0 in order to reach the maximum yields [92].

On sunny days with moving clouds, there are repeated radiations of, for instance, 1000 W m^{-2} that are reduced for short periods to, for instance, 500 W m^{-2} by the clouds. If the data are averaged over an hour, then we could obtain a value of 800 W m^{-2}. From this, we cannot see that the inverter was overloaded. This is the reason why momentary values are so important.

7.5.2 Voltage Dimensioning

In addition to the power, the voltage and the current of the solar generator must also be adapted to the inverter. As a means of presenting an overview of the relevant parameters, Figure 7.22 shows the operating region of the inverter with the example of the SMA SMC 8000TL.

First we will consider the voltage dimensioning: Each inverter has a maximum permissible voltage $V_{Inv-Max}$ that will cause it to shut down if it is exceeded. The most critical situation, for instance, would be a restart of the inverter on a cold, sunny, winter's day as the modules then possess their maximum open-circuit voltage.

If one assumes that the module temperature ϑ_M is −10 °C, then the maximum number of modules n_{Max} per string is

$$n_{Max} = \frac{V_{Inv\_Max}}{V_{OC\_M}(-10^\circ C)}. \tag{7.24}$$

The minimum number of modules n_{Min} is determined by the MPP working region of the inverter (Figure 7.22). We will, therefore, consider a summer's day during which the module temperature reaches 70 °C. In this case, the string MPP voltage must not be less than $V_{MPP-Min}$ of the inverter as, otherwise, it will not provide the maximum possible power or may even shut down. The equation is therefore

$$n_{Min} = \frac{V_{MPP\_Min}}{V_{MPP\_Module(70^\circ C)}}. \tag{7.25}$$

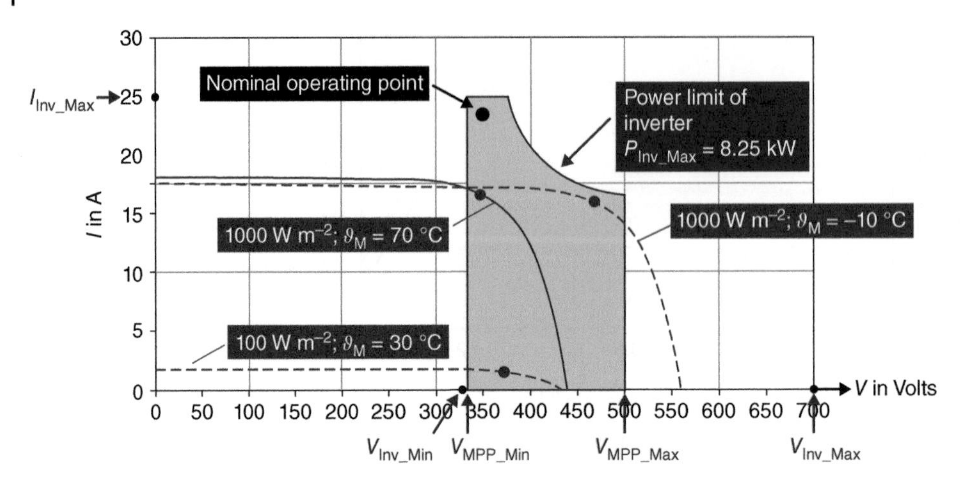

Figure 7.22 Possible operating region of an inverter using the example of the SMC 8000TL: It is limited to the left and right by the minimum and maximum MPP voltage, and at the top by the maximum current and the maximum inverter input power [22].

7.5.3 Current Dimensioning

The number of possible strings n_{String} is prescribed by the maximum current $I_{Inv-Max}$ of the inverter and the maximum string current $I_{String-Max}$:

$$n_{String} \leq \frac{I_{Inv\_Max}}{I_{String\_Max}}. \tag{7.26}$$

As a precaution, because of occasionally assumed higher irradiances than 1000 W m^{-2}, one should assume for $I_{String-Max}$ a value of 1.25 × MPP current.

Today, simulation tools are mostly used for dimensioning the inverter (see Chapter 9). However, the rules and equations given here are helpful as they are immediately comprehensible and can at least be used for critically checking the simulation results.

7.6 Requirements of the Grid Operators

Inverters must fulfill certain technical requirements in order to be permitted to feed into the power supply grid. Besides the prescriptions of the quality of the fed-in current (low harmonic portion), these are primary measures to exclude an undesired stand-alone operation of the inverter with associated danger to people. A feasible case, for instance, would be the uncoupling of a street of houses from the power supply grid in order to carry out maintenance. If the inverters were to continue to feed in, then the maintenance personnel would be in grave danger.

7.6.1 Prevention of Stand-Alone Operation

For the German electricity grid (as in most other countries), there are narrow tolerance limits for the grid voltage and the grid frequency:

$$90\% < \frac{V}{V_N} < 110\%, \tag{7.27}$$

where V_N is the nominal voltage of the power supply grid, and

$$99.6\% < \frac{f}{f_N} < 100.4\%, \quad (7.28)$$

where f_N is the nominal grid frequency.

Therefore, in a 230 V/50 Hz grid, the voltage has to be between 207 and 253 V and the frequency between 49.8 and 50.2 Hz. If there is a risk that these limits may be exceeded, control power plants have to intervene (see also Section 11.4 in Chapter 11).

In order that the inverter properly recognizes a stand-alone operation, it must continuously monitor all three phases of the mains for adherence to voltage and frequency tolerances. The VDE V0126-1-1 standard has defined these as follows:

$$80\% < \frac{V}{V_N} < 115\%, \quad (7.29)$$

$$47.5\ \text{Hz} < f < 50.2\ \text{Hz}. \quad (7.30)$$

If one of these requirements is not fulfilled, then the inverter must switch itself off from the grid within 0.2 s. However, the 50.2 Hz upper limit has been shown to be a problem. If it is exceeded, then all the photovoltaic inverters suddenly shut themselves off from the grid, which could lead to instability in the supply grid. For this reason, the new low voltage guideline VDE-AR-N 4105 mandates that the inverters must power down more and more between 50.2 and 51.5 Hz in order to prevent sudden power jumps. They are only fully switched off from 51.5 Hz. With these measures, the "50.2 Hz problem" could be rendered ineffectual.

Theoretically, the case would be conceivable that the loads in a separated grid region use just as much power as the connected inverters generate. In this case, the inverters would not recognize the undesired stand-alone operation and possibly feed in for a longer period. However, this case is extremely improbable as the smallest deviation between supply and demand would generate an asymmetry in the three-phase grid that would immediately be recognized.

An alternative to recognizing the undesired stand-alone operation is the self-acting disconnection unit with grid impedance measurement. For this purpose, the inverter increases the fed-in current for a short period by an amount ΔI and measures the resulting voltage increase ΔV (Figure 7.23). The ratio of the two sizes is then the grid impedance Z_{Grid}:

$$Z_{Grid} = \frac{\Delta V}{\Delta I}. \quad (7.27)$$

If a value of $Z_{Grid} > 1\,\Omega$ is measured, then the inverter must switch off from the grid within 5 s. A great advantage of this method over the voltage/frequency measurement is that, according to the standard, only the feeding-in phase need be checked. For single-phase inverters, no three-phase cable must therefore be laid up to the inverter.

Why do we recognize with grid impedance whether we are dealing with a stand-alone or regular grid?

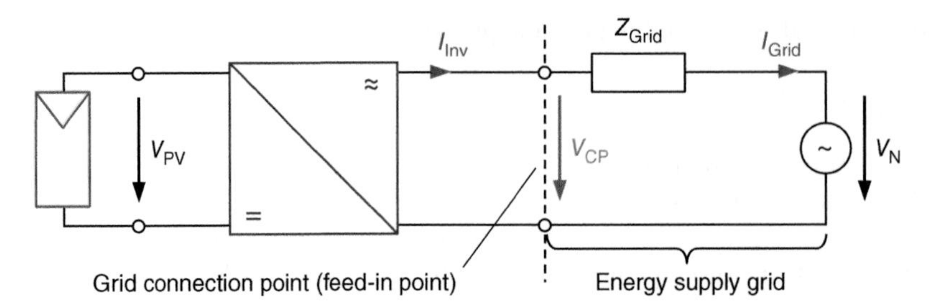

Figure 7.23 Simple equivalent circuit diagram of the energy supply grid built up of an ideal voltage source V_N and grid impedance Z_{Grid}: Depending on the size of the fed-in current, the voltage V_{CP} increases at the grid connection point compared with the nominal voltage V_N.

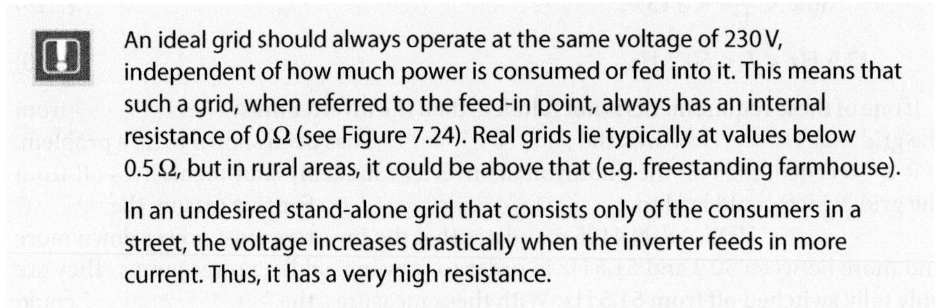

An ideal grid should always operate at the same voltage of 230 V, independent of how much power is consumed or fed into it. This means that such a grid, when referred to the feed-in point, always has an internal resistance of 0 Ω (see Figure 7.24). Real grids lie typically at values below 0.5 Ω, but in rural areas, it could be above that (e.g. freestanding farmhouse).

In an undesired stand-alone grid that consists only of the consumers in a street, the voltage increases drastically when the inverter feeds in more current. Thus, it has a very high resistance.

7.6.2 Maximum Feed-in Power

In order to prevent an unbalanced load in the three-phase power system, the feed-in power from the single-phase inverter is limited to 4.6 kVA. If several single-phase

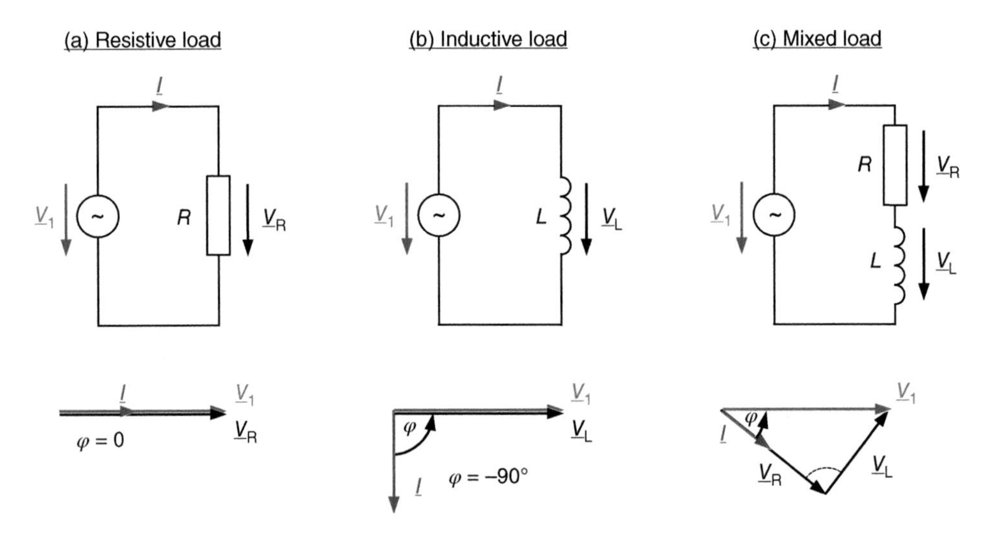

Figure 7.24 Depiction of different load cases: Whereas with a resistive load, the voltage and the current are in-phase, an inductive load induces a lagging of the current with respect to the voltage.

inverters are used, then they must be distributed over individual phases so that there is no unbalanced load greater than 4.6 kVA [93]. However, it is more sensible to use three-phase inverters directly if possible.

Before connecting a photovoltaic plant to the grid, we must check whether it can accept the planned power. The measuring limit here is that the voltage at the connection point must not permanently rise by more than 3% over the nominal voltage. For plants below 30 kW, the grid operator is required to extend the grid correspondingly if necessary. For larger plants, the plant owner must lay a suitable line up to a suitable connection point (Section 5 and Section 9 of the EEG 2009).

Since 2012, the German Renewable Energy Law requires a remote-controlled shut-down facility (so-called feed-in management) for all photovoltaic plants. The background is the fact that the solar plants installed in Germany now produce so much power that on certain days, there can be an excess supply of electrical power. The grid operators must, therefore, have the possibility to shut down the power of photovoltaic plants. The solution consists, for instance, of special electronic ripple control receivers that sent shut-down commands to the inverters.

Owners of smaller plants (<30 kW), however, have an (not really convincing) alternative. Instead of participating in the feed-in management, they can underdimension their inverters by 30% from the start. In this way, the typical midday peak on a sunny day is mitigated.

7.6.3 Reactive Power Provision

The voltage in the grid increases slightly as soon as a photovoltaic plant feeds into the power supply grid. With the increasing number of photovoltaic plants, it often occurs that nonpermissible increases in voltage take place in the low- and medium-voltage grids. An alternative to a grid expansion (larger cable cross-sections, additional lines, etc.) is reactive power provision. In this case, the inverters, in addition to active power, also feed in inductive or capacitive reactive power. This can partly moderate the generated voltage rise caused by active power feeding.

Why is it possible to reduce the voltage with reactive power feed-in? Anyway, what actually does reactive power feed-in mean?

To answer this, we have to go into greater detail. First, we will make clear what is meant by reactive power. After that, we explain how it can be used to stabilize the voltage in electric grids.

Figure 7.24 shows three different loads that are connected to an alternating voltage source. In case (a), a resistive load is connected. In this case, a voltage increase at the source directly leads to an increase in the current through the resistor. This is also illustrated by the phasor diagram below the circuit.

In a phasor diagram, the vector pointing to the right represents a voltage at the time $t = 0$. In the case of an alternating voltage with a frequency of 50 Hz, the vector would actually have to rotate continuously around its origin. Let us now assume that we as an observer rotate with the vector and therefore see the vector (called "phasor") always in the same position. The underlines below the current and voltage symbols mean that they are complex dimensions, for which both the amount (length of the arrow) and the phase angle (direction of the arrow) have to be specified.

In the case of a resistive load (Figure 7.24a), the current vector points in the same direction as the voltage vector. Hence, in this case, the phase delay φ between voltage and current is zero. It is also said that "voltage and current are in phase."

Case (b) in Figure 7.24 shows an inductive load. In the case of the coil, a voltage rise causes a delayed current rise as first a magnetic field has to be established in the coil. For a sinusoidal voltage, this results in the current through the coil always flowing 90° delayed to the voltage. Thus, the current shows a phase shift of $-90°$ with respect to the applied voltage (see phasor diagram).

A mixture of both cases can be seen in Figure 7.24c. Here, the resistor and coil are connected in series. Owing to the series connection, the current in both devices must be the same. Also here it flows delayed with respect to the voltage; the phase shift now lies between 0° and $-90°$. The voltages across the two devices are also depicted. In the case of the resistive load, voltage and current have to be in-phase; therefore, the vector V_R lies in the direction of the current I. In contrast, the voltage V_L across the coil runs 90° in front of the current, and consequentially V_L lies in a right angle to V_R.

The current in case (c) is reduced with respect to the first two cases, as the voltage source sees a total impedance Z consisting of the series connections of resistor and coil:

$$Z = \sqrt{R^2 + X^2}, \tag{7.28}$$

where X denotes the reactance of the coil, which results from the inductivity L and the grid frequency f as

$$X = 2 \cdot \pi \cdot f \cdot L. \tag{7.29}$$

Not shown in Figure 7.24 is the case of a capacitive load. In a capacitor, the current leads the voltage, and the phase shift between voltage and current is now $+90°$. As a reminder, one can imagine a discharged capacitor. For a voltage source, this will represent a short circuit in the very first moment. Only when the capacitor has taken some current can a voltage gradually be established.

In the case of a resistive load (Figure 7.24a), at the resistor a power of $P = U \cdot I$ is transformed, which can lead, for example, to heating up of the resistor. Here, we speak of active power. Totally different is the case of the inductive load. At the beginning, energy is transported from the voltage source to the coil to build up a magnetic field there. However, this field is again decomposed in the further course of the grid period, through which the released energy runs back to the voltage source. Thus, the energy oscillates back and forth periodically between source and sink so that, on average, no power is transported. In this case, we speak of reactive power.

We have seen that with reactive power, no "real power" can be transported. Nevertheless, it can be used to influence electrical grids. An example is shown in

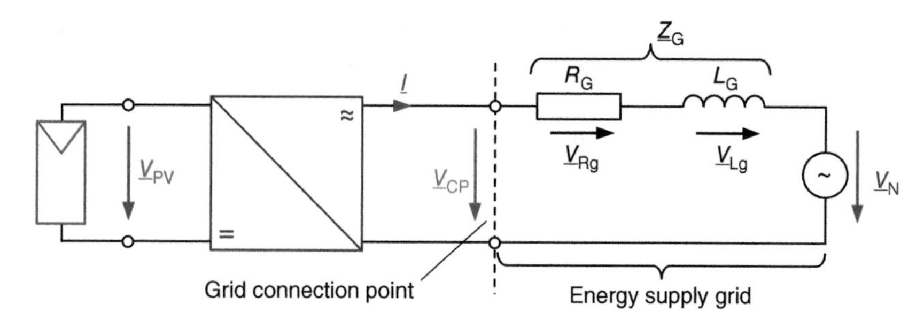

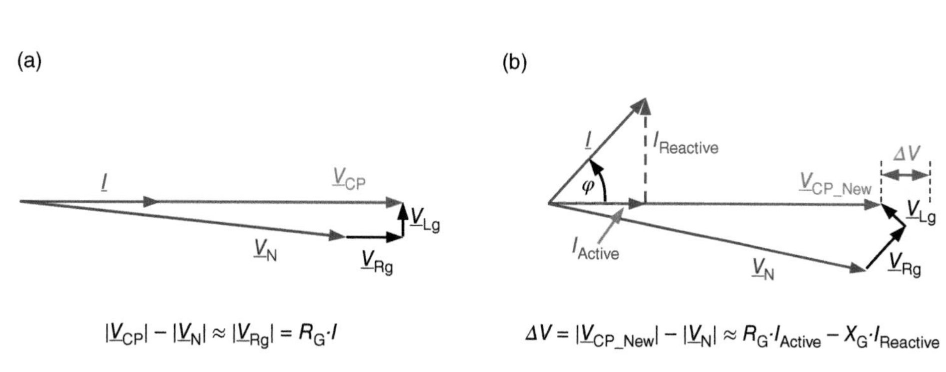

Figure 7.25 Effect of reactive power feed-in: The voltage drops V_{Rn} and V_{Ln} are turned in their phase in such a way that a voltage reduction at the grid connection point is achieved. Source: After [40].

Figure 7.25. Here, the energy supply grid is again modeled with a series connection of an ideal AC source and a grid impedance. In a low-voltage grid, V_N for example is the open-circuit voltage on the secondary side of a medium-voltage transformer. The grid inductance L_N comprises the inductance of the transformer windings and the line inductance (e.g. of a ground cable). The grid resistance R_N is mainly caused by the line.

In case (a) in Figure 7.25, the inverter feeds active power into the grid, as voltage V_{CP} and current I are in-phase. The resulting voltage V_{CP} increases clearly against V_N through the voltage drop at the grid impedance. From the phasor diagram, it can be seen that the voltage increase is caused mainly by the grid resistance R_N. It can be roughly determined using the following equation:

$$|\underline{V}_{CP}| - |\underline{V}_N| \approx |\underline{V}_{Rg}| = R_G \cdot I. \tag{7.30}$$

This voltage rise can lead to the fact that the inverter has to switch off to avoid damage to the devices in the grid.

However, there is an even better solution. This is made clear with case (b) in Figure 7.25. Here the inverter, in addition to the proper active power, also feeds in capacitive reactive power. The phase angle follows a value between 0° and +90°. The current vector I now consists out of two components: The active current I_{Active} with $I_{Active} = I \cdot \cos\varphi$ and the reactive current $I_{Reactive}$ with $I_{Reactive} = I \cdot \sin\varphi$.

The resulting voltage V_{CP_New} at the grid connection point is clearly smaller than V_{CP} in the case of a pure active power feed-in. The trick is that the two phasors V_{Rn} and V_{Ln}, being perpendicular to each other, are turned in such a way that V_{CP} becomes smaller.

The voltage reduction ΔV in case (a) can be roughly determined by considering only the active components (horizontal components) of the voltage drops at the grid resistance and the grid inductance:

$$\Delta V = |\underline{V}_{CP\_New}| - |\underline{V}_N| \approx R_G \cdot I_{Active} - X_G \cdot I_{Active}, \quad (7.31)$$

where X_N is the grid reactance.

The low-voltage guideline VDE-AR-N 4105 mandates that the grid operator can prescribe a fixed power factor ($\cos\varphi$). With larger inverters, it is also possible to prescribe a power factor/active power characteristic curve so that $\cos\varphi$ can be dynamically changed as a function of the currently fed-in power. Nowadays, all newly installed inverters in Germany fulfill this requirement. More details about reactive power provision can be found in, for example [40].

7.7 Safety Aspects

The most important aspects of safety of on-grid plants will be briefly discussed in this section.

7.7.1 Earthing of the Generator and Lightning Protection

For the protection of persons, supports and module frames must be earthed except when they are classed as protective class 2 or the open-circuit voltage of the PV generator is less than 120 V. However, this earthing should also be carried out for inverters without transformers for modules of protective class 2. The reason is that there is a high capacity between the module frame and the cell strings, and it is possible that high voltages exist at the module frame due to the changing generator potential [40].

The earthing of the plant also has advantages for lightning protection, as lightning currents can be diverted to Earth. The earthing should be carried out with a massive copper line (at least 6 mm^2, better 16 mm^2) in the shortest path to the potential compensation rail. Basically, it can be said that lightning protection is not a requirement for private residences. If, however, a lightning protection system exists, then the photovoltaic plant must be connected to it [94]. A very comprehensive discussion on lightning protection of photovoltaic plants can be found in, for instance [40]. Contrary to constant rumors, it should be mentioned here that a photovoltaic plant on the roof of a house does not increase the risk of a lightning strike [94].

7.7.2 Fire Protection

In recent years, there have been repeated reports of fire protection hazards in connection with photovoltaic plants. On the one hand, a photovoltaic plant represents an extended electrical plant that, like other electric plants, can trigger a fire in case of error. There are known cases, for instance, of scorched junction boxes on modules of BP Solar caused by poorly soldered terminals [95]. However, these problems are special cases that can be prevented by means of better quality management on the part of the manufacturer.

Far more critical is the fact that a photovoltaic plant on a burning house can be a great hazard for firemen. This also arises with normal mains installations, but that can be stopped relatively simply by switching the house off from the grid. In contrast, in the case of a solar power plant, even after switching off the inverter, there is still a voltage of several hundred volts in the solar generator. Added to this is the fact that the possibly arising electrical arcs are not self-extinguishing. The first proposals for a solution, such as placing the string lines on the outer wall of the house or in metal tubes inside the house, lessen the danger only slightly. There must be a permanent solution that will safely short-circuit each module in case of fire or that uncouples it from the string line. Conceivable would be a self-conducting semiconductor switch (e.g. MOSFET) in the module junction box that will cancel the short circuit only when it receives a signal from the inverter. If the inverter switches off in case of a fire, then this signal is missing, and all modules are short-circuited. It is to be hoped that this type of solution will become mandatory in the foreseeable future in order to spare the fire brigade unnecessary danger and the photovoltaic industry from damage to its image.

Section 10.4 gives concrete examples of photovoltaic plants that describe the construction, the components used, the dimensioning of the inverters, and the operating results.

8

Storage of Solar Energy

The question of storing electricity is, nowadays, a frequent subject in conferences, in discussion rounds, and in pub conversations. To achieve a complete transition of the energy supply on to renewable energies, storages will be indispensable. Thereby a wide range of storage capacities is thinkable: Small storages in private households to enhance the self-consumption of solar energy, medium storages to support the local power grid, and large storages to store the electrical energy over days, weeks, or even months.

We want to focus in particular on the use of storages in direct proximity to the PV plant. Solar electric energy, which is used or stored at the place of generation, must not be transported over long grid lines. At the same time, decentralized distributed storage units lead to a large reliability against blackout.

For this, we will first have a look on the principle of solar storage with grid-coupled plants. Afterward, we learn to know different battery types together with their modes of operation. Finally, at concrete plants, it is considered, how the self-consumption rate can be enhanced with storages and what advantages the decentralized storages offer from the grid point of view.

The chapter is rounded off with the view on photovoltaic stand-alone systems. Besides the description of solar home systems and hybrid plants, the dimensioning of stand-alone systems is carried out by means of a concrete example.

8.1 Principle of Solar Storage

The principal build-up of a grid-coupled PV plant with solar storage is shown in Figure 8.1. The generated solar energy can be fed into the public grid, consumed in the house, or stored in the battery. In the latter case, it is available at a later time for the domestic users.

In case of a DC coupling of the storage (Figure 8.1a), the battery is connected via a charge controller (see Section 8.2.2) and a DC/DC converter for voltage adaption directly to the DC line of the solar generator. The energy delivered from the solar generator can therefore immediately charge the battery. An alternative is shown in Figure 8.1b – with the AC coupling, the transfer of the energy only happens on alternating current side. To charge the battery, first the direct current is converted

Photovoltaics – Fundamentals, Technology, and Practice, Second Edition. Konrad Mertens.

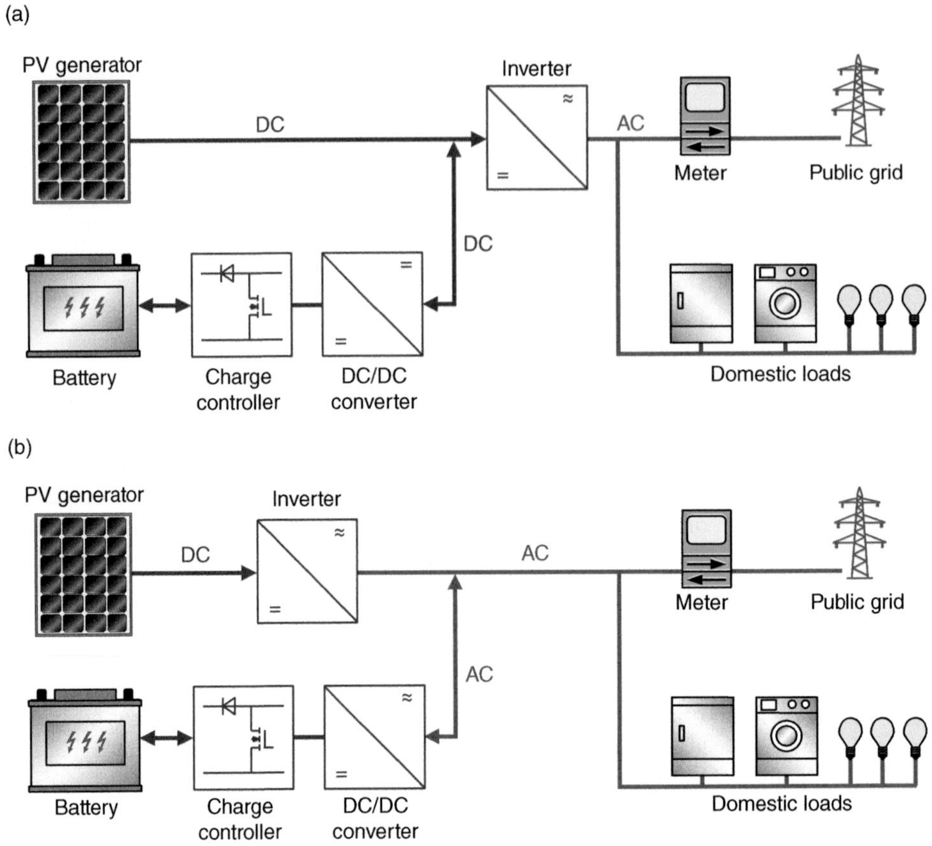

Figure 8.1 Principal build-up of a grid-coupled PV plant with battery storage: (a) DC-coupling of storage, (b) AC-coupling of storage.

to alternating current and then finally again to direct current. This rather leads to higher losses, however, considering the high efficiencies of modern inverters, carries no weight. More important is the main advantage of AC coupling: The storage system can be upgraded and expanded at any time.

8.2 Batteries

Batteries are used as storages for electrical energy in the low power range, mainly electrochemical storages. In the case of small electronic devices, these are often primary batteries, which are not rechargeable. Secondary batteries in contrast are repeatedly rechargeable. The expression *accumulator* is used synonymously. In the following, we will use the expressions battery and accumulator likewise for a rechargeable electrochemical storage.

A battery mostly consists of the interconnection of several electrochemical cells. In the single cell, a chemical reaction is used to generate an electrical current. There two spatially separated electrodes are connected with each other by an ion conducting electrolyte. At the electrodes, a redox reaction, which is split into two half reactions, takes place. Thereby energy in the form of electrons flowing into the outer circuit is set free, which can be used electrically. During the charging of the cell, the opposite redox reaction takes place.

Meanwhile, different battery types as solar storages are on the market. We will look at the most important types and will discuss their peculiarities as well as their pros and cons.

The lead accumulator thereby takes a special position for it exist extensive experiences out of, meanwhile, 150 years of development history.

Therefore, using the example of the lead battery, we will consider the most important properties of electrochemical storages together with their modes of operation and charging strategies. Afterward, we will learn to know the further battery types with their respective characteristics.

8.2.1 Lead-acid Battery

8.2.1.1 Principle and Build-up

Figure 8.2 shows the principle of the lead-acid battery (or "lead battery" for short). It is filled with an electrolyte of diluted sulfuric acid (H_2SO_4). The negative electrode consists of lead, while the positive electrode is of lead oxide (PbO_2).

Let us consider first the discharging process: The lead at the negative electrode reacts with the electrolyte under the loss of electrons (oxidation) to lead sulfate ($PbSO_4$):

$$Pb + SO_4^{2-} \rightarrow PbSO_4 + 2e^-. \tag{8.1}$$

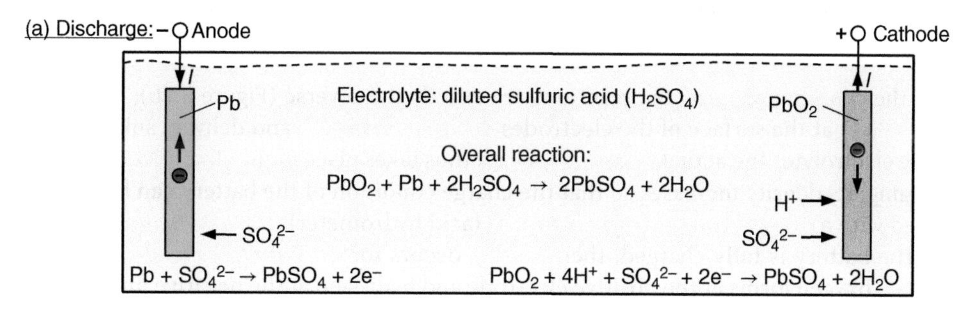

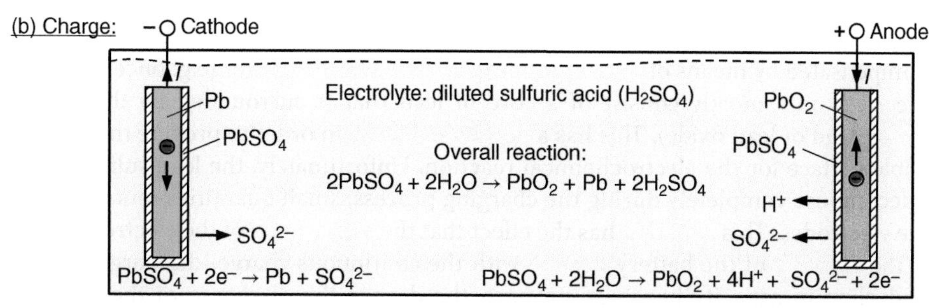

Figure 8.2 Principle structure of a lead-acid battery: When discharging, a layer of lead sulfate ($PbSO_4$) forms at both electrodes that is decomposed again during charging (Nomination of electrodes: See Section 8.2.3.2).

At the same time, the lead oxide of the positive electrode reacts with the sulfuric acid under the take-up of electrons (reduction) to lead sulfate and water. This causes a layer of lead sulfate to grow on both electrodes.

$$PbO_2 + 4H^+ + SO_4^{2-} + 2e^- \rightarrow PbSO_4 + 2H_2O. \tag{8.2}$$

Hereby, at both electrodes, a layer of lead sulfate grows. As the total reaction of the battery cell, we get

$$PbO_2 + Pb + 2H_2SO_4 \underset{\text{Discharging}}{\overset{\text{Charging}}{\rightleftharpoons}} PbSO_4 + 2H_2O. \tag{8.3}$$

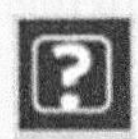

Why do we speak of "reduction" when the lead oxide takes up an electron? The number of electrons has not been reduced anyway, but instead increased.

The nominations reduction and oxidation are founded historically. Originally, oxidation describes the reaction of a substance with oxygen. If, for example, carbon reacts with oxygen to form carbon oxide (CO_2), then the four-valent carbon atom gives four electrons to the double bonds with the two oxygen atoms ("oxidation"). Nowadays, the expression oxidation generally describes a reaction where one substance gives electrons to a reaction partner. On the contrary, the reaction partner takes up electrons (in our example, each oxygen atom takes two electrons from the carbon atom, thus the oxygen is reduced). As reduction and oxidation always happen coupled, one speaks of redox reaction.

In the charging process, the reactions are exactly the reverse (Figure 8.2b): Now the lead sulfate at the surface of the electrodes decomposes again and delivers sulfate ions to the electrolyte. The actual storage of energy thus takes place in the electrolyte. During charging, its density increases so that the charge condition of the battery can be determined with a density-measuring instrument (acid hydrometer).

If the battery is fully charged, then gassing occurs for exceeding the end-of-charge voltage: oxygen forms at the positive electrode and hydrogen at the negative one, which together can form explosive oxyhydrogen gas. For this reason, it must be ensured that the room has sufficient ventilation. Gassing also eventually leads to a loss of water that must be compensated by means of regular topping up with distilled water (e.g. once a year).

The electrodes mostly consist of a core of lead that is surrounded by the active material (lead or lead oxide). This has a porous structure in order to provide the largest possible surface for the electrochemical reaction. Unfortunately, the lead sulfate does not decompose completely during the charging process; small quantities remain stuck to the electrodes. This sulfation has the effect that the active mass of the electrodes and thus the capacity of the battery decline with the continuous charge–discharge cycles. This decline in capacity becomes stronger, the deeper the discharge of the battery. Frequent deep discharges thus result in an extreme reduction in the length of life of a lead battery. The lifespan is defined as the time at which the battery falls below 80% of its original nominal capacity [96].

Why is that the end of its life? Surely, the battery is still usable, or is this not the case?

Actually, the battery can still be used, but only with reduced and further sinking capacity. At some time, the lead sulfate deposits of the positive and negative electrodes touch each other and cause a short circuit. Figure 8.3 shows the linkage between depth of discharge and maximum achievable number of cycles for various types of batteries.

8.2.1.2 Types of Lead Batteries

Which type of batteries are to be used in a stand-alone system depends to a great extent on the particular requirements. First, there is a difference between the method of operation of lead batteries in buffer operation and cycle operation. A normal car battery (starter battery), for instance, is operated in buffer mode. For most of the time, it is fully charged, but occasionally must deliver short-term high currents for starting the engine. Things are different with a forklift battery: It is fully charged overnight and almost fully discharged on the next day and thus goes through whole charge and discharge cycles. This type of operation is called cycle operation, which tends also to be the case with solar plants (especially stand-alone plants). The different requirements affect the method of construction of the different types of batteries.

A starter battery needs to provide high currents for short periods, which is why there are many electrodes close together in the form of plates that offer a large surface. With a stand-alone plant with typical cycle operation, the starter battery would be unusable in a few weeks due to progressive sulfating and corrosion.

A car battery is not suitable as storage in a stand-alone solar plant as it would become defective in a short period due to the cycle operation.

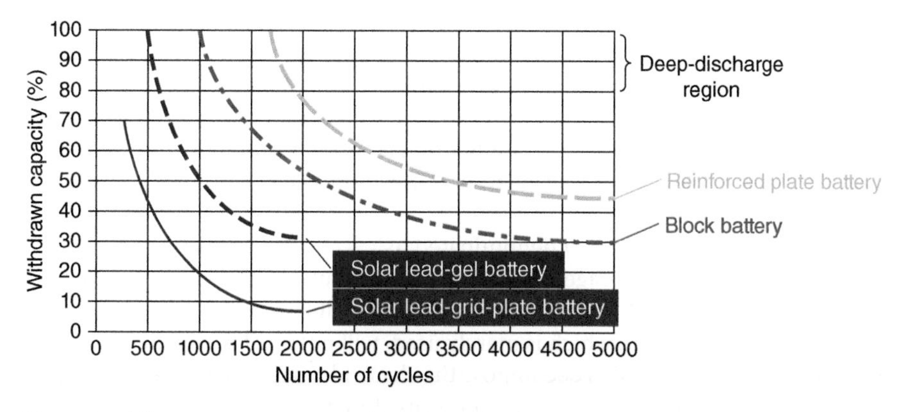

Figure 8.3 Lifespan of various types of batteries: The deeper the battery is discharged, the fewer number of cycles it can carry out until it reaches the end of its life [22].

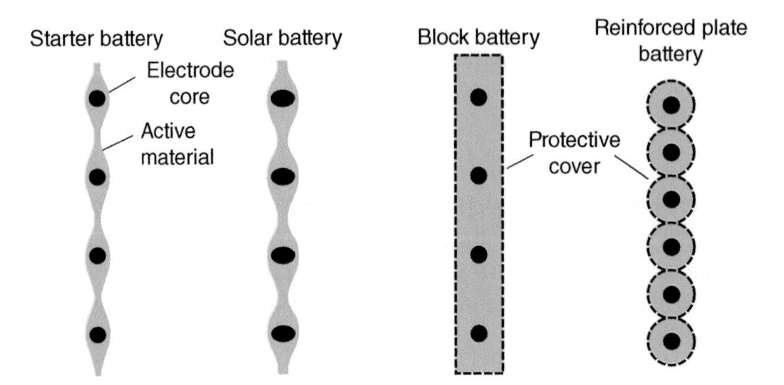

Figure 8.4 Principle arrangement of the electrodes of various types of batteries: Block and reinforced plate batteries keep the active material in place by means of an electrolyte-permeable protective sleeve.

Solar lead grid-plate batteries are modified starter batteries in which thicker plates with larger spacings are used, and the lead plates are hardened with an antimony additive (Figure 8.4). They only reach a lifespan of 300 cycles if they are discharged to a maximum of 30% of their capacity (depth of discharge 70%) (see Figure 8.3). They can reach 1000 cycles for a discharge of only 20%. Thus, they are only suitable for sporadic use such as in weekend cottages.

In solar lead-gel batteries, the electrolyte is thickened to a gel by additives. Here the battery can be completely sealed and is thus leak-proof. In addition, no gas leaks, so no distilled water needs to be topped up. A special charge controller that ensures adherence to the end-of-charge voltage is important, as otherwise, the battery will dry out due to gassing. In contrast to this, an occasional gassing is desirable for standard batteries in order to mix the electrolytes. Depending on the method of operation, the length of life of solar lead-gel batteries is almost double that of the classic solar batteries (see Figure 8.3).

If one wishes to achieve a continuous operation over 15–20 years, then stationary reinforced plate batteries are the best choice. These are available in two variants: The stationary reinforced plate with a special separation and fluid electrolyte battery and the sealed stationary reinforced plate battery. They are normally used for battery-supported emergency power systems and cost two to three times more than simple solar batteries. The positive plate consists of lead rods that are individually surrounded by tubes. These keep the active material together and prevent premature loss of mass (Figure 8.4) [40].

A middle path finally is the block battery (also called the stationary grid plate block). Here, several lead rods are surrounded by a common protective sleeve. They are therefore cheaper than reinforced plate batteries and yet achieve relatively high lifespans.

The lifespan presented here is also denominated as cycle lifespan, as it specifies the length of the usability of the storage not as a time span but as a number of cycles. It describes in a way that part of the aging process, which results from of the real use of the storage. However, also a battery, which is merely used, will not last forever. Reasons can be chemical decay processes, which decompose the electrode materials or the electrolyte in the course of time. This influence can be described by the calendrical lifespan. For lead-acid batteries, it lies around 10 years.

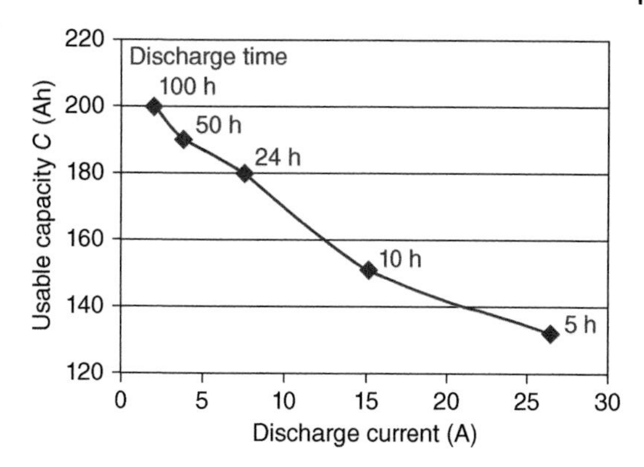

Figure 8.5 Dependency of the capacity of a reinforced plate battery (*block solar power* 200) on the amount of the discharge current.

8.2.1.3 Battery Capacity

The capacity C of a battery is not a fixed amount but depends on the discharge current. If the battery is discharged with a low current, then the sulfate ions can penetrate deep into the active mass of the electrodes and can convert into lead sulfates. With greater current draw off, the first stored sulfur molecules block the penetration of the following molecules so that the active mass cannot be fully utilized. For this reason, the nominal capacity of a battery is always mentioned in connection with a certain nominal charge current. Mostly, the nominal capacity is referenced to a nominal charge current I_{10}, a current that discharges the battery in 10 h.

This is shown for the sake of clarity in Figure 8.5 on the basis of a reinforced plate battery of the Hoppecke Company: If the battery is discharged within 5 h ($I_5 = 26.5$ A), then only a capacity of C_5 of 132 Ah is usable. The other extreme case would be a discharge over 100 h, which gives a capacity of C_{100} of 200 Ah at a current of $I_{100} = 2$ A. The type designation of the battery is called the *block solar power 200*, but actually, the nominal capacity C_{10} is only 150 Ah. This example emphasizes the need to always check on the datasheet as to which discharge current was used for the nominal capacity.

The temperature also has an effect on the usable capacity of a lead battery. Anyone who has not been able to start his car in winter knows that the capacity is less at low temperatures. It lowers the capacity, for instance, at 0 °C to about 80% of the capacity at 20 °C. At much lower minus degrees, the electrolyte freezes and limits the functionality of the battery completely.

8.2.1.4 Voltage Progression

The nominal voltage of an individual battery cell is 2.0 V. Typical is a series connection of six individual cells, so that, depending on the charge condition, there is a battery voltage of 12.0–12.7 V. Figure 8.6 shows at the left the progression of the voltage in the discharge of the battery. Starting from the open-circuit voltage, the voltage is reduced up to the end-of-charge voltage of 10.8 V. If current continues to be drawn off, it will reach deep-discharge that can damage the battery.

The progression of the charging can be seen at the right in Figure 8.6. A voltage greater than the open-circuit voltage must be applied to press the sulfate ions into the electrolyte again. Usually, the so-called *I/V* charging method is used: First, the battery is charged with a constant current. When the end-of-charge voltage has been reached, then there is a change to constant charge voltage. The charge voltage of 13.8–14.4 V (depending on

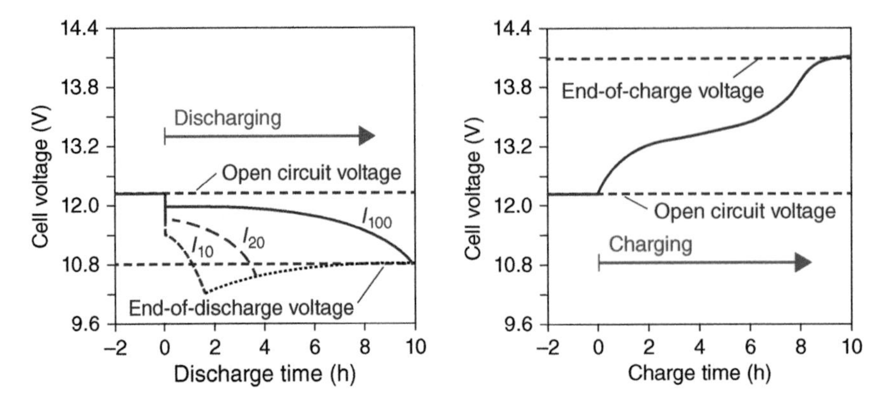

Figure 8.6 Progression of voltage of a 12 V battery when discharging and charging: The lower permissible limit is the end-of-discharge voltage, whereas the upper permissible limit is the end-of-charge voltage [96].

temperature) should not be exceeded in order not to overload the battery and to avoid gassing.

8.2.1.5 Summary

Lead-acid batteries are in use since decades and are, therefore, a proven technology. For different applications, different variants are available. However, the lifespan with about 10 years and 2000–3000 cycles is clearly restricted. Due to the relative high specific weight of lead, only moderate energy densities of 30–40 Wh kg^{-1} are attained.

8.2.2 Charge Controllers

We now want to consider the tasks and operating modes of charge controllers. For a better understanding, we start with simple systems consisting of a solar module, a battery, and simple DC loads. In the case of grid-coupled systems, the principle is still the same. However, then high DC voltages are present at the charge controller so that many battery modules have to be connected in series possibly combined with a DC/DC converter (see Figure 8.1).

Meanwhile, we have learned that batteries require care in handling in order to enjoy their use for a long time. This depends on a series of tasks for the charge controller:

- Overload protection
- Deep-discharge protection
- Prevention of unwanted discharging
- State-of-charge monitoring
- Adjusting to battery technology (electrolyte/gel)
- Voltage conversion (possibly)
- MPP tracking (possibly).

8.2.2.1 Series Controller

The arrangement of a classic series controller is shown in Figure 8.7. The control electronics constantly measure the battery voltage, and when the end-of-charge voltage

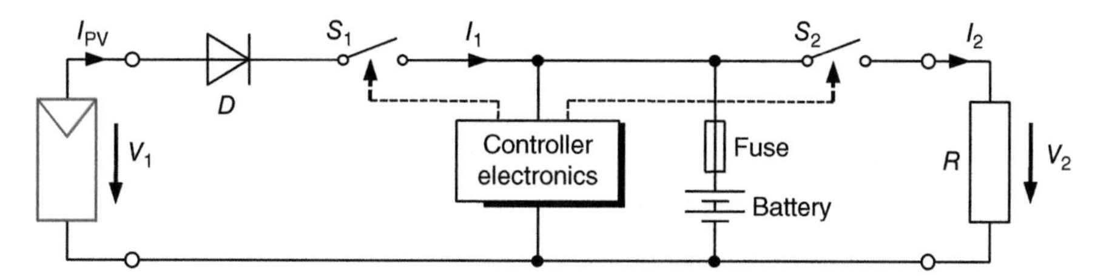

Figure 8.7 Principle of a series charge controller: Switch S_1 interrupts the charge current when reaching the end-of-charge voltage; switch S_2 serves for the deep-discharge protection of the battery.

has been reached, turn the switch S_1 (mostly a MOSFET) off. Switch S_2 separates the load from the battery in the case of falling below the end-of-discharge voltage and thus ensures deep-discharge protection. The diode at the input of the charge controller is meant to prevent the battery from being discharged at night by the inactive solar generator. A problem can arise when the battery is deep-discharged at night, and there is insufficient power available to operate the control electronics. As switch S_1 is possibly opened, the battery cannot be reloaded again despite the Sun's radiation the next morning.

8.2.2.2 Shunt Controller

An alternative to the series controller is the shunt controller (parallel controller). Here the transistor is connected in parallel to the solar module (Figure 8.8). As soon as the transistor conducts, it short-circuits the solar generator and interrupts the loading of the battery. An advantage of this concept compared to the series controller is the fact that the unavoidable (but relatively small) voltage drop at the switched-through MOSFET incurs no losses during charging.

The second advantage is that the MOSFET blocks without voltage signal at the gate, and thus the battery can be charged in the morning after the deep discharge case described previously. These advantages result in the shunt controller being mostly used today.

8.2.2.3 MPP Controller

Naturally, MPP tracking makes sense also in stand-alone systems in order to obtain the maximum energy from the solar generator. Figure 8.9 shows a charge controller that realizes the MPP tracking similar to that described in Section 7.1.3 via a DC/DC converter. This is mostly a buck converter that, for instance, reduces the input voltage from up to 48 V to a system voltage of 12 or 24 V.

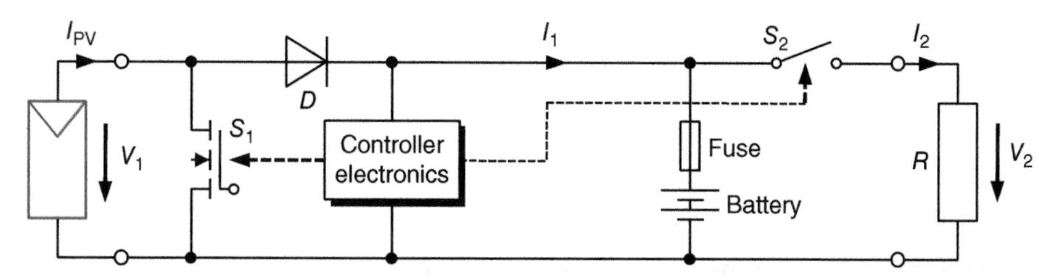

Figure 8.8 Principle of the shunt controller: If the charge current is to be interrupted, then the transistor S_1 short-circuits the solar generator.

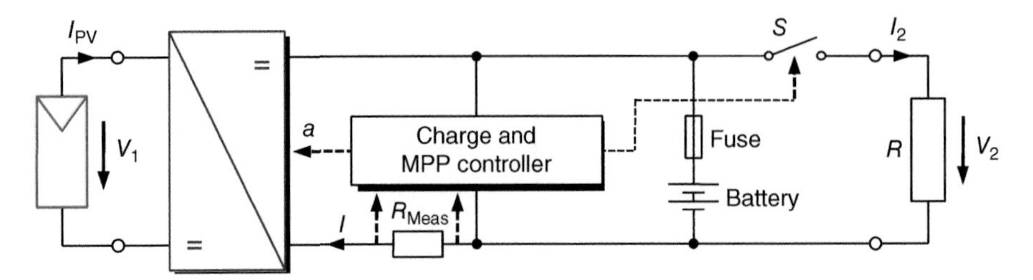

Figure 8.9 Principle of a charge controller with MPP tracker: The voltage of the DC/DC converter is varied by varying the duty factor a and thus the MPP of the solar generator is reached.

Figure 8.10 Solar charge controller of the Steca Elektronik Company: The PR 0505 shows the actual condition of the battery relatively coarsely via two LEDs, while the PR 3030 has a graphic display showing an intelligent charge monitoring.

8.2.2.4 Examples of Products

Figure 8.10 shows two solar controllers of the Steca Elektronik Company: The technical data is listed in Table 8.1. The PR 0505 is a simple controller for a module up to 78 Wp.

The maximum input voltage is 47 V and the maximum current is 5 A. The room temperature is determined by means of an internal sensor in order to determine the correct end-of-charge and end-of-discharge voltages. The controller is arranged as a series controller and controls the current in PWM operation with a switching frequency of 30 Hz.

The PR 3030 is much more powerful; it is suitable for solar generators up to 900 Wp. It works as a shunt controller and offers various additional functions such as an external temperature sensor that is fixed directly to the battery. A further feature is the intelligent charge condition determination with which current, voltage, and temperature are continuously monitored during charging and discharging. This makes possible a very exact representation of the actual battery charging state on the graphic display.

Both controllers make use of a variation of the already mentioned I/V charging process: The current prescribed by the solar module is charged up to the end-of-charge voltage. Then the electronics control the MOSFET by means of a PWM signal such that the end-of-charge voltage is maintained.

8.2.3 Lithium Ion Battery

Already in the year 1909, Thomas Edison proposed to use lithium as electrode material for rechargeable batteries. It lasted though about 80 years, until in 1991 Sony brought the first commercial lithium battery on the market. It was a lithium–cobalt-dioxide battery

Table 8.1 Data for two different solar charge controllers [www.steca.de].

Designation	Steca PR 0505	Steca PR 3030
System voltage (V)	12	12 (24 V)
Maximum charge current (A)	5	30
Connectable PV power (Wp)	78	900
Controller principle	Series controller, PWM-operation	Shunt controller, PWM-operation
Temperature compensation	Yes, with internal sensor	Yes, optional with external sensor
PWM control	Yes	Yes
Overload protection	Yes	Yes
Deep discharge protection	Yes	Yes
Polarity reversal protection	Yes	Yes
Charge condition determination	Via, voltage, and temperature	Via current, voltage, and temperature
Charge condition display	2 LEDs	Graphic-LCD-display
Features	—	Menu-controlled operation

for a Hi8 video camera with a capacity of 1200 mAh. Since then, the technology was continuously further developed so that lithium batteries, nowadays, are typically used in many portable devices (mobiles, laptops, cameras, etc.). In larger construction forms, they are used in electric cars and, also, in stationary solar storages.

8.2.3.1 Principle and Build-up

The first variants or lithium batteries were lithium metal batteries. This means that they use metallic lithium (thus a lattice of lithium atoms) as electrodes. However, massive safety problems encountered. The reason lies in the fact, that metallic lithium, as one of the alkali metals, is highly reactive. It does not only react with oxygen but also with nitrogen, which is also present in air. If lithium gets in contact with a reaction partner like water, it ignites by itself. Therefore, today's lithium batteries no longer use any metallic lithium. Instead, the lithium is used in the shape of ions in which it is no more able for reactions. Thus, the correct denomination of the current battery type should be "lithium ion battery." In the following, for simplification, we, however, will still sometimes use the term *lithium battery.*

Figure 8.11 shows the principal build-up of a lithium ion cell. On the left side, we see the negative electrode made of graphite. Graphite is a modification of carbon (besides diamond and the fullerenes). The carbon atom like silicon is tetravalent and, therefore, has four valence electrons (cf. Chapter 3). In graphite, the atom forms a stabile bond with each three neighbor atoms. Thus, a layered horizontal structure is formed, built out of many hexagons. Each fourth free valence electron ensures that in the horizontal layer a good electrical conductivity is established. Thus, graphite, in contrast to diamond, is a good electrical conductor.

The different layers are only weakly tied to each other and, therefore, have a relative large distance of about 335 pm from each other. In this gap now, the lithium ions can be

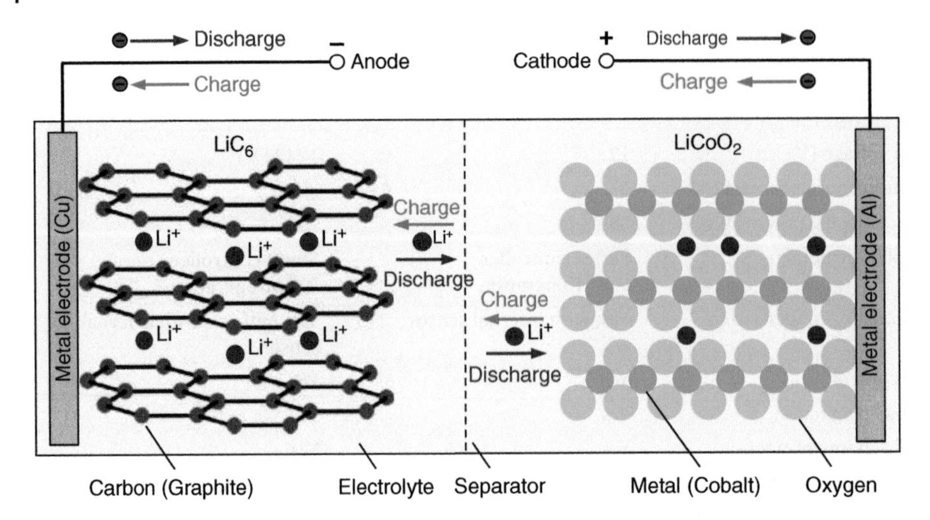

Figure 8.11 Build-up of a lithium ion cell: During the charging, the lithium ions are embedded in the graphite lattice. During discharging, the ions move out of the graphite into the cathode material [97].

embedded, which only have a diameter of 68 pm. Per six carbon atoms each one lithium ion can be embedded. The positive electrode consists of transition metal oxides, which also can embed lithium ions. A standard material for device batteries is, e.g. lithium cobalt oxide ($LiCoO_2$). The separator similarly to the lead acid battery is used to avoid a direct contact of the two electrodes. At the same time, it has to facilitate a current flow through ion migration. As a material, extremely thin (e.g. 30 μm) porous polymer foils are deployed, e.g. polyethylene (PE) or polypropylene (PP). As electrolyte, typically organic solvents are used, in which a conducting salt (e.g. $LiPF_6$) is solved.

The process of embedding ions in the respective crystal lattices is also called intercalation (from Latin: *intercalare* = insert). Accordingly, the taking out of ions is called deintercalation.

8.2.3.2 Reactions During Charging and Discharging

During charging, electrons are drawn out of the crystal lattice of the positive electrode (oxidation). As compensation, positive-charged lithium atoms migrate out of the lattice and dissolve in the electrolyte:

$$2LiCoO_2 \rightarrow 2Li_{0.5}CoO_2 + Li^+ + e^-. \tag{8.4}$$

At the negative electrode, electrons move into the graphite (reduction). As a result, the Li^+ ions being present in the electrolyte are attracted and embedded in the graphite lattice:

$$Li^+ + e^- + 6C \rightarrow LiC_6. \tag{8.5}$$

The summation formula shows that each of the six carbon atoms is needed to intercalate one lithium atom.

During discharging, the described reactions run the other way round. As overall reaction at the lithium ion battery, we get

$$LiC_6 + 2Li_{0.5}CoO_2 = \underset{\text{Discharge}}{\overset{\text{Charge}}{\rightleftarrows}} C_6 + 2LiCoO_2. \tag{8.6}$$

In Figure 8.11, the negative electrode is termed as anode and the positive electrode as cathode. In other books, it is often the other way round. Which terminology is correct?

In fact, the terminology changes, depending on whether the charging or the discharging of a battery is considered. The anode is always that electrode at which an oxidation takes place, while at the cathode a reduction happens. During charging, therefore, the positive electrode is named anode and the negative electrode the cathode, thus exactly opposite to the captions in Figure 8.11. In our terminology, we always assume the discharging case. This, however, is strictly speaking only correct for primary batteries, which cannot be recharged. The best thing is to only speak of positive of negative electrode, then you can do nothing wrong.

8.2.3.3 Material Combinations and Cell Voltage

Besides lithium cobalt oxide and graphite as active materials (intercalation materials), there exist a large number of possible substances. For the cathode, metals such as manganese, nickel, or iron are available. For the anode, e.g. lithium silicate (LiSi) can be used instead of graphite. Depending on the chosen material combination, there will be another cell voltage due to the different chemical potentials of the lithium atoms in the respective electrode material. To make the electrode potential comparable, it is typically measured with reference to a standard electrode of metallic lithium, whose potential is defined to the value zero. The cell voltage of a concrete battery cell then simply results out of the difference of potentials of the two involved electrode materials.

To illustrate this, Figure 8.12a shows the electrode potentials of different intercalation materials. If, for example, the already mentioned material combination $LiCoO_2$ and

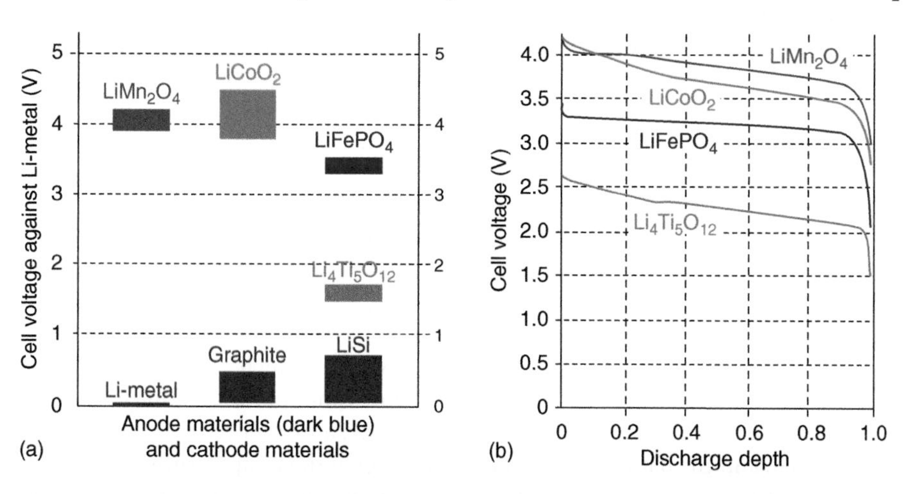

Figure 8.12 Voltage situation at the lithium ion cell: (a) Electrode potentials of different anode and cathode materials and (b) discharge curve in dependence of the discharge depth [98, 99].

graphite is used, the maximal voltage of the call is calculated to

$$V_{\text{Cell}} = V_{\text{Cathode}} - V_{\text{Anode}} \approx 4.5\ \text{V} - 0.4\ \text{V} = 4.1\ \text{V}. \tag{8.7}$$

This voltage, however, only applies for the full-loaded lithium ion cell, as it similarly to the cell of the lead battery dips with decreasing state of charge. This is depicted in Figure 8.12b for different cathode materials (each with graphite as anode material). An unfavorable curve shows $LiCoO_2$, as the voltage reduces down to about 3.5 V at a discharge level of 80%. Furthermore, in the batteries with $LiCoO_2$ cathode, only a part of the lithium can be used, as otherwise the layer structure would be unstable. More drawbacks of the material are the toxicity and limited availability of the cobalt.

Titan dioxide (e.g. $Li_4Ti_5O_{12}$) is a particularly long-living anode material. However, the batteries have only reduced energy densities due to their small voltage (see Figure 8.12b). This is in particular a disadvantage for the deployment in the automotive sector; yet for stationary storages, they can be used.

A relatively new material is lithium iron phosphate ($LiFePO_4$). It has a lower price than $LiCoO_2$ and is nontoxic. The batteries made of this material offer a high energy density, because with this intercalation material almost the whole lithium can be used for storage. Another advantage is the very constant cell voltage at different discharge depths (see Figure 8.12b).

A special variant of the lithium ion battery is the lithium ion polymer battery, which is often shortened and designated as "LiPo battery." Instead of the liquid electrolyte, it contains a gel-like or solid foil on polymer basis. This has the benefit that the batteries can be built in almost arbitrary housing forms. Therefore, an adaption to all possible electronic devices is possible (even in only 1 mm thickness in chip carts). The batteries show a high power density and are the favorite battery types especially for model builders. Applications in electric cars and stationary storages are also common.

8.2.3.4 Safety Aspects

Repeatedly, articles about safety problems of lithium batteries are shown in the newspapers. Besides others, fires in laptops, cars, and even planes were reported. The main reason is the high reactivity of the metallic lithium, already mentioned above.

As described earlier, during charging of the battery, the lithium ions are intercalated in the graphite lattice of the anode (Figure 8.11). If the battery is overcharged, the ions push further in the graphite lattice and press it apart, which can lead to an irreversible damage. During further charging, the ions start to build up an irregular lattice at the surface of the anode. This tree-like crystal is also named dendrite (from Greek: *déndron* = tree).

The dendrite, growing in the direction of the cathode, can penetrate the separator and thus lead to a short circuit and, finally, to the destruction of the cell.

As a remedy, special porous polymer separators are used, which soften above about 130 °C. This softening provokes a sealing of the foil pores and thus a stopping of the ion current is achieved (so-called *shutdown*). With this trick, a further charging of the cell can be prevented, and the battery is protected. However, if the temperature rises above 150 °C, a complete melting of the separator can follow, which leads to a short circuit of the cell ("thermal runaway") [99].

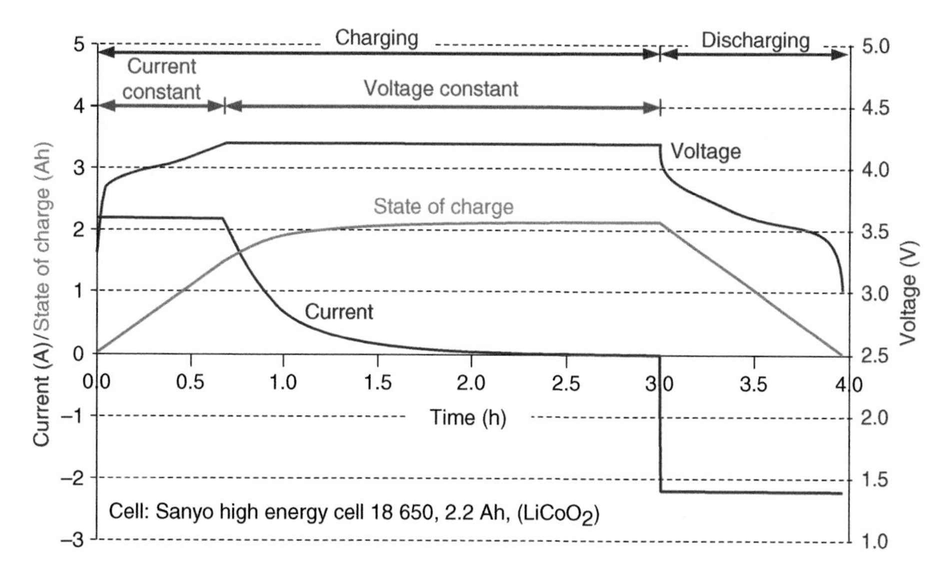

Figure 8.13 Charging characteristics of a lithium ion battery: If after charging with constant current the charging end voltage is reached, a constant voltage is applied to achieve full charge. (Source: After [97].)

8.2.3.5 Charging Procedures

The standard charging procedure for lithium ion batteries is the I/V-charging already known from the lead battery (Section 8.2.1.4). It is also named CCCV-procedure (constant current/constant voltage). Figure 8.13 shows exemplary course of current and voltage during charging and discharging of a lithium ion cell of a capacity of 2.2 Ah.

A constant current of the dimension $I_1 = 2.2$ A charges the cell until the charging end voltage of 4.2 V is reached. Afterward, the battery is fully charged by applying a constant voltage.

8.2.3.6 Battery Design

The internal build-up of a standard lithium ion cell is shown in Figure 8.14a. To reach a good volume utilization, the sandwich out of cathode, separator, and anode is winded up. An additional separator foil provides the electrical isolation of the different sandwiches. Lithium cells contain an overpressure valve as well as a safety device (current interrupt device, CID) to switch off the current in case of a short circuit.

Figure 8.14b shows different battery designs. Most common is the round cell, and for larger capacities, the prismatic design (rectangular structure) is used. This one has the advantage of a better heat dissipation; however, this is dearly bought by a smaller energy density.

Figure 8.15 shows how separate cells are joined to battery modules, and these again are combined to a complete battery system (battery pack). As the overcharge of cells represents a high risk, the continuous monitoring is necessary. A microprocessor-controlled battery management system (BMS), therefore, surveillances temperature and charge state of the single cells. Unequal cell and module voltages are detected and balanced; this serves as well to prohibit damage as well to optimally use the cell capacity. With

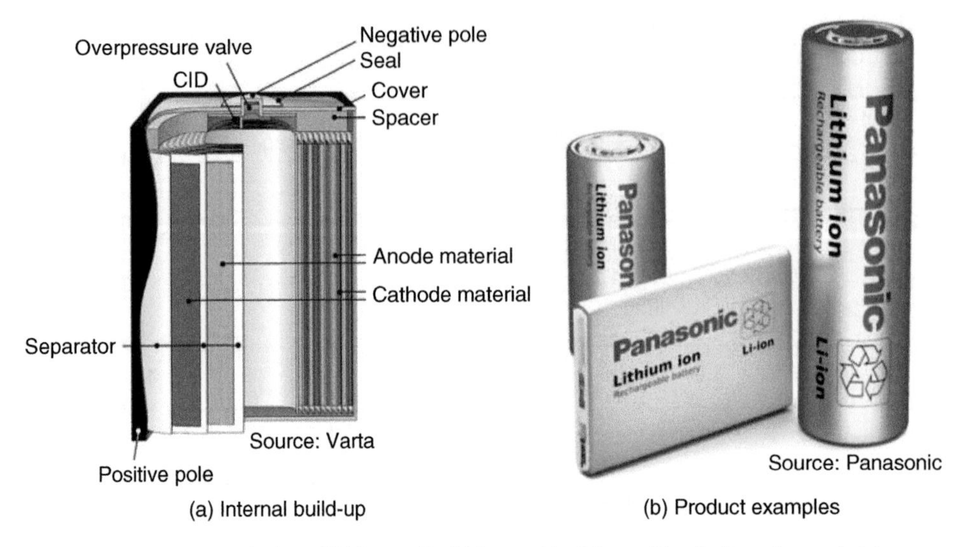

Figure 8.14 A concrete design of lithium cells: (a) Internal build-up with winding of a sandwich structure and safety facilities. (b) Examples of round cells and prismatic cell.

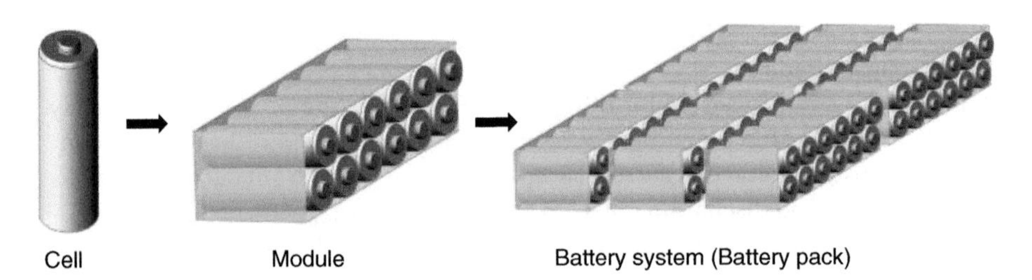

Figure 8.15 From the single cell over the battery module to the complete battery system. (Source: After [97].)

targeted "bypassing," the charge current is guided past the already charged cells, while weaker cells obtain a higher charge current (details see, e.g. [100]).

8.2.3.7 Lifespan

As for the lead battery, also the lifespan of lithium ion batteries strongly depends on the mode of operation. The calendrical aging mainly is determined by the storage temperature and state of charge, while for the cycle stability, also the level of the charging and discharging current and the discharge depth decisive. In the case of deep discharging, strong volume changes are caused in the intercalation material through the in and out swapping of the lithium ions. This finally leads to a breakup of the single layers and thus to a loss of electrochemical active material.

Nevertheless, the lifespan of lithium batteries is typically clearly above the one of the lead batteries. Some manufacturers promise a lifespan of, e.g. 20 years; a number of cycles of 5000 and a discharging depth of 90% (see also Section 8.3.1.3).

8.2.3.8 Application Areas

Due to their high power density and high lifespan, lithium ion batteries, nowadays, are used in electronic devices, electric cars, and solar home storages. Even several large storages in the megawatt region have been realized. Thus, for example, the energy supplier Wemag AG in Northern Germany operates a 5 MW storage, which is build up out of 25 500 lithium manganese oxide cells. It has a capacity of 5 MWh and serves to moderate short-term feed-in fluctuations of wind and solar energy in order to stabilize the grid frequency (see also Chapter 11).

8.2.3.9 Summary

Lithium ion batteries have made an impressive development in the last years. They show lifespans, which are drastically higher with respect to those of lead batteries. This also holds for the energy density, which attains up to 250 Wh kg^{-1}. As the development of new materials for this battery type is still not finished, more improvements and areas of application are to be expected [100].

8.2.4 Sodium Sulfur Battery

8.2.4.1 Principle and Build-up

In comparison to the two battery types already considered, the sodium sulfur battery (NaS) is a real exotic. It only works at high temperatures, and the electrodes are made of liquid material, and the electrolyte again is a solid body.

Figure 8.16 shows the principal build-up of the NaS battery as a total view as well as a detailed view to understand the electrochemical processes. In the center of the cell, the negative electrode is situated. At the operating temperature (about 300 °C), it is surrounded by melted sodium, which represents the active material of the anode. During

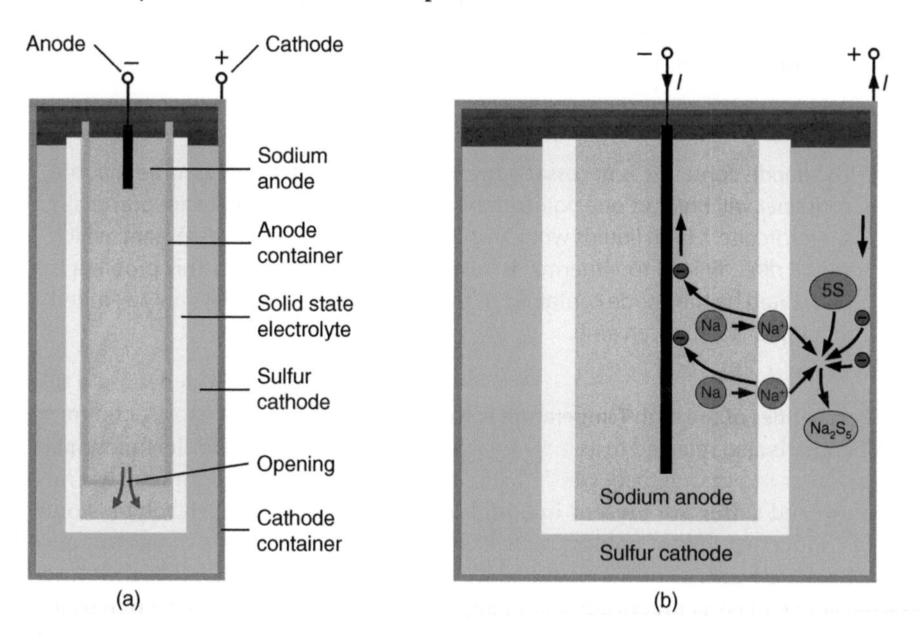

Figure 8.16 Principal build-up of the "exotic": (a) Total and (b) detailed view to depict the electrochemical reaction during discharging. (Source: After [99].)

discharging, the liquid sodium is oxidized at the electrolyte:

$$2Na \rightarrow 2Na^+ + 2e^-. \tag{8.8}$$

The electrolyte consists of doped aluminum oxide ("β-aluminate"), a ceramic-like, ion-conductive solid-state body. The generated sodium ions now migrate through the electrolyte to the cathode, which is again formed by graphite felt soaked with liquid sulfur. There they react with the sulfur and take up electrons (reduction):

$$2Na^+ + 5S + 2e^- \rightarrow Na_2S_5. \tag{8.9}$$

As total reaction, we get

$$2Na + 5S = \underset{\text{Discharging}}{\overset{\text{Charging}}{\rightleftarrows}} Na_2S_5. \tag{8.10}$$

The conductive graphite felt in the cathode is necessary because sulfur is a nonconductor. Moreover, the felt structure helps the Na_2S_5 (sodium-pentasulfide), which is produced during discharging, will not deposit at the cathode surface. This ensures that further on liquid sulfur can be delivered to the reaction zone.

In the charged state, the open-circuit voltage of the cell is 2.08 V. It stays at that level, even in case of more than 50% discharging. With further discharging, the sodium and sulfur content in the range of the electrolyte declines. This leads to the fact that instead of Na_2S_5, less complex compounds like Na_2S_4 and Na_2S_3 are formed. Thereby, the voltage gradually sinks down to 1.78 V. A further discharging should be avoided, as otherwise thermal losses or damage of the cell become probable [99].

In Figure 8.16, it is to see, that the liquid sodium is surrounded by an "anode container." This, however, has an opening at the bottom. Couldn't we just as well omit this container?

The anode container is necessary due to safety reasons. Let us assume that the container will break at one point. In this case, anode and cathode are practically short-circuited. Both liquids would react under strong heat development, which would directly lead to a thermal runaway (see Section 8.2.1.4). This problem is prohibited by the anode container, as through its opening, always only a limited amount of sodium can emit.

8.2.4.2 Peculiarities of the High Temperature Battery

The NaS battery is also referred to as high temperature battery. In fact, for its functionality an operating temperature between 270 and 350 °C has to be kept continuously. Only then, sodium and sulfur are present in liquid forms; moreover, the electrolyte is only capable of conducting ions at not less than 300 °C.

The particular advantage of the liquid active materials lies in the fact that the structure of the active mass is practically not changed during the charging and discharging process. Thus, there is almost no loss of the amount of actually usable mass in contrast to the lead battery. At the same time, the formation of dendrites like in the case of the lithium ion battery is relatively unlikely. Out of this follows a high lifespan of 4500 cycles at a maximum discharge depth of 90%. The calendric lifespan is in the order of 15 years [101].

To ensure that the heating system spends as little energy as possible, the battery should be well thermally insulated. During the active use of the battery (repeated charging and discharging), the reaction heat normally is sufficient to maintain the necessary operating temperature. In case of long downtimes, however, the battery has to be externally heated.

It is clear that the NaS batteries are not suitable as persistent storages due to their heat losses. Instead, they can be well used as highly effective short-term storages, e.g. to compensate strong load variations in the grid or to buffer large wind or PV plants.

8.2.4.3 Sodium Sulfur Batteries in Practice

A pioneer in the development and the deployment of the NaS battery is the Japanese company NGK Insulators. The company offers battery modules with a power of 50 kW as shown in Figure 8.17a. Each module contains 230 cylindrical cells. In comparison to the battery cells considered so far, the NaS cells with their height of 54 cm and a diameter of 9 cm are extremely large. Over 4 h, the cell delivers a current of 90 A with a cell voltage of about 2 V. Between the cells dry, pressed sand is located to mechanically stabilize the cells and at the same time facilitate a good heat conduction.

The battery modules can be connected to larger blocks. In Figure 8.17b, an example of such a block system is shown, which is used by the company Younicos situated in Berlin. It comprises 20 modules with 50 kW each so that a total power of 1 MW is attained [101].

A showcase project of NGK is a 34 MW system, which serves to moderate the power fluctuations of a 54 MW wind park. The system can store a total energy of about 240 MW.

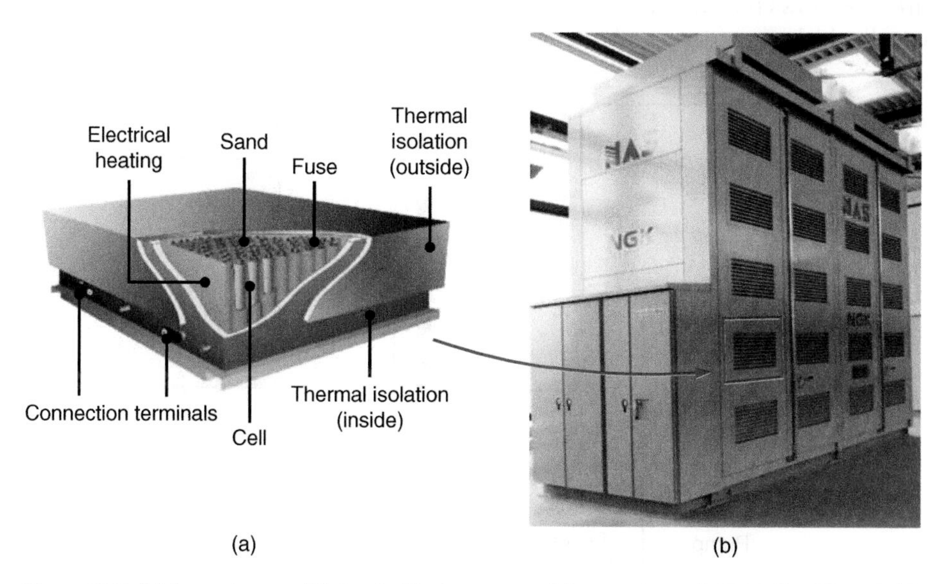

Figure 8.17 (a) Concrete use of NaS cells: The battery module contains each 320 cells with a total power of 50 kW. (b) 20 battery modules form the 1 MW storage block. (Photos – NGK Insulators, Younicos AG.)

In 2011, a fire outbreak happened during the installation of a 2 MW system of NGK. Out of a defective cell, liquid electrode material was leaking and caused a short circuit in a battery module. As no fuses were built-in, the module inflamed and finally, the whole storage block burned. After this incidence, the safety schemes were significantly enhanced: Fuses between the cells and fire protection plates between the cell blocks, nowadays, shall ensure that such an occurrence will not repeat.

8.2.4.4 Summary

Sodium sulfur batteries feature the use of low-cost materials, a high energy density of more than 200 Wh kg^{-1}, and a high number of cycles. The high operating temperatures and the losses connected with that recommend the use as large stationary energy storages.

8.2.5 Redox Flow Battery

The redox flow battery, also named liquid battery, is another exotic after the NaS battery. The peculiarity here is that two energy-storing electrolytes circulate each in separated loops. The connection is then achieved over the ion exchange in the membrane of an electrochemical cell (galvanic cell).

The fundamentals for redox flow cells were already laid in the 1950s through the University of Braunschweig. About 30 years later, concrete patent applications and realizations were done in Australia and the United States, based on the transition metal vanadium. Since in 2006, the patents expired, different research groups and companies worldwide work on the further development and commercialization of the technology.

8.2.5.1 Principle and Build-up

Figure 8.18 shows the build-up of the redox flow battery. In the center, we see the galvanic cell, which contains a membrane and two electrodes. The electrodes mostly consist

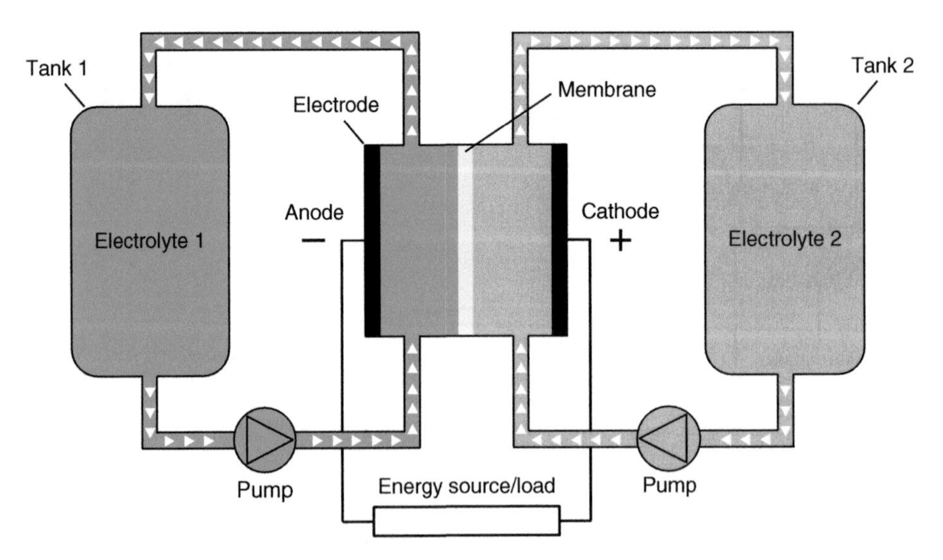

Figure 8.18 Build-up of the redox flow battery: The electrolytes circulate in two separated loops that are only connected through an ion-conducting membrane.

of graphite or graphite felt, respectively, to offer a largest possible surface. They do not take part at the electrochemical reaction but only serve as current conductors for the electrons. The actual reaction instead occurs at the membrane where the electrolytes form the reaction partners. In case of the redox flow battery, they are, therefore, quasi "active material." As electrolytes, typically organic or inorganic salts solved in solvents are used. With the help of the pumps, the electrolytes are transported from the tank to the actual cell.

The most commonly used cell variant is the vanadium redox flow battery (VRF battery). This is due to a peculiarity of the vanadium: In solution, it can exist in four different oxidation states. The ions can be charged two- to fivefold: V^{2+}, V^{3+}, V^{4+}, and V^{5+}. Thus, in contrast to the usual two electrochemical active materials, we only need one in this battery type.

First, in an empty battery system, diluted sulfuric acid (H_2SO_4) together with vanadium sulfate is filled. This results in an electrolyte, where V^{3+} and V^{4+} ions are dissolved to equal portions. During the electrical charging process, the penetration of the protons (H^+ ions) through the membrane ensures that the ions in the left cell half are reduced to V^{2+} and V^{3+}, while in the right part, they are oxidized to V^{4+} and V^{5+}. Starting from this charged state, we now want to have a closer look at the processes in the cell during discharging (Figure 8.19).

At the left side of the cell, the sulfuric acid dissolves in the aqueous electrolyte solution. This means the sulfuric acid molecule decomposes into its ions and releases protons, which migrate through the membrane on the cathode side. For compensation, at the anode side, the bivalent vanadium (V^{2+}) donates an electron to the electrode and

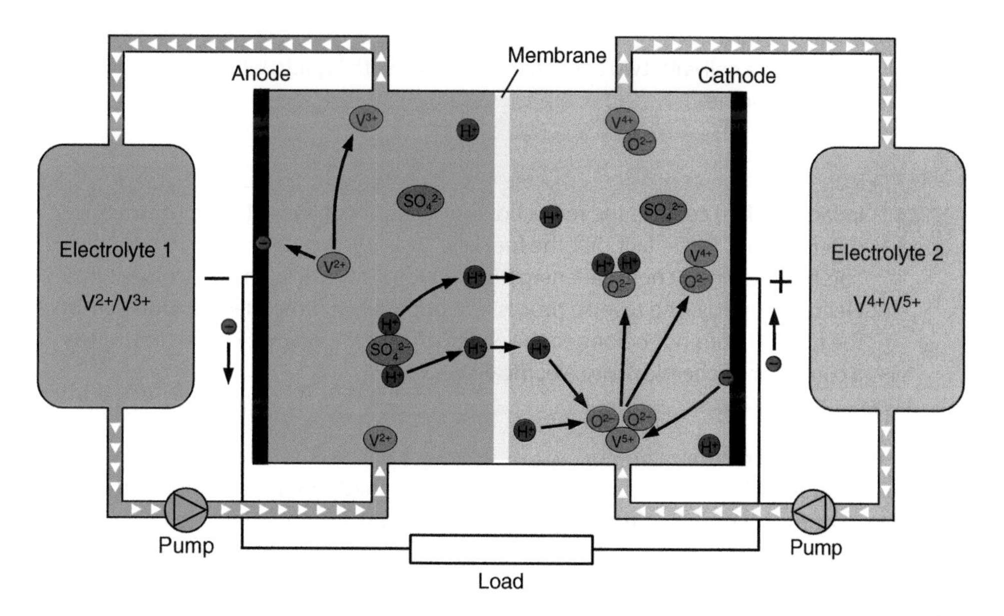

Figure 8.19 Discharging process in the vanadium redox flow battery: At the anode side, V^{2+} ions oxide to V^{3+} ions, at the cathode side, however, V^{5+} ions are reduced to V^{4+} ions (by different intermediate steps). (Source: After [99].)

becomes a V^{3+} ion:

$$V^{2+} \rightarrow V^{3+} + e^{-}. \tag{8.11}$$

At the cathode side, the pentavalent vanadium ions (V^{5+}) coalesce with the oxygen (VO_2^+). Subsequently, they take up an electron and then form with the protons delivered from the membrane water molecules (H_2O) and the oxygen compound VO^{2+}:

$$VO_2^+ + 2H^+ + e^- \rightarrow VO^{2+} + H_2O. \tag{8.12}$$

Thus, as desired, pentavalent vanadium ions (V^{5+}) were reduced to tetravalent ions (V^{4+}):

$$V^{5+} + e^- \rightarrow V^{4+}. \tag{8.13}$$

In case of charging, the reactions accordingly run the other way round. As total reaction, we get:

$$V^{2+} + VO_2^+ + 2H^+ \underset{\text{Discharging}}{\overset{\text{Charging}}{\rightleftharpoons}} V^{3+} + VO^{2+} + H_2O. \tag{8.14}$$

If only the vanadium ions are considered, the total reaction can also be expressed in the following way:

$$V^{5+} + V^{2+} \underset{\text{Discharging}}{\overset{\text{Charging}}{\rightleftharpoons}} V^{4+} + V^{3+}. \tag{8.15}$$

Thus, the discharging continues until all V^{5+} ions were reduced to V^{4+} ions, and all V^{2+} ions were oxidized to V^{3+} ions. With that, the initial state is restored.

Is a redox flow battery actually the same as a fuel cell? Also, that one, after all, consists of a cell with two compartments filled with liquids, which are separated by a membrane.

In fact, the fuel cell and the redox flow cell are built up very similar. The decisive difference lies in the fact that the fuel cell only knows one conversion direction: The conversion of chemical energy (e.g. in form of hydrogen and oxygen) into electrical energy. The reverse process is not possible in the fuel cell. Due to this, the fuel cell does not belong to the electrochemical storages; instead, it is solely a converter of chemical into electrical energy.

8.2.5.2 Behavior in Practice

Figure 8.20 shows the open-circuit voltage of the vanadium redox flow battery in dependence of the charging state. It continually rises and can therefore (in connection with measuring the temperature) be used for a simple determination of the charging state.

The electrical efficiency of the cells lies in the order of 90%. However, if we include also the losses of necessary periphery devices (especially the pumps), the efficiency is

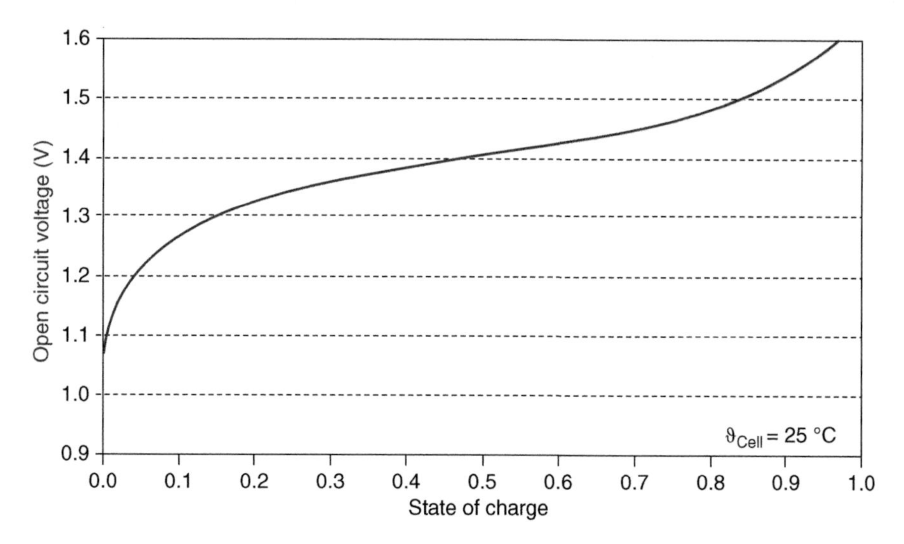

Figure 8.20 Open-circuit voltage of the vanadium redox flow battery at different charging states: It changes relatively strongly and can therefore be used for a simple monitoring of the charging state. (Source: After [102].)

reduced to about 70–80% [99]. The operating temperature should be possibly between 10 and 35 °C to attain a high lifespan.

Regarding the charging behavior, the cell offers pleasant properties. It is capable of deep-discharge, as there exists neither a dendrite growth nor a reduction of the active material. Therefore, cycle numbers in the order of 10 000 can be assumed. It shows only little self-discharging and is therefore well suited for high operating times.

A peculiarity of the redox flow cell in contrast to all other battery types lies in the fact that the storage capacity is not dependent on the cell size but only on the size of the tanks. Thus, storage capacity and battery power can be dimensioned separately. At the same time, the energy density of the electrolyte of 25 Wh kg^{-1} is relatively small. This is caused by the limited solubility of the V^{5+} ions in sulfuric acid. Meanwhile, research is going on with further cell variants that promise higher energy densities: Vanadium/bromine cell and zinc/bromine cell with doubled and threefold energy density with respect to the pure vanadium cell.

8.2.5.3 Concrete Applications

Resulting out of the described properties of the redox flow battery result different obvious fields of application. Thus, they are used, e.g. as large stationary storages, for load peak balancing in energy power supply grids or for buffering the electricity of wind parks and large PV plants.

The example of a commercial product is shown in Figure 8.21 with the EverFlow Compact Storage CS 5/15. It represents a storage based on the vanadium redox flow battery technology. It comprises a built-in inverter and delivers a maximum electrical power of 5 kW. The capacity is 15–45 kWh (depending on model). The company even offers a large container model with a power of 15–60 kW and a capacity of up to 200 kWh. The further technical data can be found in Figure 8.21.

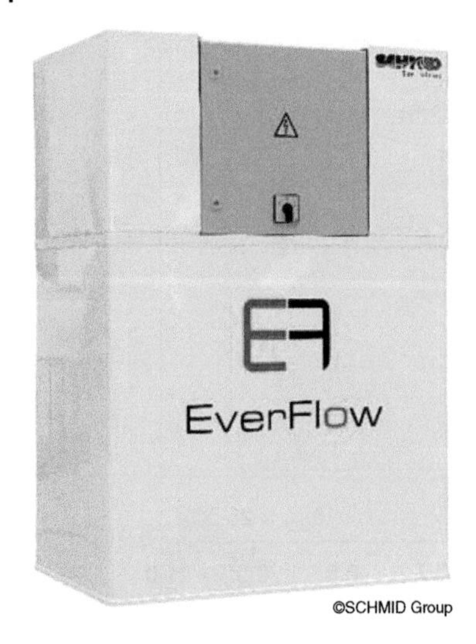

Technical data:

Name:	EverFlow Compact Storage
Type:	CS5/15
Power:	5 kW
Capacity:	15 kWh
Output voltage:	230/110 V_{AC}
Efficiency:	DC > 80%
Depth of discharge:	100%
Calendric lifespan:	up to 20 a
Cycle lifespan:	≥ 10 000 cycles
Self-discharge:	< 1% per year
Ambient conditions:	5 – 30°C
Dimensions:	1.25 m × 0.8 m × 1.74 m
Weight (full):	1500 kg

Figure 8.21 View and technical data of the EverFlow Compact Storage: The storage offers a power of 5 kW and a capacity of 15 kWh. (Photo – SCHMID Group.)

Occasionally, the deployment of redox flow batteries in electric vehicles is discussed. Here, the redox flow technology gains from the clear advantage that the charging can be done quite easy (and quick!) by refueling each of the two charged electrolyte liquids. However, it has to be kept in mind that the current density of the vanadium redox flow battery is only 1/350th of the energy density of petrol (compare Table 1.2). Thereby, an extremely large volume and large load would be required only for the storage.

8.2.5.4 Summary

Redox flow batteries have the decisive advantage that power and energy content are scalable independently of each other. Moreover, they have the largest lifespan of all considered battery types. The energy density of 15 Wh kg^{-1} is very small and leads to the fact that their deployment area will be limited to stationary storages.

8.2.6 Comparison of the Different Battery Types

After we have dealt intensively with the most important battery types, Table 8.2 presents again their respective pros and cons and lists the main fields of deployment. The large variation of the specified energy densities stem from the fact that some literature sources only consider the energy density of the active materials, while others also take into account further parts of the battery (electrodes, housing, etc.)

8.3 Storage Use for Increase of Self-consumption

Meanwhile, the feed-in tariffs for solar electric energy lie below the power consumption price of normal domestic customers (so-called Grid Parity, see Section 11.3). Therefore, it is nearby to improve the return of a PV plant by self-consuming a possibly large portion of the produced electricity. In the following, we will first look at the possibilities

Table 8.2 Comparison of the different battery technologies [99, 102, 103].

Technology	Pros	Cons	Deployment area
Lead acid/lead gel	• Cost-effective • Proven technology • Safe	• Small energy density (30–50 Wh kg^{-1}) • Small lifespan • No quick charging	• Starter batteries (vehicles) • Emergency power systems • Uninterruptible power supplies • Domestic solar storages
Lithium ion	• High energy density (110–250 Wh kg^{-1}) • High cycle lifespan • High energy efficiency (>90%) • Small self-discharging • Further development potential	• Relatively expensive • Safety problems • Charge surveillance necessary	• Electronic devices • Electric vehicles • Domestic solar storages • Buffer storages for wind and solar parks • Load peak balancing in electrical grids • Provision of control and balancing energy in electrical grids
Sodium sulfur	• High energy density (100–200 Wh kg^{-1}) • Very high cycle lifespan • No self-discharging • Relatively safe	• Danger potential in case of cell breakage • Complex thermal management • Continuous energy expenditure for temperature stabilization	• Buffer storages for wind and solar parks • Load peak balancing in electrical grids • Provision of control and balancing energy in electrical grids
Redox flow	• Suitable for large storage capacities • Low maintenance • Simple cell construction • Deep-discharging safe • Safe	• Small energy densities (10–50 Wh kg^{-1}) • Auxiliary units (pumps) necessary • Restricted temperature range	• Buffer storages for wind and solar parks • Load peak balancing in electrical grids • Provision of control and balancing energy in electrical grids

of a domestic household to enhance the self-consumption of the solar energy. As the frame conditions for commercial enterprises partly differ, we will consider them afterward separately.

8.3.1 Self-consumption in Domestic Households

Let us imagine the example of a four-person family Summer, which has installed a 5 kWp PV plant on their roof. This plant yearly yields just 4500 kWh and, therefore, in terms of figures could cover the total electricity demand of the family. However, the solar electricity offer is very unequally distributed over the year and, also, over the respective day. Therefore, always only a part of it can be used directly. For a sunny day, Figure 8.22 shows this in the upper diagram. Obviously, more energy is produced than family Summer can use. Simultaneously, there are times in the morning and in the evening, where the

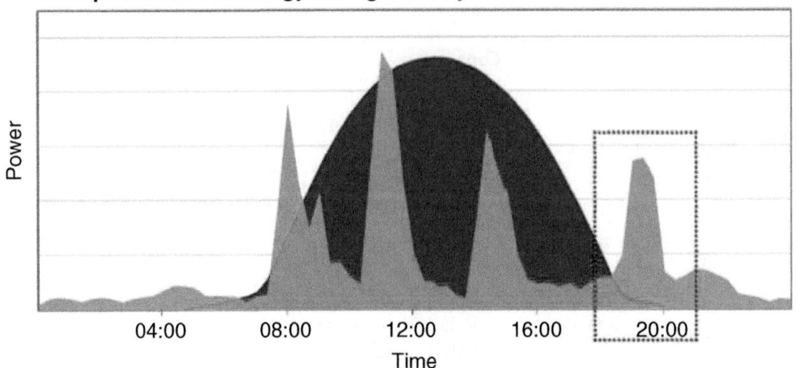

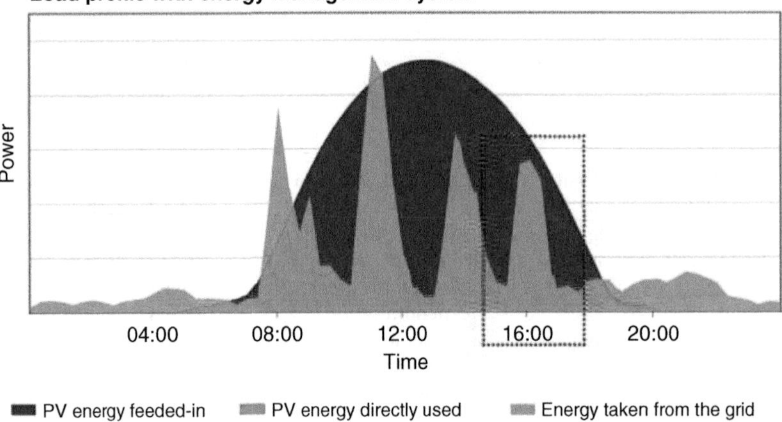

Figure 8.22 Self-consumption enhancement through load shift with an intelligent load management system: By starting the washing machine earlier, more solar energy can be used directly. (Source: SMA Solar Technology AG.)

demand is higher than the produced solar energy. Viewed over the entire year, for such a dimensioning, a solar energy self-consumption rate (or "degree of self-consumption," see Section 7.2.1) of about 30% can be expected [104].

8.3.1.1 Solution Without Storage

How this self-consumption rate can be enhanced? In the most simple cases, devices, such as a washing machine or a dishwasher, are only started during the day. Meanwhile, there are several suppliers of energy management systems, which optimize the self-consumption by automatically switching on and off (e.g. with remote-controlled sockets) domestic appliances. Then also devices, which do not continuously need electric energy (e.g. freezer), can be switched on in dependence of the solar electricity offer.

Figure 8.22 shows an example for this in the lower diagram. Here, the washing machine was started at 4 p.m. instead of at 7 p.m. and could be driven totally with solar energy. Simulations and first experiences with test households show that the self-consumption rate of systems with EMS can be increased up to about 45% [104].

8.3.1.2 Solution with Storage

If one will come to higher self-consumption rates, the use of battery storages is the right choice. For the dimensioning of the storage, first the question has to be cleared if a high degree of self-sufficiency (see Section 7.2.1) has to be achieved or if the aim is simply a most profitable solution. Basically, with the use of a large storage, it is possible to increase the degree of self-sufficiency up to 100%. This, however, stands in no relation to the costs of the necessary "several week storage." More profitable is the deployment of a small storage together with an intelligent load management.

One possibility for this offers, e.g. the Sunny Boy 5000 Smart Energy, a 5 kW inverter with integrated lithium ion battery, which can store about 2 Kilowatt-hours. This is combined with an energy management of the domestic consumers ("Sunny Home Manager"). The energy management system knows the typical daily load course of the household as well as (over the internet) the weather forecast for the following day. Based on this, it steers a part of the loads and the charging and discharging of the storage.

The exemplary result is shown in Figure 8.23. Again, the load course of family Summer is assumed. During the day, the surplus solar electricity is charged into the storage (depicted in yellow). This stored solar energy is then at disposal in the evening and morning for self-consumption (orange). With a smart combination of energy management system and storage, thus the self-consumption rate can be increased to about 55–60%.

8.3.1.3 Examples of Storage Systems

Table 8.3 lists concrete examples of domestic storage systems offered at the market. The systems 1–4 contain an inverter, which independently of the actual solar inverter feeds into the AC side (see Figure 8.1b). System 5 is the Sunny Boy 5000 SE already mentioned above. It comprises of a normal 5 kW solar inverter, which is coupled to the storage on the DC side (Figure 8.1a).

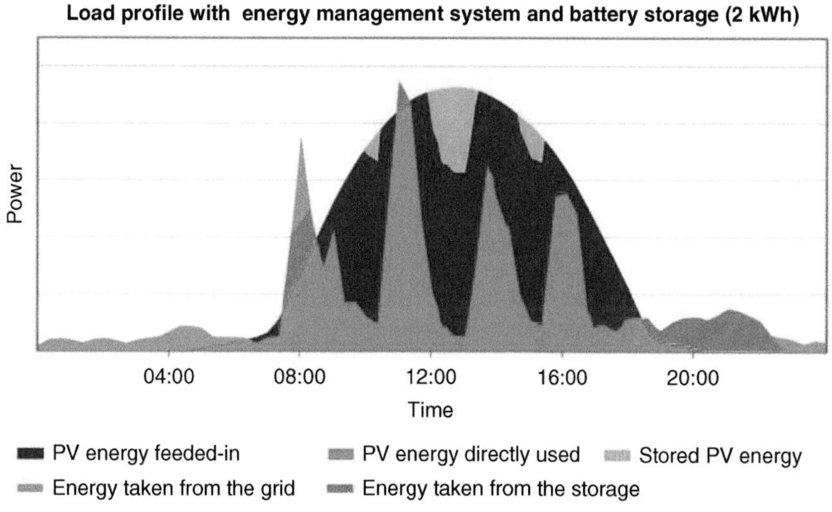

Figure 8.23 Further self-consumption increase by the use of a 2 kWh storage: The surplus solar energy is stored in the battery during the day and can be used in the evening and morning for self-demand. (Source: SMA Solar Technology AG.)

Table 8.3 Technical data of some current solar storage systems for private households.

Consecutive number	No. 1	No. 2	No. 3	No. 4	No. 5
Provider	Senec.ies	sonnen GmbH	IBC Solar	IBC Solar	SMA Solar Technology AG
Product name	SENEC.Home G2 plus	Sonnenbatterie eco 8	IBC SolStore 8.0 Pb	IBC SolStore 5.0 Li	Sunny Boy 5000 Smart Energy
Technology	Lead oxide	Lithium iron phosphate	Lead gel	Lithium ion polymer	Lithium ion
Type of coupling	AC	AC	AC	AC	DC
Gross capacity C_{Gross} (kWh)	16	8	8.0	5.0	2
Net capacity C_{Net} (kWh)	8.0	8	4.0	4.5	2
Maximum discharge power (kW)	2.5	3.3	4.6	4.6	2
Depth of discharge (DoD) (%)	50	100	50	90	100
Cycle lifespan at DoD	3200	10 000	2700	5000	>4000
Self-discharge rate	2%/month	n.s.	2%/month	1%/month	1%/month
Calendar lifespan	10–13 a	20 a	10 a	15 a	>10 a
Maximum total efficiency of the storage system (%)	86	94	85	95	96
Power of solar inverter (kW)	–	–	–	–	5
Battery voltage (V)	48	n.s.	48	48	150
Emergency power function	Yes	Yes	Yes	Yes	No
Price (net) *Cost* (€)	6.900	11.300	6.300	8.500	4.600
Specific price (net) *cost* (€ kWh^{-1})	841	1.413	1.575	1.889	2.300

Roughly speaking one has the choice between products with lead batteries and those with lithium ion batteries. The latter are typically more expensive; however, they show a higher lifespan. In particular, attention has to be paid to the specified discharging depth; it reaches from 50% with lead batteries up to 100% with lithium ion batteries. Thus, e.g. the lead battery of IBC Solar has a clearly higher gross capacity as the lithium counterpart of the same company; at the same time, the net capacity is smaller than that of the lithium battery.

8.3.1.4 How Much Cost a Kilowatt-Hour?

If a storage system is profitable for the family Summer, then this strongly depends on the acquisition costs. Further influencing factors are the lifespan and the capacity of the storage. With a rough calculation, we can calculate the costs per stored kilowatt-hour out of these parameters.

As described, the actual net capacity (also named usable capacity) depends on the allowed depth of discharge (DoD). Moreover, the storage efficiency, η_{Storage}, leads to a further reduction of the usable energy. Finally, the maximum number of cycles, N_{Cycles}, describes how often the battery can be charged and discharged. The total energy, W_{Total}, which can be stored in and got out of the storage over the whole lifespan, can be calculated with the following formula:

$$W_{\text{Total}} = C_{\text{Gross}} \cdot DoD \cdot \eta_{\text{Storage}} \cdot N_{\text{Cycles}}. \tag{8.16}$$

As an example, let us consider System 4 from Table 8.3, the lithium ion storage IBC SolStore 5.0 Li:

Example 8.1 ***Costs of the storage System 4.***

Costs	$\text{Cost}_{\text{Storage}}$	8500€
Gross capacity	C_{Gross}	5.0 kWh
Depth of discharge	DoD	90%
Storage efficiency	η_{Storage}	95%
Maximum number of cycles	N_{Cycles}	5000
Calendric lifespan	T_{Lifespan}	15 a

■

Over the lifespan, the system can store and take out again the following amount of energy:

$$W_{\text{Usable}} = 5.0\ \text{kWh} \cdot 0.9 \cdot 0.95 \cdot 5000 = 21375\ \text{kWh}$$

With the investment cost given above, the specific costs for a kWh result to:

$$\text{cost}_{\text{Storage}} = \frac{\text{Cost}_{\text{Storage}}}{W_{\text{Total}}} = \frac{8500€}{21375\ \text{kWh}} = 39.8\ \text{cent}\ \text{kWh}^{-1}.$$

This means that each kWh, which has been stored during the lifespan of the storage, costs about 40 cent for the owner! This is clearly more than the difference between the current feed-in tariff and the domestic electricity price. Therefore, this system is a no good bargain for the owner of the system.

The shown cost formula in reality is a really rough calculation. Strictly speaking, the annual interest ought to be taken into account (see Section 10.2). Additionally, the calendric lifespan for the lithium ion battery is stated 15 years from the manufacturer. If we assume for a typical solar plant with storage a full cycle number of 250 per year, then in 15 years, only 3750 cycles are used. The maximum number of cycles of 5000, therefore, can possibly not be reached at all, because already the calendric lifespan has expired earlier. In addition, it should be kept in mind that the data about number of cycles, lifespan, etc., were provided by the manufacturer. Unfortunately, there are only few independently determined data yet.

If we apply the calculation of Example 8.1 on System 1, we already get clearly better numbers of about 30 cent kWh^{-1}. This amount can further be reduced to, e.g. 20 cent kWh^{-1}, if the government support for solar storages is used (see Section 8.4.2). System 2 even ends up below 20 cent kWh^{-1} without the government support (see Exercise 8.5 at the end of this book). However, one has to trust in the specifications of the manufacturer about a cycle lifespan of 10 000.

8.3.1.5 The Smart Home

If the self-consumption rate shall be further enhanced, it is a good idea to integrate more consumers into the energy management system. Figure 8.24 shows the example of such a "smart home." Here, additionally, a domestic hot water heat pump is supplied by the electricity from the solar plant. In winter, a mini combined heat and power unit (CHP) produces the heat for domestic heating and at the same time charges the battery storage.

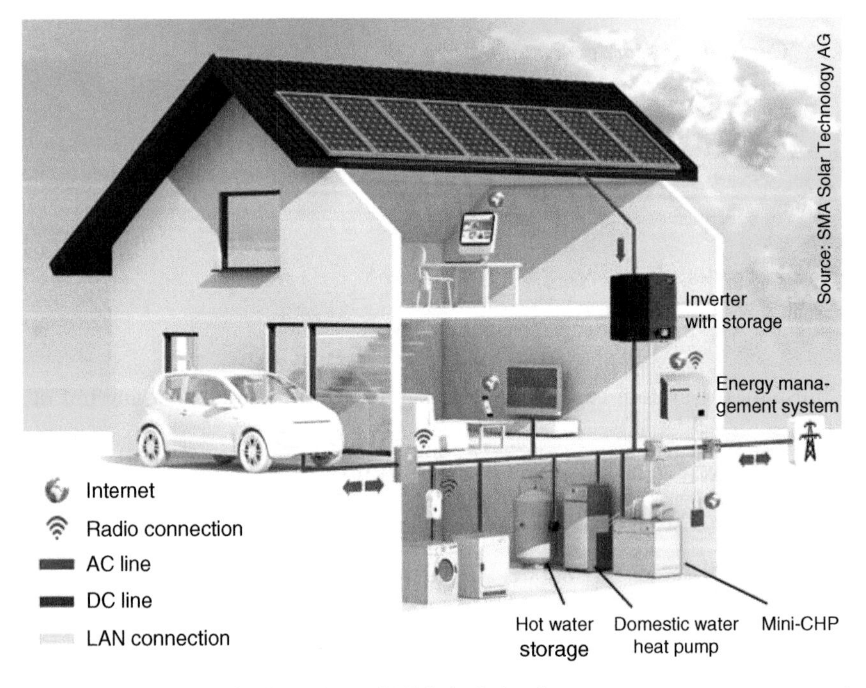

Figure 8.24 Example of a "Smart Home": With the help of an energy management system, including storage, the different loads are automatically switched on and off to enhance the self-consumption rate.

If an electric car is purchased, its battery can be used as an additional external storage. This may increase the self-consumption rate even up to 100%.

The profitability of such a solution will improve with further falling storage costs. At the same time, more and more people are willing to spend extra money for a far-reaching self-provision with eco-friendly energy. Additionally, storage systems often offer an emergency function (buffer function) that can supply the house in case of a power blackout.

8.3.2 Self-consumption in Commercial Enterprises

8.3.2.1 Example Production Factory

Commercial enterprises are especially suited for solar energy use. Often, the buildings provide large roof areas for the solar plant with at the same time a high electricity demand of the company. Furthermore, the demand typically arises at daytime. As an example, Figure 8.25 shows the load course of a production factory. Additionally, the feed-in power of a 200 kWp solar plant for a sunny (yellow) and a covered day (light blue) is depicted. On weekdays, even on a sunny day, the whole produced energy is self-consumed. Only at the weekend, the demand is clearly smaller than the solar offer due to the halted production. Considered over the full year, a self-consumption rate of about 80% can be assumed.

8.3.2.2 Example Hospital

A second example is shown in Figure 8.26. Here, the load curves of a hospital are depicted. Also in this case, the highest energy demand occurs during daytime, now however on all days of the week. Due to the very good temporal correlation of solar

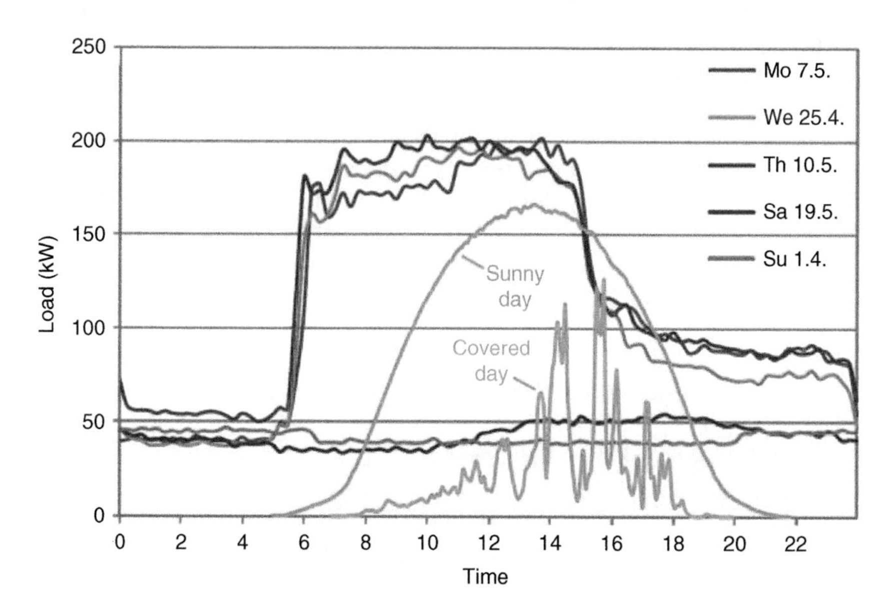

Figure 8.25 Solar energy use in a production factory: Due to the good temporal match between solar offer and demand, high self-consumption rates follow. Only at the weekend outweighs the solar energy. (Source: Load profile: Solarpraxis AG.)

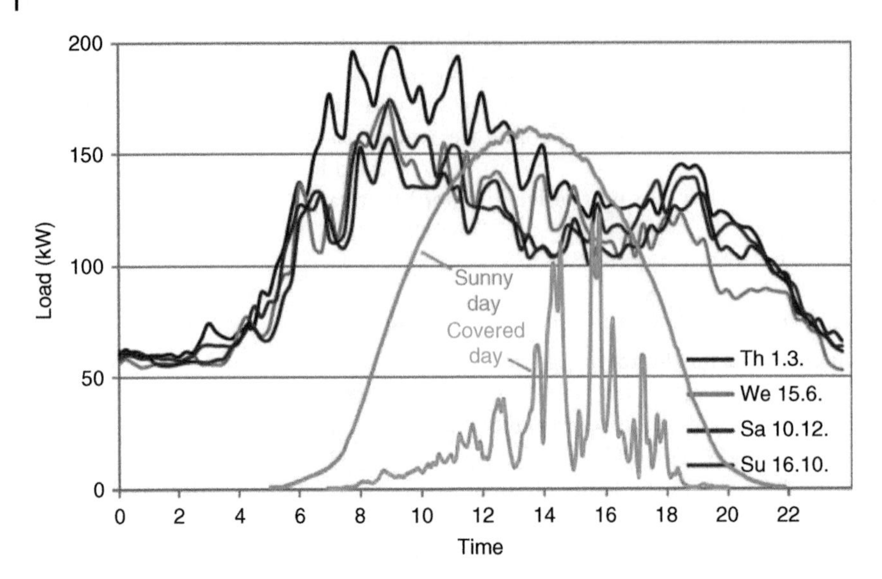

Figure 8.26 Solar energy use in a hospital: As there is a high power demand every day of the week, practically the whole produced solar energy can be self-consumed. (Source: Load profile: Solarpraxis AG.)

offer and demand here, practically the whole produced solar energy of the 200 kWp PV plant can be self-consumed.

In both examples, we can assume that the investment in a PV plant will amortize in a few years. In both cases, a storage is not necessary.

In connection with commercial enterprises, there is currently mainly only one application, where a storage can be profitable: to limit the peak power taken out of the grid. Many companies have electricity tariffs, where besides the energy price (price per kWh) a power price has to be paid. This price depends on the taken peak power during a month (or a year). Here, an energy management system can be employed, which limits the power demand to a defined value. In case of such a limitation, the consumers are provided by the battery storage. However, in a first step should be always cleared if the temporal switching-off of some loads is possible. Only if this potential is exhausted, the (relatively expensive) solution with a storage should be taken into consideration.

8.4 Storage Deployment from the Point of View of the Grid

The solutions described above all have the aim to enhance the self-consumption of the operator of the plant. The deployment of a storage, however, does not automatically lead to a grid relieve. For instance, it is possible that on the morning of a sunny day, all plants start to charge their storage with the solar energy. Afterward, it could happen that, e.g. at noon, all storages are fully loaded and suddenly, the total solar power would be fed into the grid. This would not be the favorable scenario of the grid operator.

Of course, it can be countered that this case is somewhat theoretical, as all plants have different sites, orientations, and storage capacities. Therefore, the described effects possibly cancel each other. Nevertheless, from the grid point of view, it would be sensible, that storages also will take over grid services in the future.

8.4.1 Peak-shaving with Storages

As an example, Figure 8.27a shows the case of a 5.6 kWp plant without storage on a sunny day. The red curve depicts the grid exchange power (positive: grid feed-in, negative: grid supply). Clearly visible are the large feed-in and demand peaks of almost 5 kW. This is primarily due to switching on and off of heavy users (e.g. a stove) in the household.

A better solution with respect to the grid requirements is shown in Figure 8.27b. Here the plant was complemented by a 5.5 kW storage. Besides an increase in the self-consumption rate, it shall also achieve a limitation of the maximum grid exchange power. For this, the storage is only charged in the case of exceeding the feed-in power of 1.9 kW. Thus, it can store energy production peaks over the whole day.

In the opposite case, the "intelligent storage" limits the power drawn from the grid in times of consumption peaks by providing additional power from the battery. As a result, a more even course of the grid exchange power can be seen clearly. Now the maximum demand peak only lies at 2.4 kW, which is solely half of the case without storage. This "peak-shaving" therefore leads to an equalization of the grid exchange power and thus to a grid relieve.

8.4.2 Governmental Funding Program for Solar Storages

With the aim of an accelerated deployment of solar storages, the German government has created a funding program. The funding consists of low-interest loans and repayment grants. The declared intent of the program is a support of only those systems that in the long run will facilitate the relieving of the grid. This is mainly achieved by the following requirement:

> The maximum admissible power output of a PV plant at the grid-connection point is limited to 50% of the plant's nominal power.

How will this requirement affect the behavior of a plant in the direction of the grid? To make this clear, Figure 8.28 shows, in the upper diagram, the example of a "conventional storage" on a sunny day. In the morning, first, mainly the battery is charged. Around 11 a.m., the storage is fully loaded, and the PV power is suddenly fed into the grid. As already described, this leads to stronger power variations as it would be in the case of a PV plant without storage. Therefore, this build-up cannot be a model for the future.

Entirely different is the case for a "grid-optimized storage" (lower diagram in Figure 8.28). Here, the maximum feed-in power is limited to 50% of the plant's nominal power. The result (with a suitable charging strategy) is a slowly increasing feed-in power in the morning. As soon as 50% of the nominal power is attained, this feed-in power is held constant. The surplus energy is loaded into the storage, or it is consumed by switchable loads. In contrast to the peak-shaving solution of the preceding section, here though the production and load peaks are not totally suppressed. Nevertheless, the very simple "50% method" ensures a very significant stabilization of the feed-in with a positive effect on the grid loading and the grid stability.

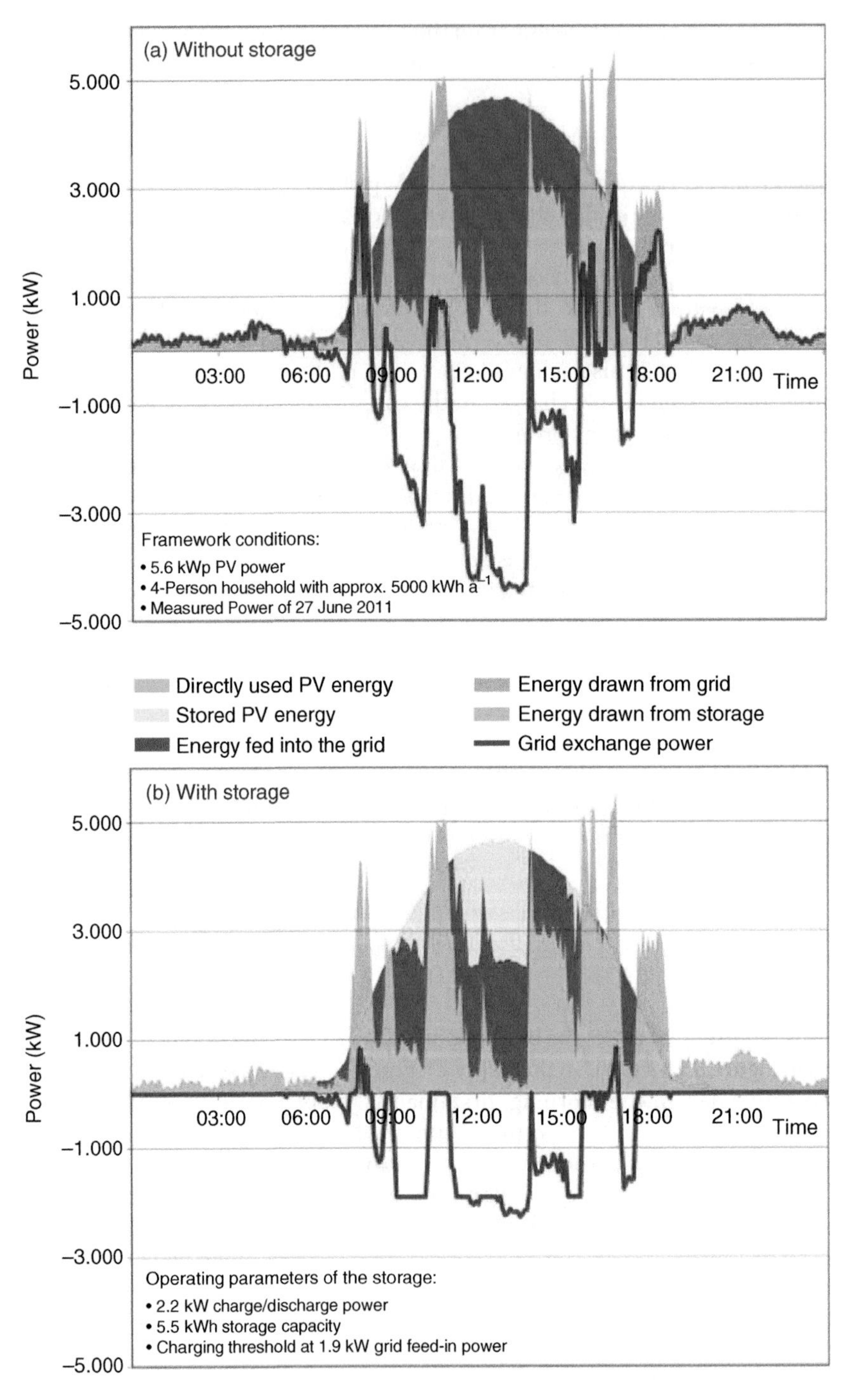

Figure 8.27 Peak-shaving using an intelligent storage: The high grid exchange power in the upper diagram is strongly reduced by the energy management system including battery as shown in the lower diagram. (Source: Solar Technology AG.)

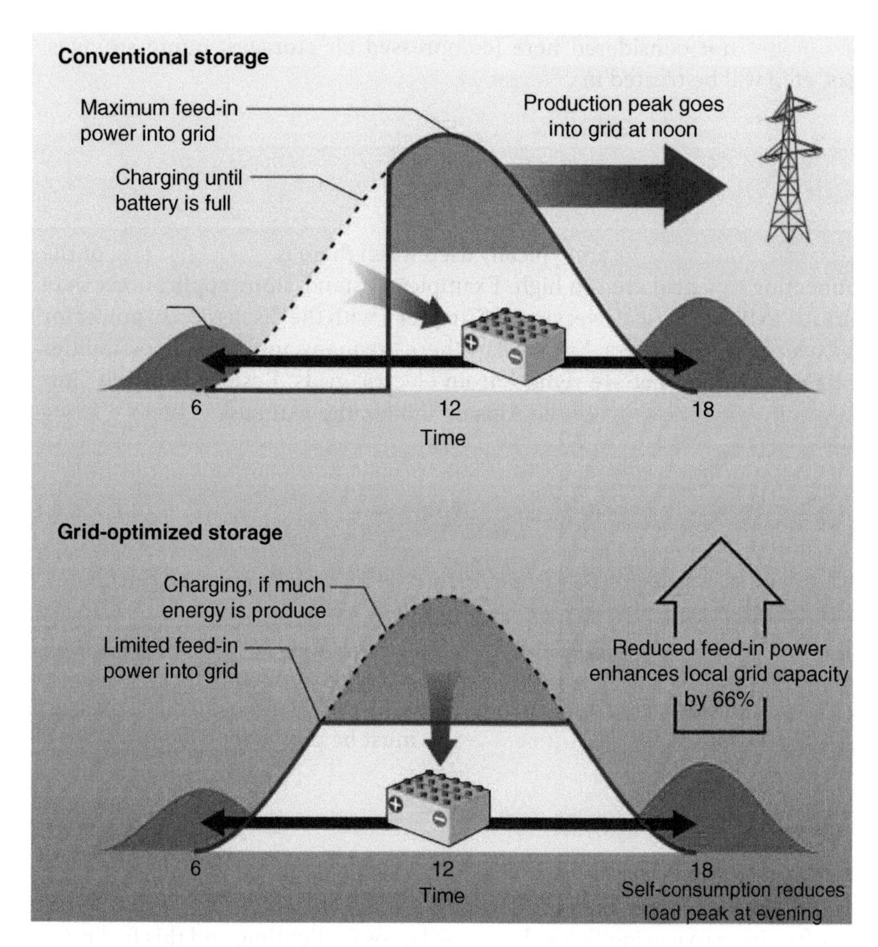

Figure 8.28 A comparison of different storage operating strategies: While the conventional method leads to strongly varying feed-in powers, this can be moderated with the grid-optimized strategy. (Source: German Solar Association.)

An additional requirement of the funding program is the installation of an interface at the inverter. This shall facilitate the remote control of the PV plant along with the storage by the grid operator. Up to now, this remote control is hardly used; however, it is already a preparation on the grid of the future (smart grid). Despite the strongly fluctuating feed-in power of renewable energies, this grid can show a high stability due to intelligent decentralized battery storages.

Summarizing, we can state that battery storages are an interesting add-on to PV plants. Due to the current prices, up to now they are profitable only in exceptional cases. To enhance the degree of self-consumption, therefore, first an energy management system should be implemented. The next step from the economic point of view is the use of the solar electricity to operate a heat pump. The produced warm water can be directly used for domestic hot water and domestic heating or can be stored in a hot water storage.

However, with the falling prices, battery storages are getting more and more attractive. Moreover, they can offer additional benefits like emergency function or grid relieve.

The large storages not considered here (compressed air storages, pump storages, power-to-gas, etc.) will be treated in Chapter 11.

8.5 Stand-alone Systems

Photovoltaic stand-alone systems are typically used when there is no electric grid or the costs for connecting to a grid are too high. Examples of stand-alone applications were discussed already in the historical overview in Chapter 1 with the discussion of power for satellites and telephone amplifiers. Even today, there are many application possibilities for photovoltaics in places that are remote from electric grids. Besides relatively limited applications in mountain huts in the Alps or similar, the main use is in developing countries.

8.5.1 Principal Structure

Figure 8.29 shows the block diagram of a simple stand-alone system using photovoltaics. An important element is the storage, which typically is a lead or lithium ion battery. This is protected from overloading by a charge controller, which contains an in-built deep-discharge-proptection. DC loads, for instance, are energy-saving lamps, radios, or also water pumps. If besides the direct current loads also alternating current loads are used, then a special additional stand-alone inverter must be provided.

8.5.2 Examples of Stand-alone Systems

8.5.2.1 Solar Home Systems

There are billions of people in the world without access to an electric supply grid. Instead, they often have to use inefficient kerosene lamps for lighting, and this fuel must be transported over wide distances. Batteries are often the only source of energy for radios, and so on. Here photovoltaics is an ideal alternative for making a considerable contribution for improvement in quality of life and for environmental protection.

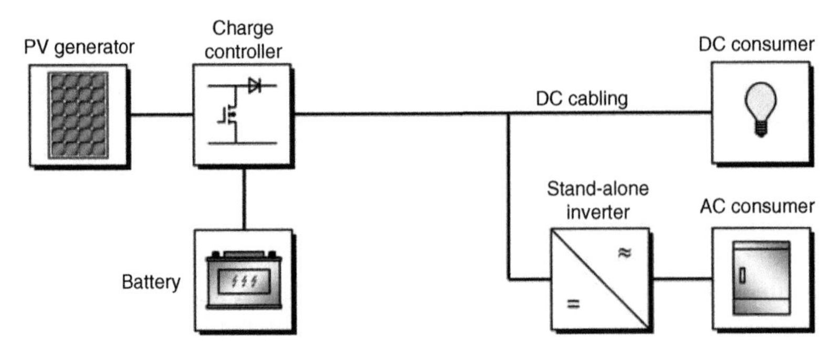

Figure 8.29 Block diagram of a photovoltaic stand-alone system: The battery fed by the module over a charge controller makes the power available for the DC consumer. An additional stand-alone inverter is provided in the case of AC consumers (e.g. refrigerator).

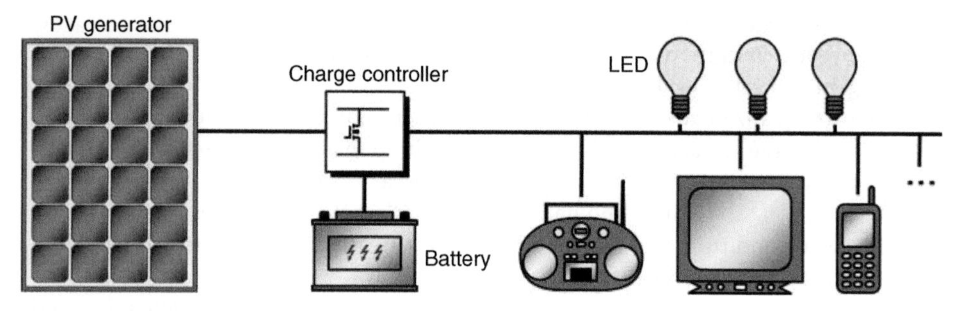

Figure 8.30 Typical arrangement of a solar home system: The loads are typically DC-operated lamps, radios, mobiles, and TV sets.

The expression solar home systems (SHS) denominates mini solar systems of one or two modules that provide the required electric energy *for* a house in developing countries. Typical loads are DC-operated energy-saving lamps, radios, and TV sets (Figure 8.30).

The main problems in the spread of solar home systems are the high initial costs of the plants. Here micro-credit models, for instance, are a help with which customers can make a down-payment and then monthly payments for their SHS. A pioneering role in this field is the *Grameen Shakti* (translation: "village energy") undertaking, a subsidiary of the Grameen Bank famous for its micro-credits. This undertaking is responsible for financing more than 500 000 solar home systems in Bangladesh (Figure 8.31). As the installations are erected by specially trained technicians and servicing of customers is organized through service offices, the result is high quality and a long life. If a system should fail, then it is frequently due to the bridging of the charge controller by the owner of the installation [105, www.gshakti.org].

The concept pioneered by Grameen Shakti now has many imitators. For instance, there is a similar project in Ecuador that is sponsored by the government via a fund for rural electrification (FERUM – Rural and Marginal Urban Electrification Fund). The PV system consists of a 100 W module, a charge controller, and a 105 Ah battery with three DC LED lamps. In addition, systems with inverters and PV power of up to 800 Wp are also being installed [106].

Figure 8.31 Photos of solar home systems in Bangladesh: (a) Arrival of the components by river and (b) installation of the plant on a hut. (Photos: Grameen Shakti; Microenergy International.)

8.5.2.2 Hybrid Systems

If a whole year's continuous supply of electric energy is to be ensured, then the solar systems quickly reach their limits.

In order to equalize bad weather periods and even more seasonal fluctuations, high PV power and large storage capacities must be built up that lead to unreasonable costs. In this case, the solution is hybrid systems, which are a combination of various methods of generating electricity. The combination with other renewable sources of energy such as waterpower, wind turbines, and biomass is obvious. In addition, the use of a diesel generator makes sense in order to bridge gaps in supply.

From an economic point of view, hybrid systems with a PV, a diesel generator, and a large battery storage are often cheaper than plants in which only diesel generators are used. Reasons for this are the high servicing effort, short life span, and very poor part-load efficiency of diesel generators [106].

Hybrid systems can be organized as pure DC systems, mixed DC/AC systems, or also as pure AC systems. For larger plants, the trend is definitely pure AC systems as they are very flexible and can be easily extended. Figure 8.32 shows an example of such a system that is equipped with inverters by SMA.

The heart is the Sunny Island battery inverter that generates a stable alternating current grid. The batteries serve as the basis for the energy, and these again are charged by a PV generator. Other plants feed into the AC grid, for example, a further PV generator via a normal grid inverter (Sunny Boy) or a wind turbine via a special wind energy inverter (Windy Boy).

If less power is used in the AC grid than is generated, then the Sunny Island feeds the excess energy into the batteries. If these are already full, then other generators must be turned off. For this purpose, SMA uses the SelfSync® process, which works in a similar manner to the power controls of large power stations in the energy supply grids. If the Sunny Island reduces the frequency of the AC grid, then this is a command for the other inverters to reduce the fed-in effective power. At the same time, the amount of voltage of the AC grid determines the quantity of the fed-in reactive power.

If, on the other hand, more power is consumed than can actually be generated, then the Sunny Island accesses the batteries. If this is still insufficient, then the device automatically starts the diesel generator or will switch off certain loads.

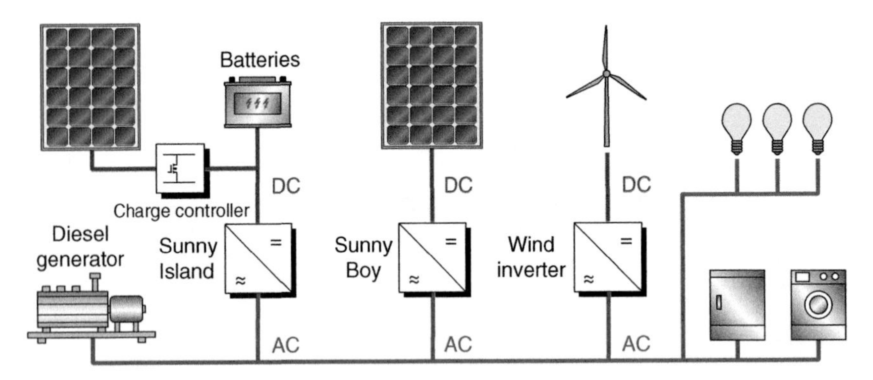

Figure 8.32 Arrangement of a hybrid system with pure AC coupling of generators and consumers: The Sunny Island generates a stable AC grid and can start the diesel generator or switch off loads in case of demand bottlenecks [107].

It is clear that this system is very flexible as further generators can simply be connected to the alternating current grid. Besides SMA, other producers also offer similar systems (e.g. KACO new energy).

If a whole village is connected to the hybrid system, this is also called a micro-grid. If, again, several micro-grids are interconnected, this is known as a mini-grid. Thus, it is clear that the technologies for developing countries offer a great opportunity to decentrally electrify rural areas in an evolutionary manner and then to interconnect to ever larger grid units.

8.5.3 Dimensioning Stand-alone Plants

Dimensioning is relatively simple in the case of plants connected to the grid as one assumes that the public grid will be able to absorb the power generated at all times. This is much different for stand-alone systems: Here, the consumption must be determined as accurately as possible and then the PV generator and storage must be dimensioned in accordance with the radiation conditions at the plant location.

In the following, we will become acquainted with a simple scheme for planning a stand-alone system that works with tables. The scheme was mostly taken over from the guidelines for photovoltaic plants of the Deutsche Gesellschaft für Sonnenergie e.V. (German Association for Solar Energy) [22]. This method of calculation is naturally not nearly as accurate as an analysis with a stand-alone plant design program, but it can provide a good first estimate.

8.5.3.1 Acquiring the Energy Consumption

The best way of determining energy consumption is by means of a table in which the nominal consumption of the individual consumers and the daily operating time is listed. As an example, we will assume a small holiday cottage near Munich that is to be used throughout the year (see Table 8.4). It is advisable to list the summer season (May–September) and the winter season (October–April) separately.

A special case is the refrigerator: In order to save energy especially in winter, it is assumed that the refrigerator is switched off in winter. Also one usually does not know

Table 8.4 Consumption balance of a small holiday cottage.

Consumer	Nom. power P_N (W)	Daily operating time t (h)		Daily consumption W (Wh)	
		Summer	Winter	Summer	Winter
Three lamps in living room	3 × 12 = 36	1	3	36	108
Two reading lamps in bedroom	2 × 7 = 14	1	2	14	28
One outside lamp with motion detector	10	0.1	0.5	1	5
One TV	50	2	3	100	150
One refrigerator	50	Unknown	Switched off	200	Switched off
Total	160			351	291

the daily operating times of a refrigerator, but the data sheet usually states the daily energy consumption (here 0.3 kWh).

After determining the energy consumption, we can turn our attention to the dimensioning of the PV generator.

8.5.3.2 Dimensioning the PV Generator

First, we must find out what the yield of a solar module is in the respective months. Basic data is shown in Table 2.2 in Chapter 2, which gives the daily radiation H on a horizontal area for various places. Now, we will look up in which month in the summer season, the radiation is weakest. In the case of Munich, this is September with an radiation of 3.53 kWh $(m^2\ d)^{-1}$.

Instead of the radiation, we will simplify by considering Sun full-load hours (again – see Chapter 2). The Sun requires 3.53 full-load hours to generate the whole radiation on a September day. A 100 W module lying flat on the ground on this day will then generate energy of 3.53 h·100 W = 353 Wh. If we slant it, then the yield is increased, which we can coarsely determine by means of a correction factor, C_{Slant}, according to Table 8.5. In the case of south alignment and a tilt angle of 30°, we get a day's yield of 353 Wh 1.25 = 441.3 Wh.

Finally, we should take into account the effects of temperature on the module yield; this is done using a correction factor, C_{Temp}, that is listed in Table 8.6. As is to be expected, the warm climate of the south-oriented site makes a difference to the yield.

Added to the given correction factors there are also loss factors. First, we will assume a fixed amount of up to 6% for the electrical losses in the direct current lines. These we will take into account with the line loss factor $V_{Line} = 0.94$.

Table 8.5 Correction factor, C_{Slant}, for deviations from the horizontal in Germany and other sites [22, 40].

Site	Direction	Slant angle	Jan	Feb	Mar	Apr	May	Jun	Jul	Aug	Sep	Oct	Nov	Dec
Germany	South	30°	1.67	1.54	1.31	1.16	1.07	1.01	1.02	1.10	1.25	1.38	1.58	1.68
		45°	1.88	1.70	1.37	1.15	1.02	0.95	0.96	1.08	1.28	1.46	1.76	1.93
		60°	2.02	1.77	1.36	1.09	0.93	0.84	0.86	1.00	1.25	1.48	1.85	2.07
	Southwest	30°	1.55	1.36	1.21	1.08	1.02	0.97	0.96	1.05	1.16	1.29	1.42	1.48
		45°	1.73	1.46	1.24	1.06	0.98	0.91	0.90	1.02	1.17	1.35	1.54	1.61
		60°	1.83	1.49	1.22	0.99	0.89	0.82	0.81	0.94	1.13	1.34	1.56	1.68
	West	30°	1.10	0.98	0.98	0.93	0.92	0.90	0.88	0.93	0.96	1.02	1.01	0.95
		45°	1.12	0.97	0.95	0.87	0.86	0.83	0.81	0.87	0.92	1.00	0.99	0.93
		60°	1.10	0.93	0.90	0.80	0.78	0.76	0.74	0.79	0.86	0.95	0.96	0.91
	Southeast	30°	1.35	1.36	1.19	1.12	1.05	1.02	1.05	1.08	1.16	1.20	1.35	1.45
		45°	1.43	1.46	1.21	1.11	1.01	0.97	1.01	1.05	1.17	1.22	1.45	1.61
		60°	1.47	1.48	1.18	1.05	0.93	0.89	0.94	0.98	1.12	1.18	1.46	1.68
	East	30°	0.87	0.98	0.95	0.97	0.96	0.98	1.00	0.96	0.95	0.90	0.94	0.95
		45°	0.83	0.98	0.91	0.93	0.90	0.93	0.96	0.91	0.91	0.86	0.92	0.93
		60°	0.80	0.94	0.85	0.86	0.83	0.85	0.90	0.84	0.84	0.80	0.86	0.89
Marseille	South	60°	1.80	1.43	1.20	0.97	0.82	0.76	0.79	0.91	1.11	1.35	1.65	1.89
Cairo	South	60°	1.39	1.21	1.00	0.82	0.68	0.62	0.65	0.75	0.93	1.15	1.34	1.44

Table 8.6 Correction factor, C_{Temp}, for various sites in case of on-roof installations [40].

Site	Jan	Feb	Mar	Apr	May	Jun	Jul	Aug	Sep	Oct	Nov	Dec
Berlin	1.06	1.05	1.02	0.98	0.94	0.93	0.92	0.93	0.96	1.00	1.04	1.06
Marseille	0.98	0.98	0.95	0.93	0.91	0.89	0.87	0.88	0.91	0.93	0.97	0.98
Cairo	0.93	0.92	0.90	0.88	0.86	0.84	0.84	0.84	0.86	0.87	0.90	0.93

In addition, a battery can never completely output the full amount of energy that was put in during charging as the electro-chemical conversion processes incur losses. Experience values for temperate regions are 10% and in hot climates approximately double. For our cottage in Munich, we will therefore calculate with a conversion loss factor, V_{Conv}, of 0.9.

Finally, stand-alone plants to 500 Wp are usually produced without an MPP controller. This we will account for with the adaptation loss factor, V_{Adapt}, also of 0.9.

Now we can finally calculate the required module power with the following equation:

$$P_{PV} = \frac{W}{N_{Sun} \cdot C_{Slant} \cdot C_{Temp} \cdot V_{Line} \cdot V_{Conv} \cdot V_{Adapt}}, \tag{8.17}$$

where

W = Energy consumed per day;

N_{Sun} = Sun full-load hours.

For summer operation in the case of a south-oriented PV generator at a 30° inclination angle, we get

$$P_{PV} = \frac{351 \text{ Wh}}{3.53 \text{ h} \cdot 1.25 \cdot 0.96 \cdot 0.94 \cdot 0.9 \cdot 0.9} = 108.8 \text{ Wp}. \tag{8.18}$$

A 110 W module would just about be good enough to cover the daily demand.

If we do the same calculation for the winter half-year (with December as the critical month), then we get a required generator size of

$$P_{PV} = \frac{291 \text{ Wh}}{0.79 \text{ h} \cdot 1.68 \cdot 1.06 \cdot 0.94 \cdot 0.9 \cdot 0.9} = 271.7 \text{ Wp}. \tag{8.19}$$

Winter thus clearly dominates the number of modules required.

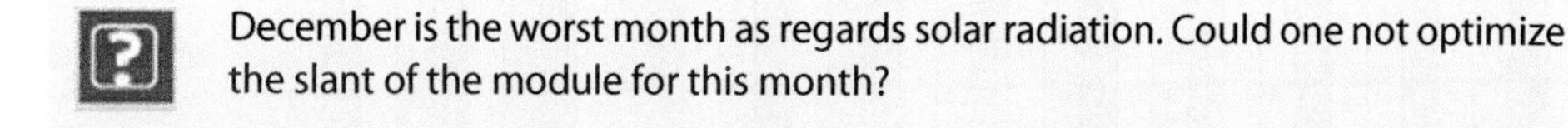

December is the worst month as regards solar radiation. Could one not optimize the slant of the module for this month?

This is really a good idea! According to Table 8.5, an inclination of 60° is the most advantageous for December. This results in a required PV performance of only 220.5 Wp. For instance, we could thus use two 120 W modules connected in parallel as a solar generator, which would then still be 10% overdimensioned.

8.5.3.3 Selecting the Battery

Two aspects are especially important when selecting the battery:

1. The number of desired autonomy days.
2. The increase of the battery life for lesser discharge depth.

The term *autonomy days* N_A is understood to be the number of days on which the consumers can be further operated even in bad weather.

A rule-of-thumb value for autonomy days in Germany is Summer: 3.5 days; winter: 5.5 days.

In order to achieve a high battery lifespan, we make the decision that batteries must never have a state of charge of less than 30%.

In the end, the user must decide what supply safety he wishes to achieve with the stand-alone system.

Accounting for these two conditions leads to the following equation:

$$C_N = \frac{W \cdot N_A}{0.7 \cdot V_N} \tag{8.20}$$

where V_N = system voltage, here assumed to be $V_N = 12$ V.

In our case, this results in summer:

$$C_N = \frac{351 \text{ Wh} \cdot 3.5}{0.7 \cdot 12 \text{ V}} = 146.3 \text{ Ah} \tag{8.21}$$

Naturally, the requirements are higher in winter:

$$C_N = \frac{291 \text{ Wh} \cdot 5.5}{0.7 \cdot 12 \text{ V}} = 190.5 \text{ Ah} \tag{8.22}$$

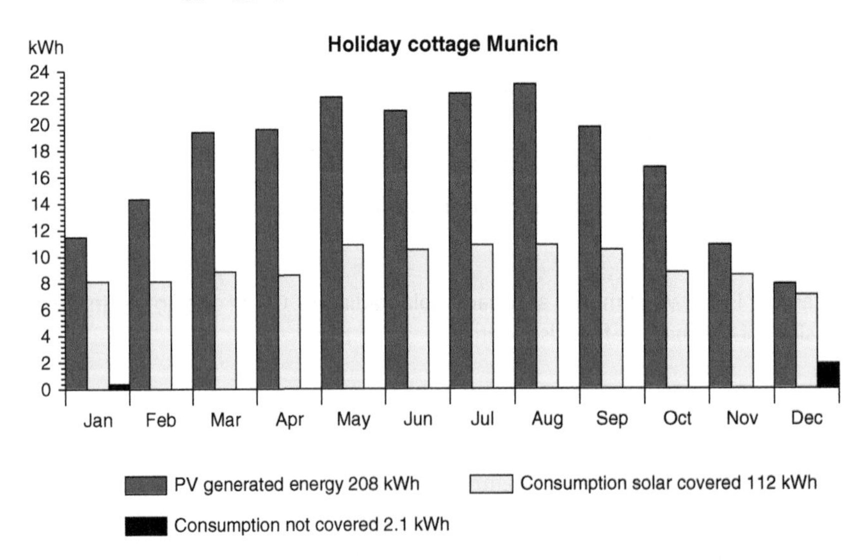

Figure 8.33 Simulation of the holiday cottage in Munich: With the design of the plant for the winter half, only about half the generated electricity can be used (simulation with PV-Sol Expert).

Here, for instance a 12 V battery with 200 Ah capacity would be a good choice. As a result, we will therefore select two solar modules each 120 W and a battery with a capacity of 200 Ah.

On the website *www.dgs-berlin.de*, there is an Excel tool with which the calculation according to this scheme can be comfortably worked out for a series of other sites in the world. Further, the tool assists in the dimensioning of the DC line cross section. There are, of course, many other simulation programs with which stand-alone plants can be calculated (see also the overview in Chapter 9).

This example shows that in summer, there will be a large surplus of energy that can normally not be used sensibly. Despite this, it is conceivable with the selected dimensioning that no full supply can be ensured in bad weather periods. If, for instance, one simulates this holiday cottage with the PV-Sol Expert simulation tool, then one actually obtains a solar coverage of 98% over the year (Figure 8.33). However, to cover the last 2 kWh with solar energy, one would have to increase the battery capacity by 50%. This is in no relationship to the additional costs. Instead, a small emergency generator would be recommended that in winter would occasionally recharge the battery. In this case, one could even halve the battery capacity, for instance to 100 Ah, and still achieve a solar coverage of 93%.

9
Photovoltaic Metrology

Trust is good – control is better. This saying also applies to photovoltaics. That is why this chapter is devoted to presenting the most important principles for acquisition of solar radiation and for performance and quality analysis of photovoltaic plants.

9.1 Measurement of Solar Radiation

It is necessary to use radiation sensors in order to determine the radiation behavior at various sites as accurately as possible. This measurement deals mostly with the determination of global radiation but sometimes also with the separate acquisition of direct and diffused radiation (see Chapter 2). For this purpose, we will look at various types of sensors and discuss their special features.

9.1.1 Global Radiation Sensors

9.1.1.1 Pyranometer

The most accurate sensor for measuring global radiation is pyranometer. The name is derived from the ancient Greek of *pyr*: "fire" and *ouranós*: "heaven"; to a certain extent, the instrument measures the strength of the Sun's "fire" from the sky. The structure of a pyranometer is shown in Figure 9.1. The decisive element is the black absorber surface. The Sun's rays cause this to heat up compared to the environmental temperature so that the temperature difference $\Delta\vartheta$ is a measure of the incident irradiance E:

$$\Delta\vartheta = \vartheta_1 - \vartheta_A = \text{const} \cdot E \tag{9.1}$$

with

ϑ_1 = temperature of the absorber;

ϑ_A = temperature of the ambience.

The temperature difference is determined very accurately by means of a thermopile, which is understood to be a series connection of several thermal elements.

The two glass domes have two tasks: First, they ensure that the heated absorber surface re-radiates as little as possible of the heat taken up. Second, the hemispherical form secures that the sensitivity depends on the cosine of the incident angle (so-called *cosine response*). A vertical radiation thus has a maximum effect on the absorber surface, whereas the horizontal one should have no effect at all.

Photovoltaics – Fundamentals, Technology, and Practice, Second Edition. Konrad Mertens.

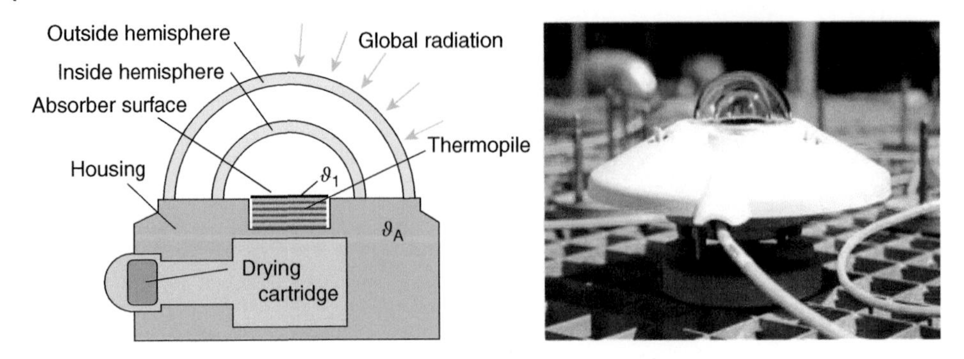

Figure 9.1 Structure and view of a pyranometer: The absorber surface is heated by the Sun's radiation so that via the temperature the irradiance can be acquired. Photo: Kipp and Zonen.

In order to prevent the glass cover from misting, many pyranometers have a drying cartridge of silica gel that absorbs the moisture. The cartridge must be replaced after about 6 months.

The photo in Figure 9.1 shows the pyranometer with the normal white plastic cover; the sunshade is meant to prevent the pyranometer housing from heating up due to solar radiation. The black absorber surface detects practically the whole of the Sun's spectrum with a constant sensitivity. However, because of the glass dome, the measurable spectrum is limited to the range of 300–2800 nm, but this is no problem as only a small portion of the solar radiation is not acquired (see Figure 9.2).

The voltage signal output by the thermopile is very small, typically around 10 mV with full Sun irradiance (1000 W m^{-2}). For this reason, the signal is best increased by means of an external amplifier to manageable voltage values such as 0–10 V.

Pyranometers are always used if as accurate as possible global radiation measurement data are required. For this purpose, various classes of accuracy are defined in the ISO Standard 9060 (see Table 9.1).

Table 9.1 shows that there are great differences between the individual classes. Special note should be taken of the peculiar naming: Best Class is not the "First Class" but "Secondary Standard"!

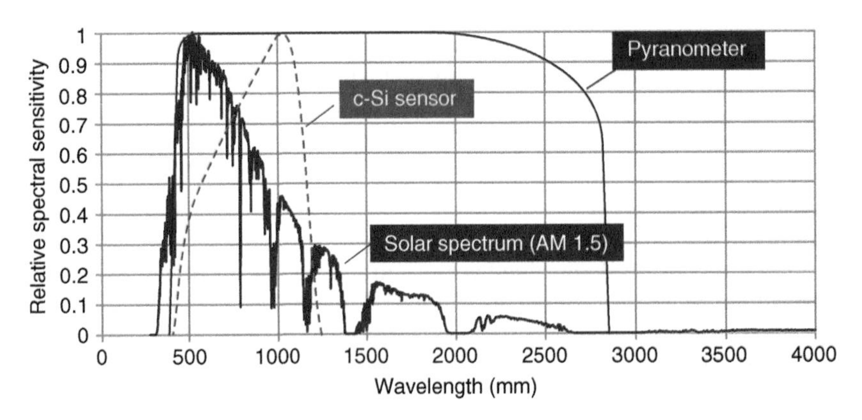

Figure 9.2 Spectral sensitivity of pyranometer and solar cell sensor: The pyranometer absorbs almost the whole of the solar spectrum, whereas the solar cell only detects a limited portion.

Table 9.1 Accuracy classes of pyranometers to ISO 9060.

Property	Second class	First class	Secondary standard
Quality	Fair	Good	Excellent
Accuracy (daily sum) (%)	±10	±5	±2
Resolution (W m^{-2})	±10	±5	±1
Long-term stability (%)	±3	±1.5	±0.8
Response time (s)	<60	<30	<15

9.1.1.2 Radiation Sensors from Solar Cells

A pyranometer costs between 600€ and 2000€, depending on the class. For this reason, radiation sensors from solar cells are an economical alternative. These are mostly small c-Si cells that have been specially encapsulated. In order to measure the irradiance, the solar cell is short-circuited with a low-ohmed shunt resistor, and the voltage drop at the shunt is measured. As the short-circuit current of a solar cell is proportional to the irradiance, a simple arrangement is possible. The temperature dependency of I_{SC} can be compensated for by building in a temperature sensor together with a downstream temperature-dependent voltage amplifier.

For a price range between 100€ and 500€, one obtains sensors with given accuracy of ±5% to ±10%. However, it must be noted that the c-Si solar cell always only measures a small portion of the Sun's spectrum (see Figure 9.2). If the sensor is calibrated to an AM 1.5 spectrum, then it will show a deviation for a low-lying Sun (e.g. AM 4). Added to this is that the flat cover sheet has a reflection that is dependent on the angle of incidence.

Therefore, solar cell radiation sensors are mainly used for continuous performance monitoring of a PV plant. For this purpose, they are mounted at the module level so that they receive the exact radiation available to the PV plant (Figure 9.3). This is why the radiation sensor should have the same technology (c-Si, a-Si, CdTe, etc.) as the modules in the plant.

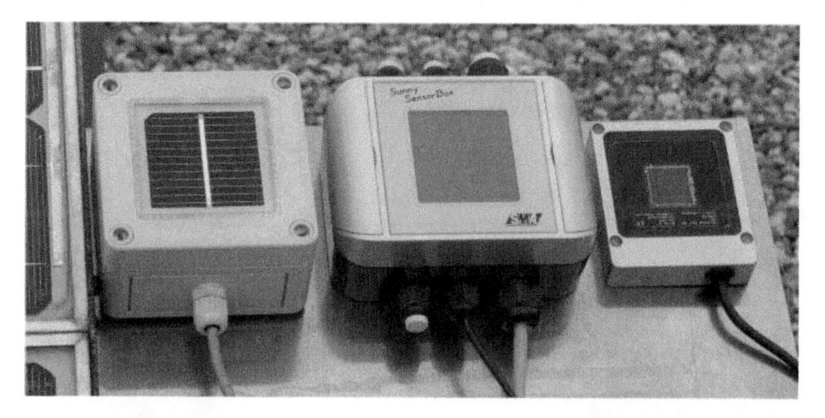

Figure 9.3 Various reference sensors of c-Si solar cells at the experimental rig of the Münster University of Applied Sciences: The modules are installed at module level to measure the incident radiation on the solar module.

Table 9.2 Comparison of pyranometers and solar cell sensors.

Pyranometer	Solar cell sensor
+ Very high accuracy	+ Economical
+ Sensitivity independent of λ	+ Behaves as a solar module
+ Hardly direction dependent	+ Small response time (<1 s)
– Expensive	– Strong spectral dependency
– Sluggish	– Strong directional dependency
Use: Measurement of global radiation for comparing various sites	Use: Measurement of radiation in module level for plant monitoring

In this case, the limited spectral sensitivity of the sensors is an advantage as they correspond exactly to that of the monitored modules. Table 9.2 summarizes the advantages and disadvantages of both types of sensors.

Nowadays, there are also sensors on the market that are sold as pyranometers with silicon photodiode. These are actually c-Si sensors that possess a glass dome with a scatter pane in order to reduce the directional dependency of the sensor.

9.1.2 Measuring Direct and Diffuse Radiation

In Chapter 2, we saw that for the most accurate yield estimate, the separation of the global radiation into direct and diffuse radiation is necessary. There are special sensors for this. For determining the direct radiation, use is made of a *pyrheliometer* (*Helios*: a Greek Sun god). A typical model can be seen in Figure 9.4: The actual sensor is situated at the lower end of the tube. This only receives light when it comes exactly in front through a pinhole at the start of the tube. For this reason, the pyrheliometer must continuously track the Sun.

In the case of diffuse radiation, one uses a normal pyranometer in which the direct radiation is carried out by a tracker shade ball (Figure 9.5). The much cheaper variant is a fixed shade ring whose height, however, must be adjusted every 10 days

Figure 9.4 Pyrheliometer for measuring direct radiation: The instrument must continuously track the Sun. Photos: Kipp and Zonen.

Figure 9.5 Sensors for measuring diffuse radiation: Whereas the shade ball variant must continuously track the Sun, the shade ring is satisfied with an occasional adjustment of the tilt angle. Photos: Kipp and Zonen.

9.2 Measuring the Power of Solar Modules

When purchasing a fairly expensive solar plant, the customer expects the power of the module to correspond to what was agreed-upon during the purchase. As we saw in Chapter 6 in the discussion about mismatching, it is not sufficient when the sum of all the power of the modules corresponds to the agreed-to nominal power; instead, the powers of the individual modules should have the smallest possible tolerance. In order to ensure this, the modules must be measured very precisely by the producer. For this purpose, almost exclusive use is made of module flashers.

9.2.1 Build-up of a Solar Module Power Test Rig

An important element of a module test rig is the module flasher (also called solar simulator). This designates a source of radiation that generates a flash of light corresponding to the spectrum of the Sun. It consists mostly of a Xenon flash lamp with a filter in front in order to approach as near as possible to AM 1.5 spectrum. During the period of the flash (e.g. 1 ms), the I/V characteristic curve is electrically measured. For this, one varies the resistance of the connected electronic load and, at the same time measures, via computer control, the voltage and current of the module (Figure 9.6).

The reason for the use of a flash of light is the fact that the module scarcely heats up in the short time period. Thus, one can assume a constant temperature. In order that the solar module is homogenously illuminated, the flash lamp should be placed at a great distance (e.g. 5 m) from the module. A laser pointer acts as an aid to direction and must be aimed at the center of the module.

The irradiance during the measurement should be exactly 1000 W m^{-2}, which is why it is controlled by a reference solar cell sensor. The sensor must again be calibrated in order to achieve a high degree of accuracy during measurement. This calibration is carried out by means of a reference module (the "golden module"), which is understood to be a module that was measured by an accredited testing laboratory (e.g. TÜV Rheinland or Fraunhofer ISE) with an error of a maximum of ±2%. After the calibration of the

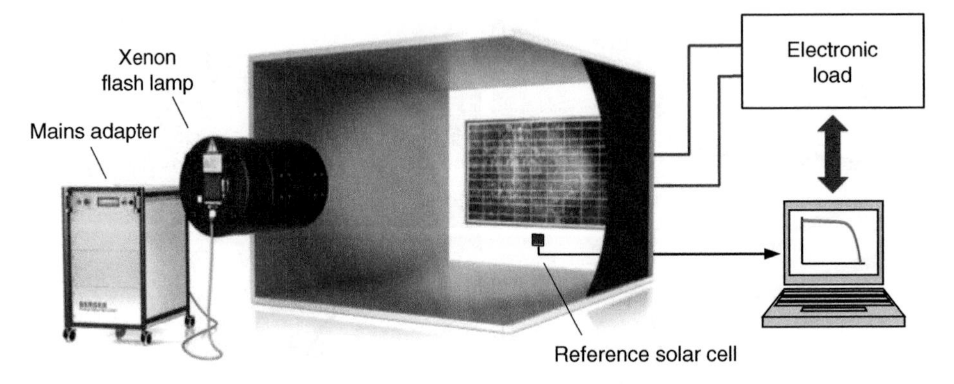

Figure 9.6 Principle arrangement of a solar module power test rig: The Xenon lamp generates a flash of light with an AM 1.5 spectrum. During the flash, the *I*/*V* characteristic curve of the solar module is measured with the help of an electronic load. Source: Berger Lichttechnik.

reference sensor, the measurement of the golden module must have the same power as the value stated on the certificate of the test laboratory.

9.2.2 Quality Classification of Module Flashers

The EN 60904-9 standard specifies the requirements that a solar simulator must fulfill with regards to spectrum as well as homogeneity and timely stability of the irradiance [108]. For this purpose, different classes are defined from which the quality of a flasher can be derived. Table 9.3 shows the conditions that lead to a division into the classes A, B, or C.

In order to evaluate the spectral adaptation, an investigation is carried out in six defined spectral regions between 400 and 1100 nm on how well the flasher simulates the AM 1.5 spectrum. In order, for instance, to achieve Class A, there must be no deviation of more than ±25% in any of the regions. Much more difficult to achieve is a homogenous illumination of the module area. The maximum deviation demanded of Class A of 2% between minimum and maximum irradiance can be reached for large measurement areas of, for example, 2 m × 2 m only with much effort. For this reason, there are a number of ABA flashers on the market. However, from Chapter 6, we know that even a reduced radiation limited to a small area can have a large effect on the power of a solar module. For this reason, one should only accept measurements that have been

Table 9.3 Classification of solar simulators [108].

Classification	Spectral adaptation to AM 1.5 (%)	Inhomogeneity of irradiance (%)	Long-term instability of the irradiance (%)
A	±25	2	2
B	±40	5	5
C	±60–100	10	10

carried out with flashers of Class AAA. In this case, the module performances should be determined with an accuracy of ±3%.

9.2.3 Determination of the Module Parameters

The most important parameters V_{OC}, I_{SC}, and P_{MPP} can be directly derived from the received I/V curve of the module under standard conditions. In addition, one can determine the shunt resistance R_{Sh} and series resistance R_S of the module from the slope in the short-circuit and open circuit point as described in Section 4.5.

However, a more accurate determination is possible with the flasher in the case of series resistance. For this, two characteristic curves must be taken at different irradiances (e.g. 1000 and 800 W m^{-2}) (see Figure 9.7). Now one determines an operating point Q_1 on the upper characteristic curve somewhat to the right of MPP and determines the current difference ΔI between I_{SC1} and the current in point Q_1. Then one determines the operating point Q_2 on the lower characteristic curve at which the current is $I_{SC2} - \Delta I$. R_S is calculated according to the following equation:

$$R_S = -\frac{V_2 - V_1}{I_{SC2} - I_{SC1}}. \tag{9.2}$$

This method was specified in the DIN EN 60891 standard in 1994 and is now somewhat out of date [109]. A follow-up standard was issued in 2010 that specifies the taking of at least three characteristic curves at different irradiances. The series resistance is defined by means of a mathematical method, the details of which can be found in Ref. [110].

In order to estimate the quality of a solar module, the weak light behavior should always be investigated. For this purpose, the module is irradiated, for instance, in sequence with 100, 200, 400, and 700 W m^{-2}, and the decay of the efficiency compared to 1000 W m^{-2} is determined (see also Section 6.1.5). The different irradiances are obtained in that an optical attenuator ("fly screen") is placed in front of the lamp.

Could we not just reduce the current of the lamp to get a lower irradiance?

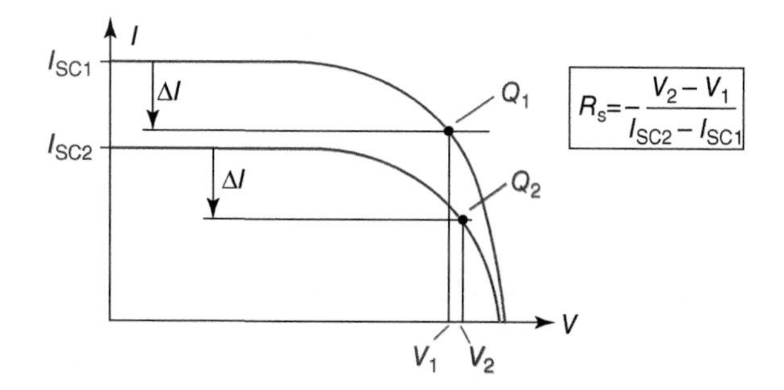

Figure 9.7 Determination of the series resistance R_S: After sketching the horizontal help lines, the two operating points Q_1 and Q_2 that lead to the determination of R_S are found [109].

Unfortunately, lowering the lamp current also lowers the lamp temperature and thus the lamp spectrum. The current should only be reduced to a maximum to 900 W m^{-2} where the change of the spectrum is still acceptable.

9.3 Peak Power Measurement at Site

When building up a photovoltaic plant, one normally assumes that the power of the purchased modules corresponds to the data sheet values. A precautionary checking of all modules with the flasher would, in any case, be too expensive. A cheaper alternative to laboratory measurements is the peak power measurement of the whole plant at site.

9.3.1 Principle of Peak Power Measurement

In peak power measurement at site, a whole string or even the whole solar generator is measured at once. The source of light is the Sun; the actual irradiance is measured with a radiation sensor that points in the same direction as the solar modules. The module temperature is measured at the same time.

The actual measurement is carried out as follows: At time point $t = 0$, the solar generator is connected to a large capacitor (see Figure 9.8). The solar generator loads the capacitor, for instance, for 1 s. In this it runs through the whole I/V curve from the short-circuit point (empty capacitor) to the open-circuit point (charged capacitor). During the charging process, the microcontroller continuously measures the voltage and current and can thus acquire the whole of the characteristic curve.

With simultaneously recorded radiation and module temperature, the determined MPP power can then be converted to the value for STC conditions (hence, the designation peak power measurement).

In the case of the module flasher, use was made of the changeable resistance in order to run through the characteristic curve. Why is a capacitor used here?

When measuring a whole solar generator, high powers of, for instance, 10 kW are easily reached. The electrical load would, therefore, have to be very large and well cooled. With the capacitor, however, almost no power loss occurs so that the device can be relatively small.

9.3.2 Possibilities and Limits of the Measurement Principle

A large disadvantage of measurements at site is the dependency on the weather. Most of the producers of characteristic curve measurement instruments admit that exact measurements are only possible with irradiance over 500 W m^{-2}. A study carried out at the Münster University of Applied Sciences (Münster UAS), however, showed that even

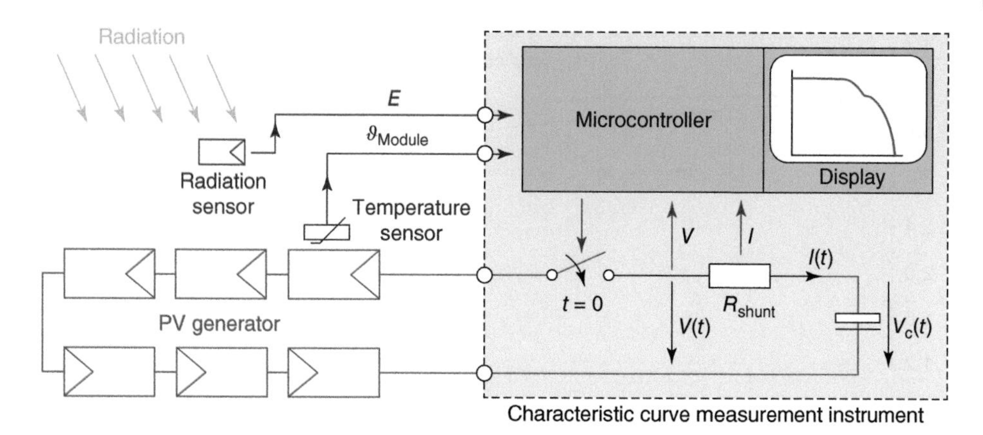

Figure 9.8 Arrangement of a characteristic curve measurement instrument: After the switch is closed, the solar generator charges the capacitor. In this, it runs through the whole *I*/*V* characteristic curve from the short circuit to the open circuit point.

in this case the peak power was still strongly dependent on the radiation and module temperature [111]. The error tolerances of some producers of ±5% are certainly very optimistic because already the radiation sensors supplied have their own tolerances of ±5%.

Besides the peak power determinations, also the reported characteristic curve is very informative. It can provide information on cabling errors or defective modules. Figure 9.9 shows two examples of characteristic curves.

The left picture shows an example of a curve of a string measurement, in which a slanted decay is noticeable in the upper region. This can be an indication of partly shaded modules. In this particular case, however, there was no shading; instead the modules were of different qualities. Because of this "proof curve," all the modules of the string were replaced by the producer. The picture on the right shows the effects of errors in the string cabling. Here a string of eight modules was connected in parallel with one string of seven modules shown by the dip in the right side of the characteristic curve.

The series resistance of the whole measured string can also be determined from this characteristic curve [14]. If this is much larger than the sum of the series resistances of the individual modules, then it points to high contact resistances at the terminals of the string cabling.

To summarize, the characteristic curve measurement at site is a very informative method with information on quality of modules and cabling. Therefore, this should be carried out as standard within the scope of plant acceptance for comprehensive quality control.

9.4 Thermographic Measuring Technology

Infrared thermography offers the possibility of carrying out a time-saving quality check of solar modules. We will first briefly deal with the principle of the function of thermography and then look at the two most important methods of bright and dark thermography.

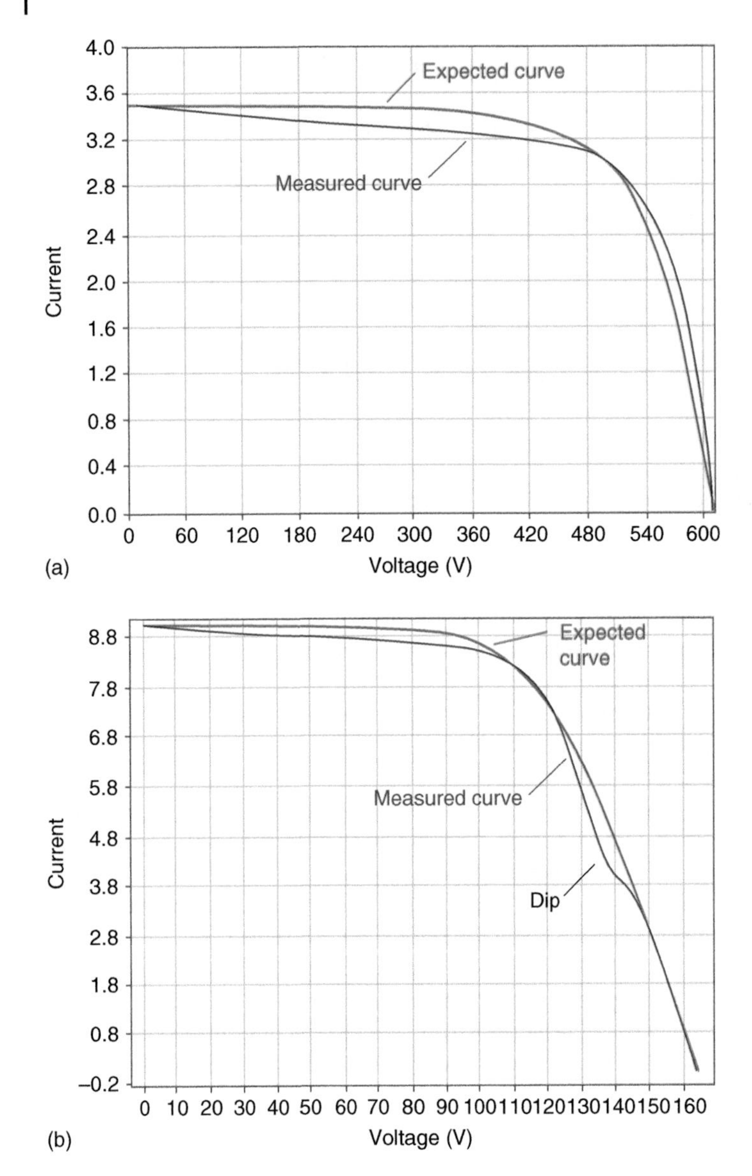

Figure 9.9 Measured measurement curves: (a) A string with modules of different quality and (b) the effects of incorrect string cabling.

9.4.1 Principle of Infrared Temperature Measurement

In Chapter 2, we learned that the Sun is a blackbody radiator. An ideal blackbody radiator radiates a characteristic light spectrum depending on the surface temperature (Planck's law of blackbody radiation; see Figure 2.2). If this spectrum is integrated over the whole wavelength region, then the result is the *Stefan–Boltzmann law*:

$$P_B = \sigma \cdot T^4 \tag{9.3}$$

Table 9.4 Emission factors of various materials [112].

Material	Emission factor	Material	Emission factor
Blackbody radiator	1	Aluminum, polished	0.18
Steel, black painted	0.96	Aluminum, oxidized	0.19
Steel, polished	0.07	Glass sheet	0.8–0.9

with

P_B = optical power of the blackbody radiator;

σ = Stefan–Boltzmann-constant: $\sigma = 5.67051 \times 10^{-8}$ W (m^2 K^4)$^{-1}$;

T = surface temperature of the blackbody radiator.

The optical power of the ideal blackbody radiator thus depends solely on a constant as well as the surface temperature of the radiator.

If the surface of the radiator is not ideally black, then only a portion of the possible power is irradiated to the outside. This is accounted for in Equation (9.3) by the introduction of an emission factor ε:

$$P = \varepsilon \cdot \sigma \cdot T^4 \tag{9.4}$$

with

ε = emission factor.

If the emission factor of a radiator is known, then one can determine the surface temperature by means of a measurement of the radiation power. Table 9.4 lists the emission factors of some materials. From the example of steel, it is clear that it is not the material in itself, but the surface properties that are decisive in the emission factor. Basically one can say that the stronger the reflective effect of a surface, the smaller the emission factor.

In simple IR thermometers, the radiation emitted by a surface is led by means of a lens to a sensor that detects the strength of the radiation. From this and the known emission factor, the surface temperature can be determined. Thermography cameras use the same principle, but here a whole array of detectors is used so that a two-dimensional temperature picture can be generated.

9.4.2 Bright Thermography of Solar Modules

Bright thermography is understood to be the temperature analysis of solar modules in sunlight. In the simplest case, one views the solar generator with a thermographic camera and searches for noticeable temperature peaks. Figure 9.10 shows an example in which an on-roof installation was investigated.

Two cells of the module at the bottom right whose temperatures are approximately 20 K above the other cells are clearly noticeable. Apparently, they deliver too little current and thus heat up (load operation, see also Chapter 6). At the same time, the region of the module connection box is clearly heated up, which indicates an active bypass diode. A characteristic curve measurement of the module confirmed that the module provided approximately 30% less power. Based on this malfunction picture, the producer replaced all modules of the installation within the warranty period.

The temperature picture of a solar module depends strongly on the actual operating condition. Figure 9.11 shows four modules, of which one is in open circuit, one in short

Figure 9.10 Example of thermographic measurement of an on-roof installation: Two cells of the module at the bottom right clearly show noticeable spots. Photos: T. Stegemann, Münster UAS.

circuit, and two are in MPP operation. The thermography picture of the module in open circuit is relatively homogenous, as here the light absorption has led to even heating. In the case of the short-circuited module, there are clear temperature differences between the individual cells; here the load operation of the weaker cells can be seen. The cell at the top right has a temperature of 75 °C; it is approximately 30 K above the temperature of the other cells. In MPP operation, the differences are not nearly so distinct.

After the short-circuit operation, the bottom-left module was changed to MPP operation and was measured by thermography again (not shown in Figure 9.11). In this case, the weak cell only showed a temperature increase of 3 K compared to the other cells.

In the short-circuit operation of a solar module, the different cell qualities are much more in evidence in the thermography picture than in MPP operation.

In Figure 9.11, the MPP modules seem to have the lowest temperatures, while the open-circuit and short-circuit modules are somewhat hotter. Is this a coincidence?

It is really no coincidence! The MPP modules give up a part of their irradiated optical power as electrical power to the inverter. This part, therefore, does not lead to heating up the module. In the two other modules (open circuit and short circuit), no electrical power is output, and therefore, they heat up more strongly. In the particular case, the difference is 2–4 K.

A module in short-circuit or open-circuit operation heats up slightly more than one in MPP operation.

Figure 9.11 Thermography picture of modules in various operating conditions: The module in open circuit shows a homogenous heating, whereas the short-circuit operation leads to different heating of the cells. Source: Photo: Münster UAS.

The knowledge of the characteristic thermo-pictures of the various operating conditions makes thermography a very effective means of quality control of photovoltaic plants. It is very easy to recognize incorrectly connected and defective modules.

If a thermography camera is too expensive, one can also do without it if need be. Figure 9.12 shows the module of Figure 9.10 taken on a cold November day with hoarfrost on the modules.

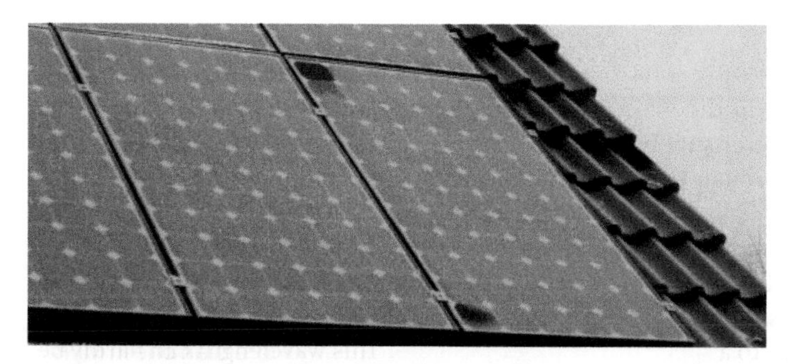

Figure 9.12 "Thermography for the poor" on a cold November day: The two cells of the modules from Figure 9.10 can clearly be seen as free from hoarfrost. Photo: T. Stegemann, Münster UAS.

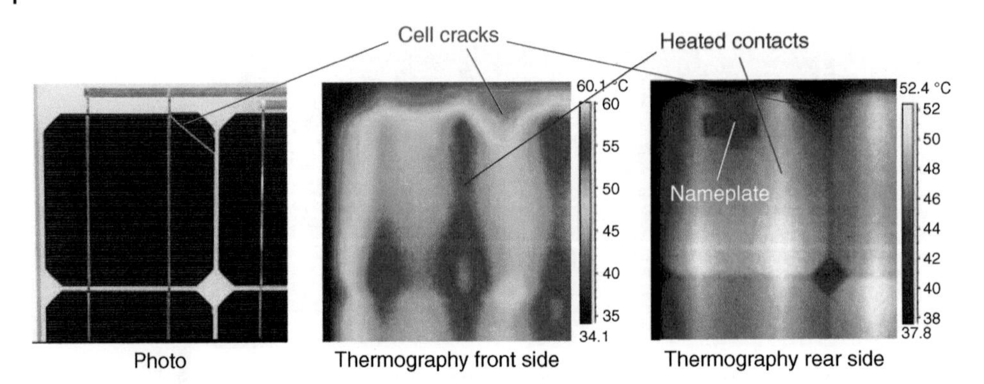

Figure 9.13 Dark thermography at the front and rear sides of a module: The inactive cell area top right can be much better recognized from the rear of the module. Photo: Münster UAS.

9.4.3 Dark Thermography

Besides thermography under illumination, measurements in the dark laboratory can also provide information on the module. In this, the module is operated in the IVth quadrant (see Figure 6.1) with a high reverse current of, for instance, double the short-circuit current.

This current leads to heating of the finger contacts on the top side of the cell. Error positions with high contact resistances are clearly seen by means of temperature peaks (Figure 9.13). At the same time, the reverse current ensures a relative even heating of the cells. Any inactive cell region can also be recognized. Measurement in the laboratory also has the advantage that the module can be viewed from the bottom. The thin rear-side foil provides much better "picture sharpness" than the thick front glass that strongly evens out the temperature picture.

9.5 Electroluminescence Measuring Technology

The last method of measuring we will look at is electroluminescence (EL measurement) measuring technology.

9.5.1 Principle of Measurement

The idea here is to light up a solar module as a large-surface LED. For this purpose, the module is operated in the reverse current region, thus in the IVth quadrant (see Figure 6.1). A strong power supply is used to drive a reverse current to the level of the short-circuit current through the module. With this, the p–n junction is operated in the forward region, and light emission is possible. However, we know from Section 3.6 that c-Si is an indirect semiconductor. As a phonon is required here for the interaction between photon and electron–hole pair, not only is the absorption of light relatively weak but, in the opposite case, the emission is too. Added to this is the fact that the bandgap of silicon leads to a spectrum of about 1150 nm. This wavelength can hardly be detected by a normal charged coupled device (CCD) camera.

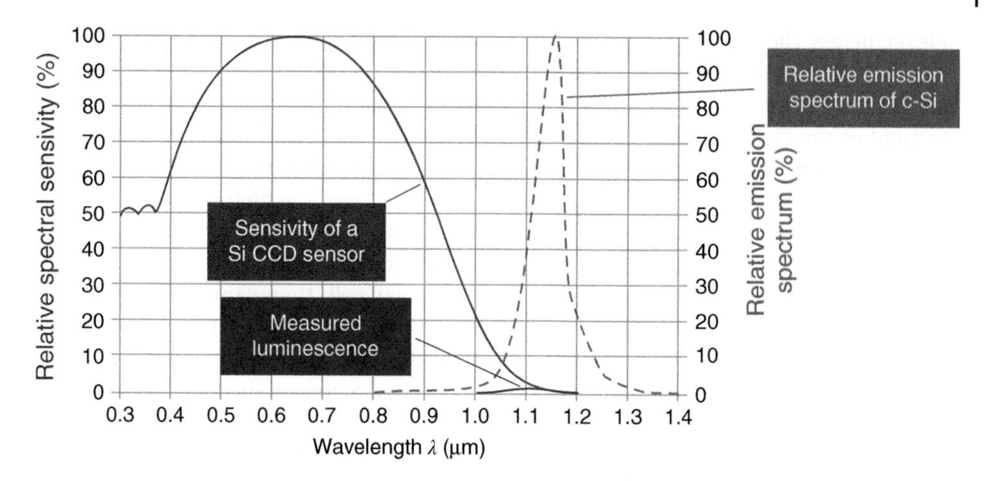

Figure 9.14 Problem with electroluminescence measurements: The emission spectrum of c-Si is at the sensitivity limit of CCD cameras and can therefore only be detected very weakly [113].

Figure 9.14 shows the problem: The spectral sensitivity of the CCD sensor made of silicon decays strongly in the region of the bandgap of silicon. The already low light emission of the solar module can thus only be photographed at the noise limit. Nowadays, special (and expensive) CCD cameras are available that can provide usable pictures.

A trick can be used to improve the quality of the photographed pictures: A picture of the current-less solar module is also taken, which is then "subtracted" from the electroluminescence picture. In this way, the inherent noise of the camera and any foreign light can be well compensated.

9.5.2 Examples of Photos

What do typical electroluminescence pictures look like? Figure 9.15 shows at the left a detailed photo of an individual cell taken with a special camera. Spider web-type microcracks can be clearly seen that have probably been caused by point mechanical loading.

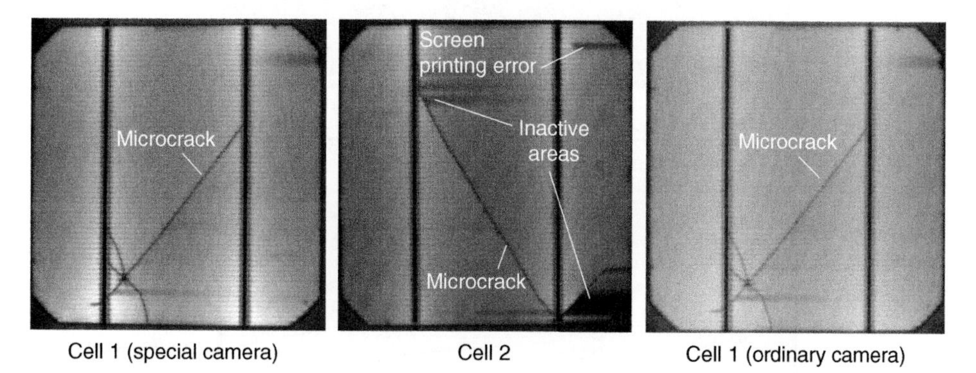

Figure 9.15 Photographed electroluminescence pictures of two solar cells: Both cells show a microcrack that does not lead to a reduction of power at cell 1. In the case of cell 2, however, dark areas can be seen that can no longer contribute to the photocurrent. The right-hand photo again shows cell 1, but here, photographed with an altered cheap digital reflex camera. Source: Münster UAS.

Nevertheless, the whole cell lights brightly as the current, as before, can be led through the two busbars and the contact fingers to all parts of the cell.

Things look different for cell 2: There is an inactive region that can be seen at the bottom right as the current cannot bridge the microcrack. There is also a dark area at the top left indicating that the microcrack runs parallel to the left busbar. The dark horizontal strip at the top right shows a missing contact finger piece, which could have been caused by a blocked screen-printing mask (see also Section 5.1.3).

The dark regions are especially critical for cell performance. Although electron–hole pairs are also formed in normal operation of the solar cell, they cannot move to the bus bar and thus recombine without being used. This causes sinking of the short-circuit current of the cell and also that of the cell string of the module.

An investigation at the Münster University of Applied Sciences showed that cheaper digital single-lens reflex (DSLR) cameras can also be used for EL photos after being altered [114, 115]. This is shown in Figure 9.15 with the same cell as before, but now as a photo with a "low-cost camera." The picture is not quite as sharp and has less contrast, yet the relevant details can be well recognized.

Figure 9.16 shows a module that seems completely undamaged from the outside. The EL photo, however, shows many microcracks and electrically inactive regions. In this case, the microcracks are due to improper transport of the module.

Naturally, the dark regions in the dotted circles from the right are particularly unpleasant. In the case of the cells with the dashed circles from the left, the microcracks have no effect on the performance of the module. But if a microcrack does exist, then it can grow in the course of time under the influence of wind, temperature changes, and snow loads

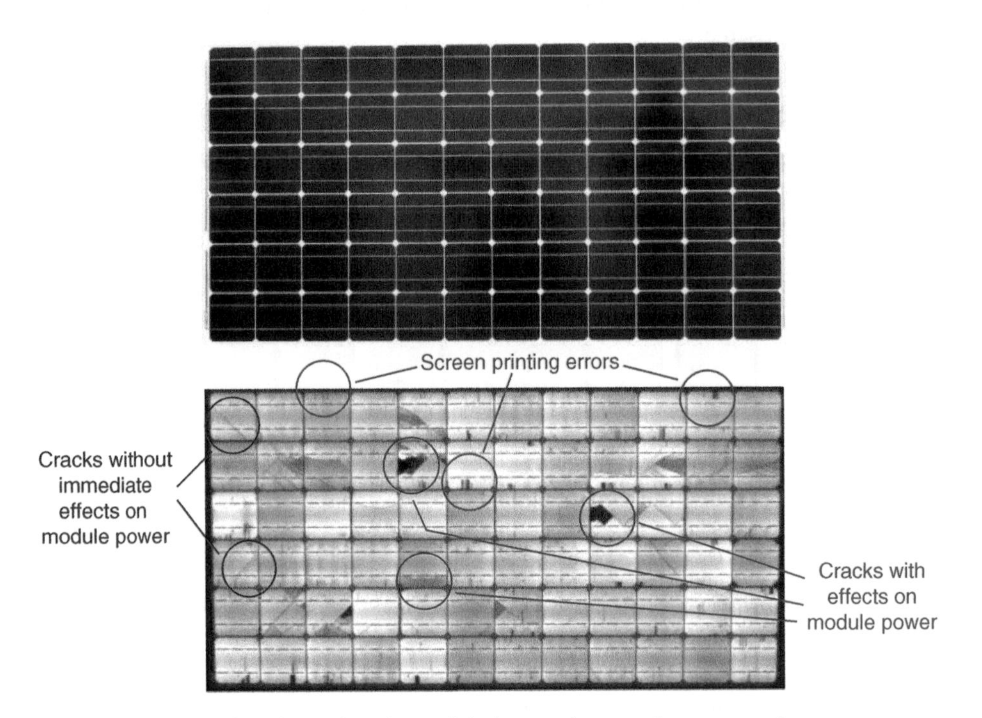

Figure 9.16 View and EL photo of a solar module that was improperly transported.

and also lead to degradation of performance [116]. Thus, by means of EL photos, one can obtain information on the damage to the module even before there are any electrical effects.

The top, solid circles in Figure 9.16 are screen-printing errors. These are not necessarily dramatic for the purchaser of the solar module as they occur during production and do not propagate themselves. The problem is rather with the producer, who has to sell the module with a lesser power than is possible.

Thin-film modules can also be investigated by means of the electroluminescence technology as is shown in Figure 9.17. This is a micromorphic module in which two dark points are noticeable. These are caused by local shunts (short circuits) that can occur during the module production in the integrated series connection (see also Section 5.2.7).

9.5.3 Low-cost Outdoor Electroluminescence Measurements

Notably the EL technique offers the possibility to get detailed information about the state of a module. Therefore, it is close at hand to use this technique also at already installed PV plants for fault detection. As we have seen in Section 9.2.5, there exist inexpensive "measurement devices" in the form of modified digital reflex cameras. With them in many cases, it is possible to detect faults at site without demounting the modules and costly taking them to the lab.

The principle of such an outdoor measurement shows Figure 9.18. With a high voltage power supply, a whole string of modules is energized to examine several modules at one time. The power supply should have at least provide 600 V and 5 A to cover the typical application cases. As well, the camera as the power supply can be remote controlled by the laptop. This facilitates, for example, the subsequent exposure of energized and nonenergized string. Both EL photos can then be subtracted from each other to reduce the disturbing influence of stray light. Additionally, the stray light should be further reduced by the use of specific infrared filters [8].

A first measurement example is shown in Figure 9.19. Here subsequently, the two strings of an on-roof PV plant were energized. This first of all serves to clear if all modules

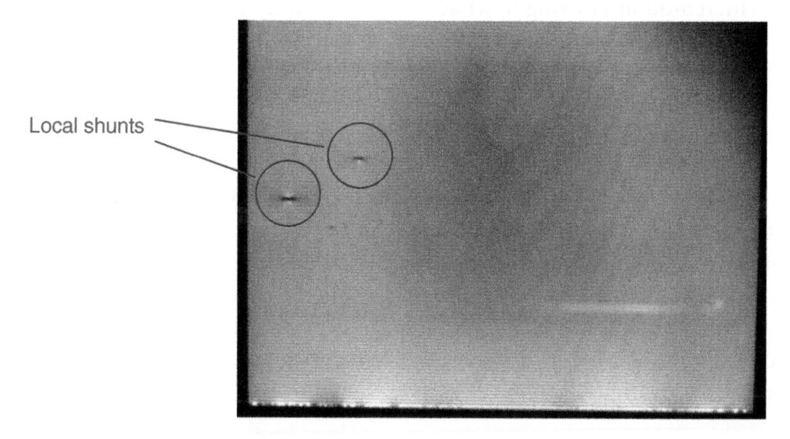

Figure 9.17 EL photo of a micromorphic thin film module: The two dark spots are caused by local shunts, which could already have been caused during module production.

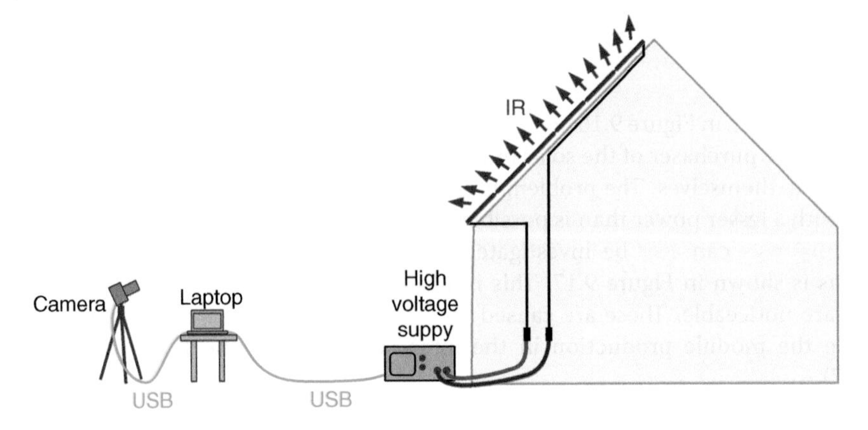

Figure 9.18 Setup for low-cost outdoor electroluminescence measurements: A whole string is energized to examine several modules at the same time.

Figure 9.19 On-site examination of an on-roof plant: Subsequently string 1 and string 2 are energized to inspect the modules for damage. Photos: M. Diehl.

are connected and which modules belong to what string. Moreover, it can be seen in the exposures that no severe module damages are present. If furthermore the modules shall be examined for microcracks, additional photos have to be taken in proximity of the modules.

Another example is to be seen in Figure 9.20. The central module contains two defective bypass diodes. These form a short circuit each, which shunts the reverse current through the two right-hand-side cell strings. Therefore, these cells remain dark. Such faults typically often are not detected over years and lead to yield losses.

Even fused, nonexistent or not connected bypass diodes can be detected with clever approaches. For this, a small reverse voltage is applied to the string under test (Figure 9.21a). If all bypass diodes are okay, they are operated in the forward direction and are therefore conductive. Per bypass diode results a forward voltage of about 0.4 V. If the string comprises, for example, 15 modules, then the following voltage drop V_{Total} results:

$$V_{\text{Total}} = V_{\text{BPD\_Complete}} \approx 15 \cdot 3 \cdot 0.4\ \text{V} = 18\ \text{V}. \quad (9.5)$$

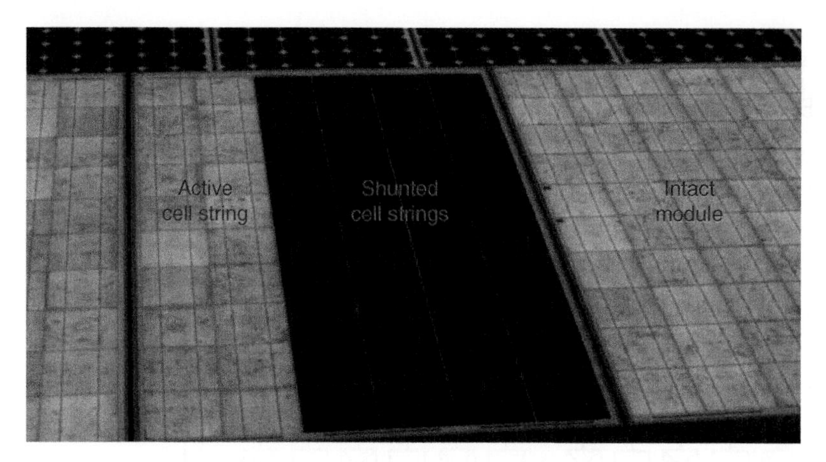

Figure 9.20 Detection of defective bypass diodes: In the central module two cell strings remain dark as they are shunted by defective bypass diodes. Photos: M. Diehl.

If now despite an applied voltage in the range of $V_{\text{BPD-Complete}}$ no current flows, then at least one bypass diode is missing or inoperable. In order to find out which bypass diode is the evildoer, the applied voltage is further increased until, for example, one tenth of the modules nominal current is reached. This current is possible through the beginning avalanche breakthrough of the cells in the affected cell string (see Section 6.1.1). As the cells are heated up due to the converted power, the concerned solar module can be easily detected by thermography. Figure 9.21 shows on the right the considerable warming up of the cells operated in breakthrough.

Taking EL photos at night is sometimes cumbersome and tedious. Significantly easier and quicker is using the video mode of the camera. With some tricks, it is possible to attain a sufficient brightness of the EL video. For details, see Ref. [18].

Summarizing, the outdoor EL technique provides a series of advantages with respect to thermography. It offers significantly higher resolutions and often gives direct evidence at the kind of the present defect. Moreover, in contrast to thermography, the EL exposures can be also done in an oblique angle to the modules. A disadvantage, of course, is the fact that solar modules under survey have to be externally energized.

It can be expected that the low-cost outdoor EL technique will become a standard measuring technology for PV plants beside thermography and peak power determination [8].

9.6 Analysis of Potential Induced Degradation (PID)

Already in Chapter 7, we shortly got to know the phenomenon of potential induced degradation (PID). Under this term, all mechanisms are summarized where solar modules show a power reduction due to a high voltage between module cable and ground potential. In case of thin-film modules, beside the power reduction even a permanent damage of the modules can occur caused by the corrosion of the transparent electrode. This is not typical for c-Si modules; here the degradation effect is (at least mostly) reversible.

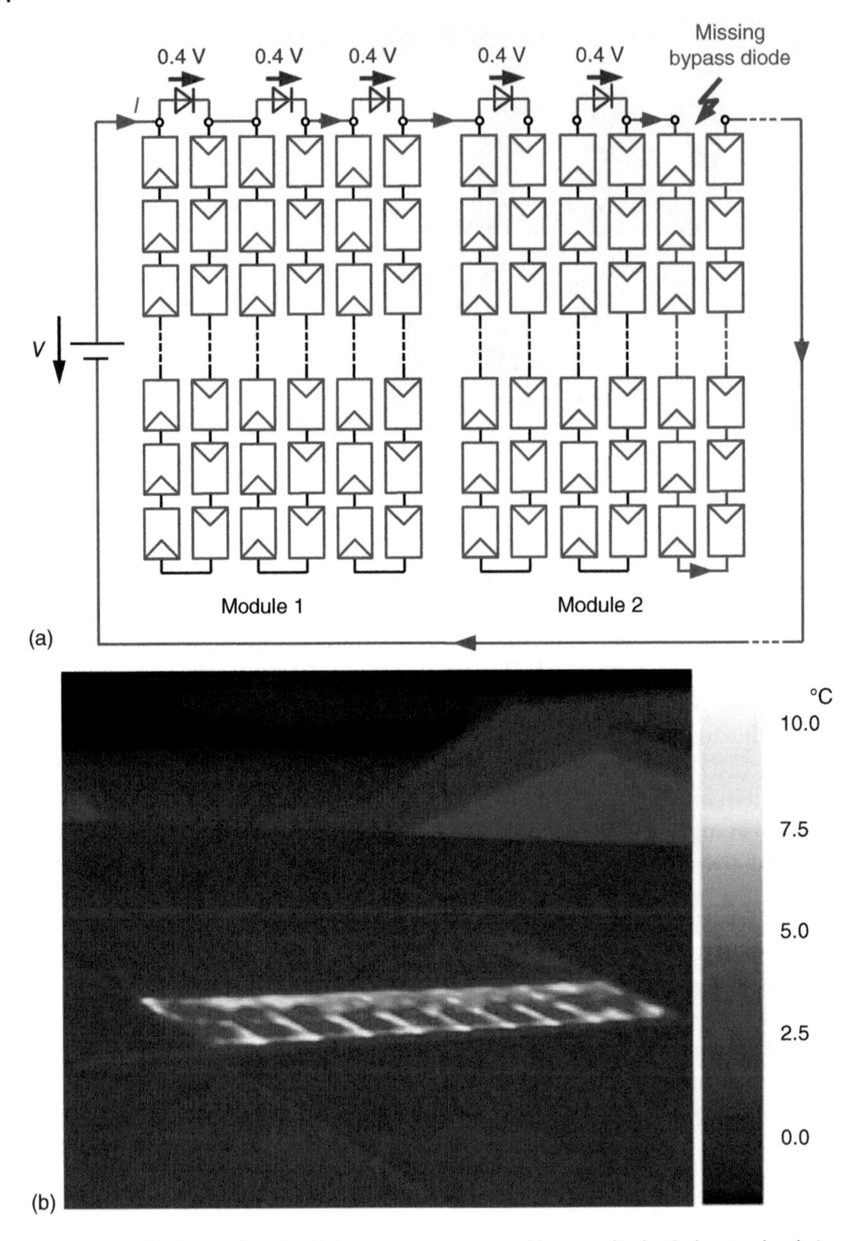

Figure 9.21 Fault case fused, missing, or not connected bypass diode: Only at a clearly increased voltage, a significant current flows. This one heats the concerned cell string, which can be easily localized by thermography. Photo: M. Diehl.

9.6.1 Explanation of the PID Effect

What now causes the power reduction? To explain this, Figure 9.22 shows the situation at the edge of a solar module with p-type cells. If the frame is grounded and, at the same time, a high negative voltage is present at the module connection cables, the potential

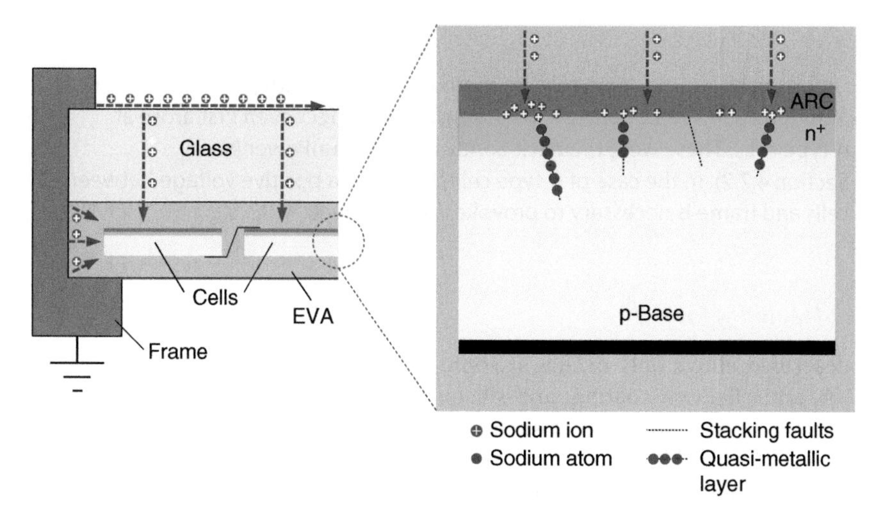

Figure 9.22 Genesis of the PID effect: Sodium ions migrate through the cover glass and the EVA to the cells. There they penetrate into existing lattice faults and provoke a local short circuit of the p–n junction.

difference can cause a migration of sodium ions through the cover glass of the module. They migrate further through the EVA and the silicon nitride antireflection coating down to the n^+-emitter. Actually, they could hardly go on then.

However, in the silicon crystal at some places, there exist so-called stacking faults. These are two-dimensional crystal faults, which extend some micrometers into the crystal and normally do no harm. In this case, though, they serve as a gateway for the sodium atoms. The Na^+-ions take up electrons of the emitter and form a razor-thin layer of sodium atoms, which reaches down into the base.

This quasi-metallic layer then locally shunts the p–n junction (PID shunting) [117]. In the IV curve of the module, this is to be seen in a drastically reduction of the shunt resistance R_{Sh} (see Figure 9.25). Moreover, the sodium layers act as recombination centers for the electron–hole pairs, which are generated by the sunlight. This leads to a further decline of the module efficiency.

The migration of the sodium ions from the frame to the cells provokes vice versa a leakage current of electrons out of the cells to the module frame. For reasons of clarity, this leakage current is not depicted in Figure 9.22.

Surveys at PV plants have shown that PID arises mainly at damp weather. Obviously, the electric conducting water film on the glass surface establishes ground potential on the whole topside of the glass. Therefore, the migration of the ions not only happens at the module edge but also on the whole module area.

Obviously, PID happens in some modules built up with p-type cells. Would it not be smarter to use modules with n-type cells instead?

Unfortunately, also n-type cells (to memorize: n-base with p^+-emitter) are not immune against PID. Concretely spoken, the PID effect even first arose at n-type cells. These were the back-contact cells of SunPower (see Section 4.7.2). In the case of n-type cells, however, a positive voltage between cells and frame is necessary to provoke the PID effect.

9.6.2 Test of Modules for PID

The effect described above only occurs at some modules. Obviously, the interaction between EVA, antireflection coating, and silicon crystal plays a decisive role. Meanwhile, the effect was even observed at such modules, which had been labeled by the manufacturer as "PID free." Therefore, in case of new module types, also the proneness of the respective module to PID is examined.

A simple testing method is shown in Figure 9.23. The module front side is totally covered with a conductive aluminum foil to simulate the "worst case." Then a voltage of $V = -1000$ V is applied for 168 h (7 days) between the shunted module cables and earth potential. In the subsequent power determination under standard test conditions, no power degradation is allowed to occur.

As an example, Figure 9.24 shows the test results for two concrete modules. These stem from a PV plant with yield losses. The testing reveals a dramatic behavior. Module A already at the beginning has a power reduction of about 6% with respect to the nominal module power P_N. After application of the voltage of −1000 V, the module power reduces within 7 days to only 62% of P_N. For module B, the degradation starts already with 13% and increases in 1 week to 33%. However, after the subsequent reverse of the voltage polarity a "healing effect" can be observed. The power of both modules attains almost the initial value again.

Also interesting is the variation of the current/voltage curve under the influence of potential-induced degradation. For this, in Figure 9.25 the characteristic curve of module A before and after the application of −1000 V is depicted. The prior damage of the module can already be seen at the irregular curve at the start. Within 3 days, the fill factor decreases from 70% to 60% and ends up on the 7th day at 50%. The main reason

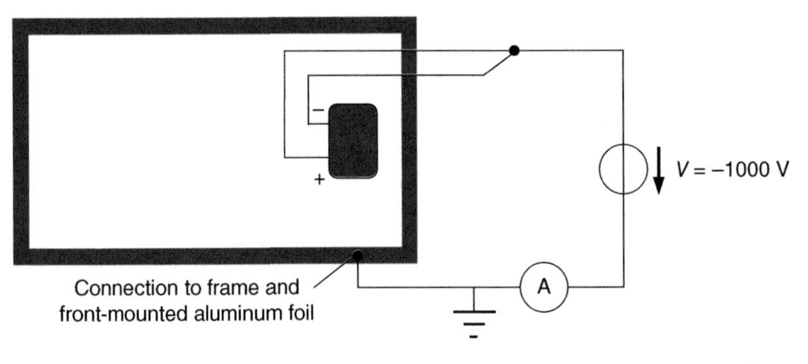

Figure 9.23 Setup to measure the PID effect: Between the cells and the module frame a voltage of $V = -1000$ V is applied. The current meter serves to determine the leakage current.

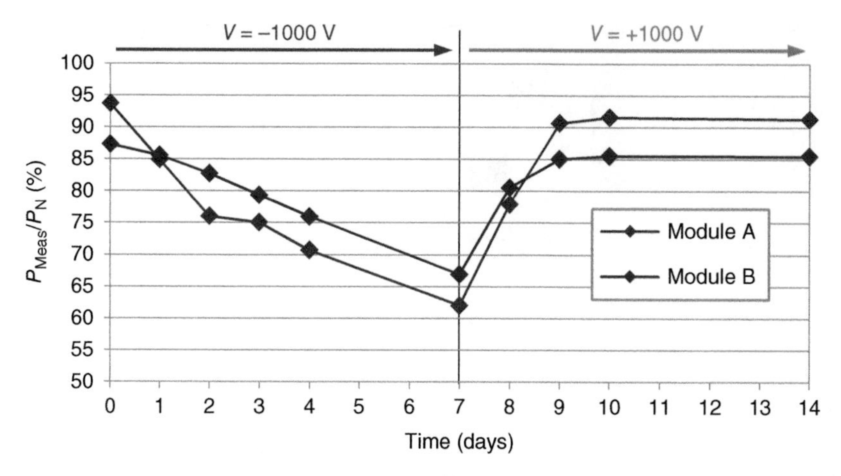

Figure 9.24 Result of a PID survey: At applied negative voltage, the modules show a considerable loss of power. After reversing the polarity, almost the original power sets in again.

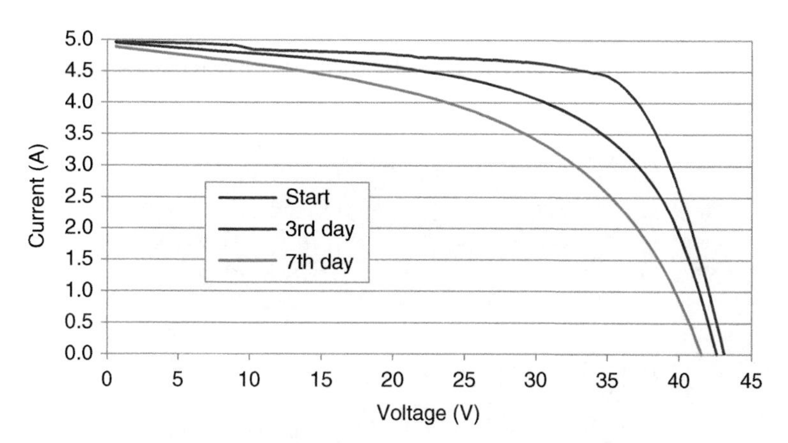

Figure 9.25 Influence of the PID effect on the *I*/*V* curve: The fill factor massively sinks, which is mainly caused by the reduction of the shunt resistance.

here is the shunting of the cells, which results in a reduction of the shunt resistance R_{Sh} (compare Figure 4.14).

9.6.3 EL Investigations to PID

Also in the case of PID, the EL measurement is very helpful. The shunting of the involved cells causes a lower brightness in the EL picture. This effect is shown in Figure 9.26. After PID influence over a week, several cells become almost totally dark (central module). Further, dark stripes and flaws appear. After applying a positive voltage over several days, practically the original picture can be seen again.

Even more relevant for practice is the PID at-site survey of modules with the help of outdoor EL measurement technique. As an example, Figure 9.27 shows EL exposures of energized strings of an on-roof plant. The bright shining modules are located at the positive end of the string. These are not affected by PID, as in normal operation almost no

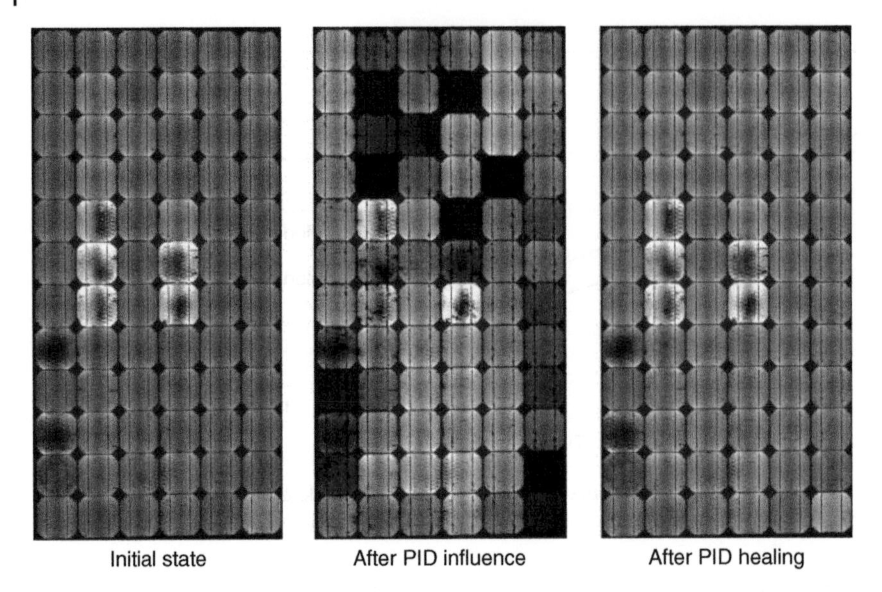

Figure 9.26 The EL exposures reveal the PID effect by clearly darkened cells, which is reversed after PID healing.

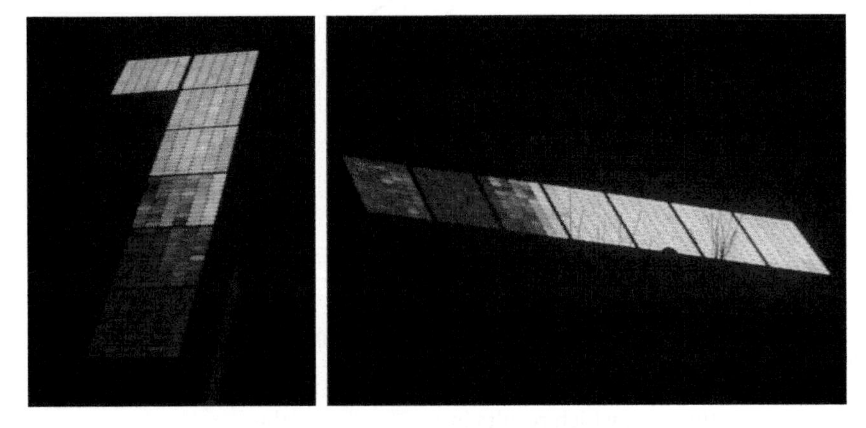

Figure 9.27 EL exposures of two strings affected by PID: The modules on the positive side of the strings show no anomalies, while the modules with stronger negative potential have the typical PID pattern. Photos: M. Diehl.

negative potential against ground is present. The situation is different for the modules at the negative side of the string. Their potential with respect to ground is (in normal power generation mode) the more negative, the more they are connected to the negative end of the string. This can clearly be seen at the modules getting darker and darker to the end.

Summarizing, it has to be admitted that the problems with potential induced degradation of solar modules are not fixed yet. Despite the progress in the explanation of the PID effect, even today PID-prone modules are still sold on the market. Here, only a consequent PID testing of newly offered module types can solve the problem. At the same time, with the outdoor EL method, a very revealing method on-site is available.

10

Design and Operation of Grid-connected Plants

This chapter deals with the steps required for erecting a photovoltaic plant. The important aspects will be discussed starting with the choice of location and suitability of the roof, through yield estimates up to choice of components, and plant installation. A further important subject is investment calculations for photovoltaic plants. Finally, we will deal with methods of monitoring plants and operating results of actual plants.

10.1 Planning and Dimensioning

This subject will place the emphasis on the selection of the site as well as the effects of shading. In addition, software tools for plant dimensioning and yield estimates will be presented.

10.1.1 Selection of Site

Before an investment decision for a photovoltaic plant is made, there should be an exact investigation of the boundary conditions at the planned site. An important criterion for the economic viability of the plant is the annual radiation at the site of the plant. As was seen in Chapter 2, this varies in Germany between 900 and 1150 kWh (m^{-2} a^{-1}). A second important factor for roof installations is the orientation of the roof and its pitch. Here, we determined that the optimum is a south-oriented roof with a pitch of approximately 35° (for Germany). Orientation of this roof to the southwest reduces the yield only by around 5%, but in the case of a west-facing roof, the losses are already almost 20% (see Table 2.4). A check must also be carried out whether shading occurs and what effect this will have on the yield of the plant.

A further cause of loss is module soiling. This seldom occurs for a module inclination from 30°, as rainwater brings with it sufficient self-cleansing. Typical losses then are between 2% and 3%. For flatter inclination angles, losses due to bird droppings, dust, and so on can easily reach up to 10%. For agricultural operations, the impairment from the dirt of stable ventilators can be even greater. In these cases, regular cleaning is recommended.

10.1.2 Shading

A shading analysis should be carried out in order to recognize possible shading and estimate its effects.

Photovoltaics – Fundamentals, Technology, and Practice, Second Edition. Konrad Mertens.

10.1.2.1 Shading Analysis

The simplest shading analysis is to stand at the installation site (on the roof) of the planned system and look to the east, south, and west to see if there are any positions of possible shading. If there is only a single object (e.g. a high tree), then its sideways position can easily be determined by means of a compass. The height angle γ_{Shade} of the shade object is calculated from the distance d and the height difference Δh to the roof center (see Figure 10.1):

$$\gamma_{\text{Shade}} = \arctan\left(\frac{h_2 - h_1}{d}\right) = \arctan\left(\frac{\Delta h}{d}\right). \tag{10.1}$$

This method is time-consuming in the case of several shading objects. A convenient aid is the Sun path indicator with transparent film (Figure 10.2a). This contains a compass with a spirit level for aligning the device as well as a wide-angle lens through which the surroundings are viewed. A marker is then used to draw the silhouette of the surroundings on the film. As the film also shows the angle of the Sun during the course of the year, a rough estimate can be made of when the shading is likely to occur. These coordinates of the objects of shading can also be incorporated into a simulation program that carries out a relatively accurate determination of the effects on the energy yield (see Section 10.1.3.2).

More elegant than manually drawing-in the shading horizon is automatic recording. In the right-hand picture, Figure 10.2 shows the SunEye device that photographs the surroundings using a fish-eye lens. The shading horizon is then determined semiautomatically. At the same time, with the use of a built-in GPS receiver, the device can determine the exact site and can thus carry out the site yield calculation directly.

Figure 10.3 shows the Sun path diagram known from Figure 2.12 with a shading horizon as an example.

Apparently, the house on the right only has a slight influence as it causes shading only during the winter months from 14:00 h. However, the tree on the left provides morning shade for almost 2 h from September to March.

10.1.2.2 Near Shading

The shading horizon is usually formed by far distant objects that often generate only a diffuse shade and do not lead to a full shading of the affected modules. Nearby objects such as roof dormers or chimneys are different as they form deep shading. A special

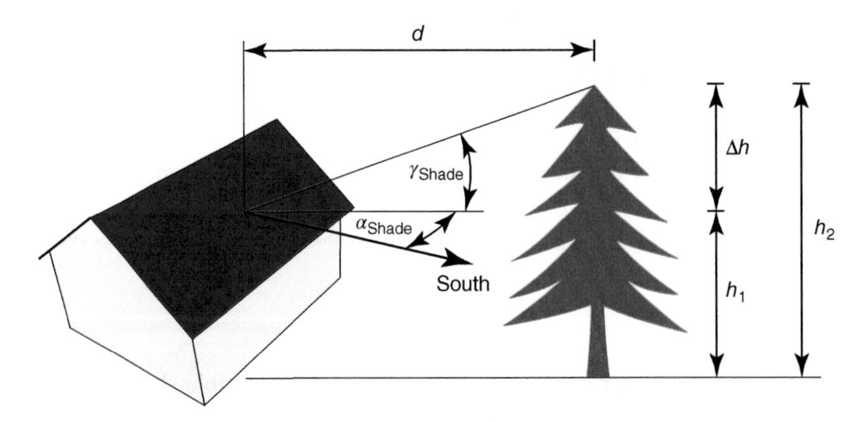

Figure 10.1 Determining the height angle of a shading object [22].

Figure 10.2 Two variants of the shading analysis: The panel (a) shows the manually operated Sun path indicator and the panel (b) shows the SunEye device that permits automatic acquisition of the shading horizon.

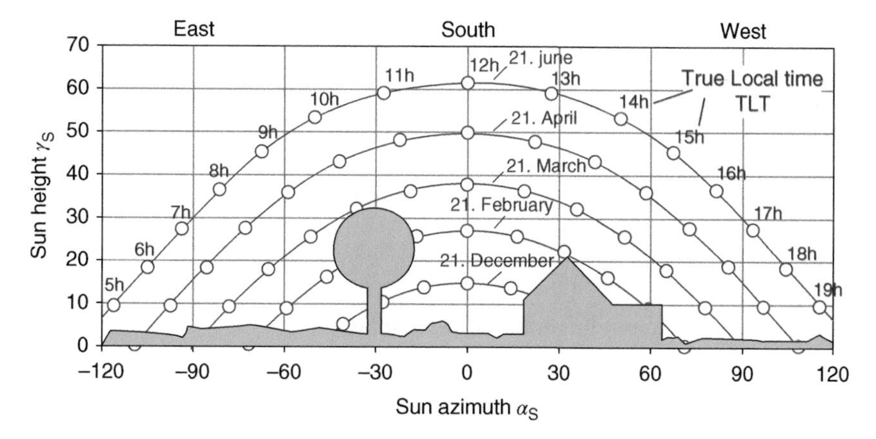

Figure 10.3 Example of a Sun path diagram with shading horizon: Whereas the house hardly has any effect, the tree provides daily morning shading from September to March.

case is caused by small objects such as aerial tubes, lightning protection rods, or overhead lines. Whether these cause deep shading depends on the respective distances. Figure 10.4 explains the correlation on the basis of a sketch.

In the upper picture, an aerial tube is positioned relatively near the solar module. The result is broad deep shading that can lead to clear yield impairment. If the aerial tube is further away, then the width of the deep shading on the module is reduced. In the lower picture, the aerial tube is just far enough away that there is no more deep shading. This minimal distance $r_{\text{Shade\_Min}}$ should be adhered to as far as possible. It can be simply calculated with the aid of the intercept theorem:

$$\frac{r_{\text{Shade\_Min}}}{d_{\text{Shade}}} = \frac{r_{\text{SE}} + r_{\text{Shade\_Min}}}{d_S} \approx \frac{r_{\text{SE}}}{d_S}. \tag{10.2}$$

With the information for r_{SE} and d_S from Table 2.1, one can then derive an approximation equation for the minimum distance $r_{\text{Shade\_Min}}$ of the shading object of solar modules:

$$r_{\text{Shade\_Min}} = \frac{r_{\text{SE}}}{d_S} \cdot d_{\text{Shade}} = \frac{149.6\,\text{M km}}{1.393\,\text{M km}} \cdot d_{\text{Shade}} = 107 \cdot d_{\text{Shade}}. \tag{10.3}$$

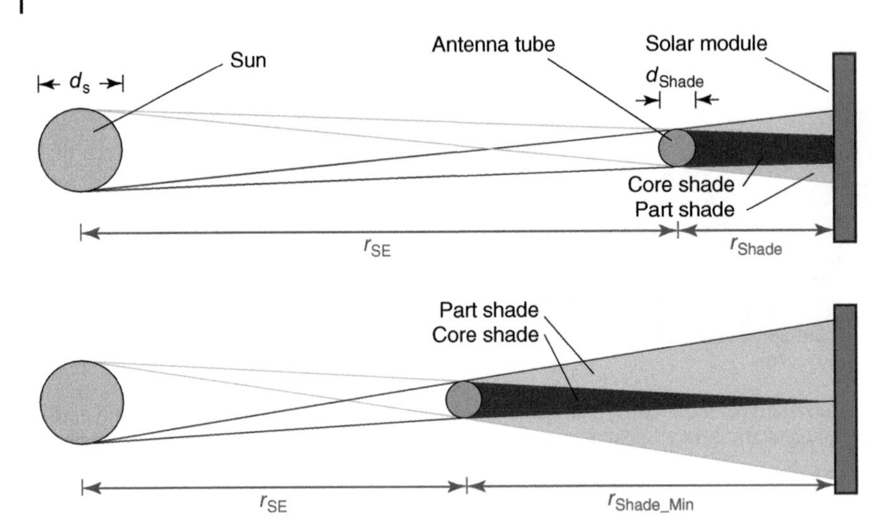

Figure 10.4 Problem of near shading: At short distances of the shading object, there is a disadvantageous deep shade. The lower picture shows the minimum distance r_{Shade_Min}, at which no deep shading will occur [22].

Example 10.1 ***Prevention of deep shading***

Lightning protection rods of 1 cm are required in front of the installation. The minimum distance for preventing deep shading is

$$r_{Shade\_Min} = 107 \cdot d_{Shade} = 107\,\text{cm}.$$

The distance of 1.07 m should not be undercut if possible. ■

10.1.2.3 Self-shading

In the case of flat-roof or freestanding plants, it is possible that the rows of modules shade each other. In order to prevent this, certain minimum spacings must be adhered to.

A rule of thumb here is that at noon on December 21 (in the Northern Hemisphere), there must not be any shading. We can see from Figure 10.3 that this requires an angle of the Sun γ_S of 15°. The case is shown in Figure 10.5 in which the minimum spacing of the

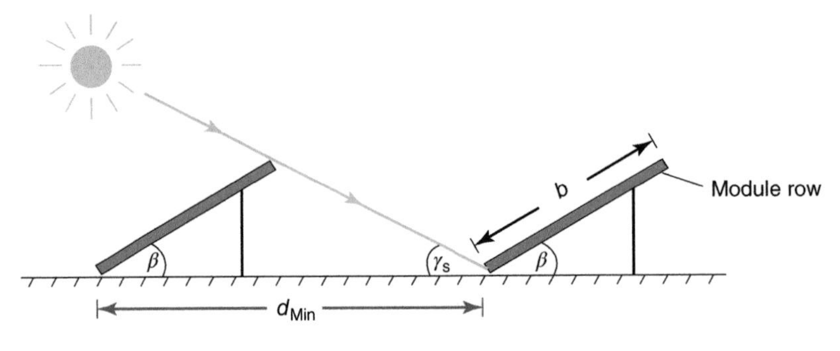

Figure 10.5 Prevention of self-shading: The spacing d_{Min} is the result of the requirement that there must be no shading even on the shortest day in the year (midday).

modules can be determined for a given angle of the Sun and defined module inclination angle.

The minimal module spacing d_{Min} in this case can be found by means of a similar calculation to that shown in Figure 2.15.

$$d_{\mathrm{Min}} = b \cdot \frac{\sin(\gamma_S + \beta)}{\sin \gamma_S} \tag{10.4}$$

with b: module width.

Unfortunately, because of this condition, only a part of the available area can be used. For this purpose, we define the area utilization ratio f_{Util} that gives the relationship of module width b and module spacing d:

$$f_{\mathrm{Util}} = \frac{b}{d}. \tag{10.5}$$

In order to accommodate as much photovoltaic power on a flat roof surface as possible, the optimum module inclination angle of 35° is often changed, for instance, to 20°.

Example 10.2 ***Area utilization ratios for various inclination angles***

Two variations of flat-roof plants are planned: For a module width of 1 m in the first case, the inclination angle is 35° and 20° in the second case. What is the area utilization ratio for an angle of the Sun of 15°? ■

In the first case, Equation (10.4) gives a minimum spacing of 2.96 m and in the second case $d_{\mathrm{Min}} = 2.26$ m. The result of this is an area utilization ratio of 34% and 44%, respectively. Therefore, in the second case, approximately 29% more photovoltaic power can be installed. The achievable energy yield per module area was only reduced by 2% (see Table 2.4).

10.1.2.4 Optimized String Connection

Section 6.2 showed that the effect of shading could be reduced by suitable string connections. We found there that simultaneously shaded modules should be connected into one string if possible. Figure 10.6 shows the application of this rule for two cases. The left picture shows a chimney, the shade of which wanders from left to right over the plant in the course of the day. Usually, only one string is affected by the shading with vertical connections of the modules. In the picture on the right, we see the slanted shading of a roof dormer that occurs in the late afternoon with the Sun low in the horizon. Here, the modules are best connected diagonally in order to limit the effects.

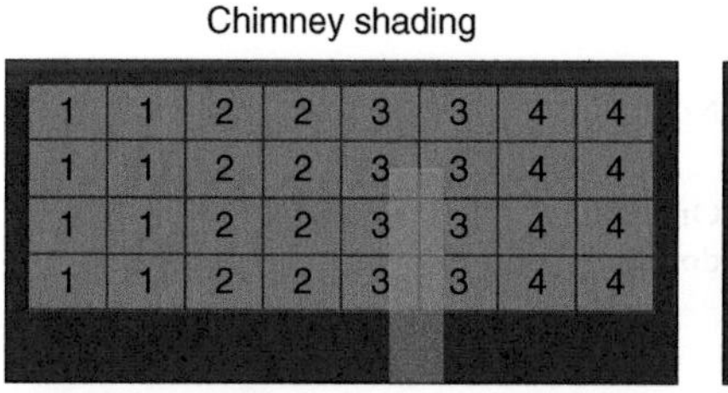

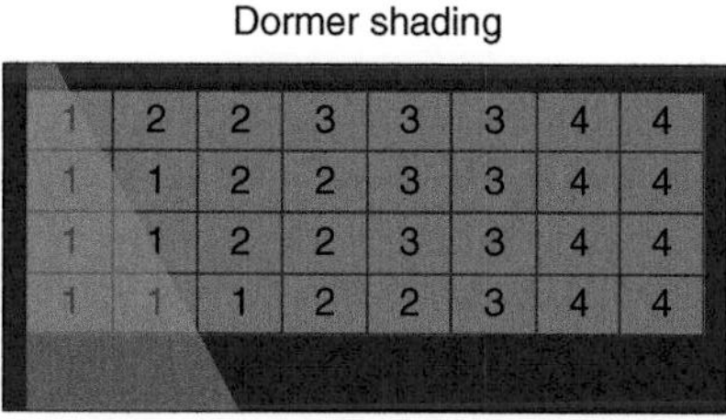

Figure 10.6 Optimized string connection for two cases of shading: Whereas the vertical chimney shading requires a vertical module connection, in the case of the slanted shading caused by the dormer, this is better carried out diagonally [118].

10.1.3 Plant Dimensioning and Simulation Programs

10.1.3.1 Inverter Design Tools

In Section 7.2, we dealt with the dimensioning of inverters and became acquainted with various rules for plant design. Nowadays, practically all inverter manufacturers offer design tools for optimum combination of solar generators and inverters. Table 10.1 lists the tools of five manufacturers.

Besides the actual inverter dimensioning, some of the programs also provide aids to cable-loss calculations. Further, some tools also comprise simple yield estimates but of different qualities. For exact yield prognosis, it is better to turn to professional time-step programs as described in the following sections.

10.1.3.2 Simulation Programs for Photovoltaic Plants

An accurate yield prognosis for an actual plant at a particular site is best achieved with a professional simulation program. The program typically works in the time-step method in which the radiation (e.g. in the three-component model, see Chapter 2) is calculated for the whole year in minute or hour grids. At the same time, the temperature coefficients and the weak light behavior of the solar modules are entered in the current and voltage calculation on the DC side. The efficiency curve of the inverters and cable losses are also taken into account in order to simulate the annual power fed into the public grid.

Table 10.2 shows the properties of a selection of the simulation programs available in the market. Besides two commercial programs, it also contains free versions.

It is important for the accurate simulation results to have a large number of weather data sets in order to represent the situation at the planned site as accurately as possible.

Table 10.1 Summary of the dimensioning tools of various inverter manufacturers.

Name	String sizing tool	PV-size	Solar configurator	Solar info design	Sunny design
Manufacturer	Kaco New Energy GmbH	Power One Inc.	Fronius International GmbH	Sungrow Power Supply Co.	SMA AG
Inverter dimensioning	Yes	Yes	Yes	Yes	Yes
Yield estimate	No	Only coarse	No	Only coarse	Yes
Cable loss calculation	No	No	Yes	Yes	Yes
Self-consumption consideration	No	No	No	No	Yes
Storage use consideration	No	No	No	No	Yes
Available as	Online tool	Download	Online tool/ download	Download	Online tool/ download
Web address	www.kaco-newenergy.com	www.power-one.com	www.fronius.com	www.sungrowpower.com	www.sma.de
Remarks	—	Cumbersome user guidance	—	—	Good user guidance

Table 10.2 Collation of simulation programs for yield prognosis.

Name	PV-Sol premium	PV Scout premium	Greenius Free	RETScreen	PVGIS
Number of weather datasets	More than 8000 worldwide	900	11	6700	From satellite pictures
Number of module datasets	13 000	50.000	25 (editable)	—	—
Time resolution of simulation	1 h, 1 min	1 min	1 h	—	—
Shading simulation	Horizon and near objects, 3D visualization	Near objects (only visualization)	Horizon	No	Horizon
Self-consumption consideration	Yes	Yes	No	Yes	Limited
Storage use consideration	Yes (including electric cars)	Yes	No	No	No
Suitable for island plants	Yes	No	No	Yes	Limited
Investment calculation	Yes	Yes	Yes	Yes	No
Price	1300€	500€	Free	Free	Free
Web address	*www.valentin-software.com*	*www.solarschmiede.de*	*freegrenius.dlr.de*	*www.retscreen.net*	*http://re.jrc.ec.europa.eu/pvgis*
Remarks	Very comprehensive	Export of part lists and connection plans	Also for wind and solar thermal	Excel-based, also for wind, water, etc.	Online-tool

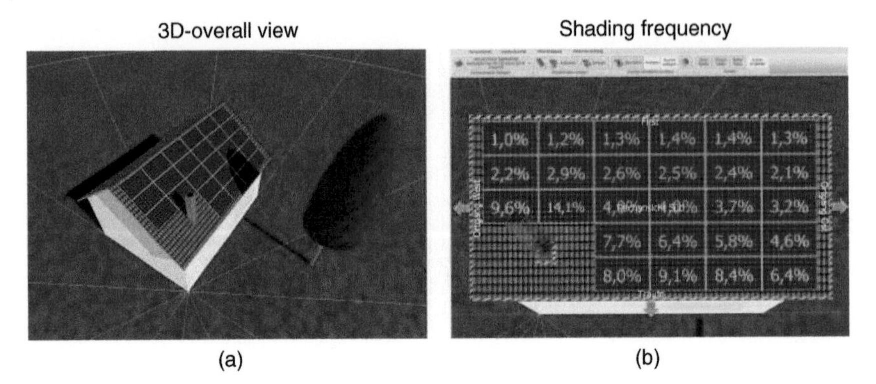

Figure 10.7 (a) Shading simulation with PV-Sol: The three-dimensional depiction permits visualization of the shading for each time of the year. The panel (b) shows the frequency of shading of the individual modules.

Besides the existing datasets, some programs provide the possibility of entering the data from additional sites.

There are many different ways of dealing with shading. Especially elegant is the three-dimensional plant depiction of the PV-Sol program with which the shading can be seen very clearly. Figure 10.7 shows on the left the three-dimensional depiction of an on-roof installation, shaded by a chimney and a tree. The example shows the shading on March 15 at 12:30 in Berlin: Because of the low position of the Sun in March, this causes heavy shading. The effect on the yield can be seen in the screenshot in the right-hand picture: Here the frequency of the shading is shown for every module over the year. The two modules above the chimney are the most affected with frequencies of 10.6% and 14.1%. This depiction can then be used to create a shade-optimized string connection and then to calculate the yield. For a simulation that is as near to reality as possible, it is able to provide information on how many bypass diodes are contained in each module per type of module.

Besides the simulation programs, photovoltaic online databases are also available for forecasting the yield of a planned installation as accurately as possible. Thus, for instance, under *www.sunnyportal.com*, yield data of approximately 50 000 photovoltaic plants worldwide are freely available. There it is possible to find average yields of plants in one's own region, which is a good estimation of the achievable yield.

Further online databases as well as links to solar maps and yield simulation programs are available in the Appendix. Besides this, there is a checklist that describes what is to be taken into account before and during the installation of a photovoltaic system.

10.2 Economics of Photovoltaic Plants

The economic operation of photovoltaic plants has only become possible in Germany with the Renewable Energy Law (Erneuerbare-Energien-Gesetz: EEG). For this reason, we will first look in more detail at the framework conditions of this law. Then we will look at a method for profit calculation with which the economics of the investment in a photovoltaic plant can be determined.

10.2.1 The Renewable Energy Law

The EEG was enacted by the German Bundestag in 2000. For the first time, it provided a country-wide regulation of cost-covering compensation of energy from photovoltaic plants. The aim of the EEG is the provision of a continuous demand for photovoltaic plants in order to achieve cost reductions through mass production. The result was very impressive: Within the first 10 years, the prices for photovoltaic plants were reduced by approximately 70%.

The amount of the compensation depends on the year of erection of the plant and remains constant for 20 years. A plant that is built in the next year receives less compensation that again remains constant for 20 years. The compensation regression of the EEG that was only 5% per year in its early years has been substantially increased as the prices of plants have been driven down overproportionally.

10.2.2 Return Calculation

If one asks various providers of photovoltaic plants what the profit is, one receives greatly varying numbers as an answer. On the one hand, this is because different annual energy yields are assumed. On the other hand, the cause is often in the definition of "return." Thus, in some calculations, possible savings in tax are taken into account. Often it is assumed that the plant is in large part financed with a credit and only the return on equity is given in which the total return only refers to the own capital that was used. We want to make it simpler for ourselves but still achieve a high degree of transparency and traceability.

10.2.2.1 Input Parameters

The input parameters of any investment calculation are the investment costs K_0 for the construction of the photovoltaic plant, the annual operation costs K_{Oper}, and the expected annual income K_{Inc}. In the investment costs and all other amounts, use is made of the net values as the VAT is only a posting that is offset with the finance department in the following year.

The following applies for the annual operation costs K_{Oper}:

The annual operating costs K_{Oper} (including service costs) are typically taken as 1.5% of the investment costs of a photovoltaic plant.

The operating costs include expenditures for insurance, meter costs, electricity costs of a data logger, and so on. Besides this, in the lifetime of a plant, one must take defects, especially of the inverter, into account. The calculated operating costs are also to be seen as reserve for repairs and should not be less than described above for the investment calculation.

The annual income depends on the amount of the feed-in tariff k_{EEG} and the achieved annual energy yield W_{Year}:

$$K_{\text{Inc}} = k_{\text{EEG}} \cdot W_{\text{Year}}. \tag{10.6}$$

10.2.2.2 Amortization Time

The simplest calculation model is the view of the amortization time. This is understood to be the time it takes to recoup the capital expended. In the years after that, one is in the profit zone. The amortization time T_{Amort} is obtained by the division of the investment sum K_0 by the annual surplus $K_{Surplus}$ that is further understood to be the difference between the annual income and operating costs:

$$T_{Amort} = \frac{K}{K_{Inc} - K_{Oper}} = \frac{K_0}{K_{Surplus}}. \tag{10.7}$$

Example 10.3 ***Amortization time of a 5 kW plant***

Susie Sunny purchases a 5 kW plant at a price of 7000€. She calculates conservatively with a specific yield w_{Year} of 850 kWh/(kWp a). She receives 12 cents/kWh as feed-in compensation. ■

The annual income is

$$K_{Inc} = w_{Year} \cdot P_{STC} \cdot k_{EEG} = \frac{850\,\text{kWh}}{\text{kWp a}} \cdot 5\,\text{kWp} \cdot 12\,\text{cent/kWh} = 510€/\text{a}$$

With the operating costs of 105€, this leads to an annual surplus of 405€. Thus the amortization time is

$$T_{Amort} = \frac{K_0}{K_{Surplus}} = \frac{7000€}{405€/\text{a}} = 17.3\,\text{a}.$$

If one calculates this example with an annual specific yield of 900 kWh/kWp, then there is a better amortization time of 16.1 years. With an operating time of feed-in compensation of 20 years, Ms. Sunny therefore has only 4 years of making a profit.

10.2.2.3 Property Return

Although the amortization time is a very clear measurement, it does not take into account the interest of the capital used. For Ms. Sunny, the 510€ that she receives after the first year is to a certain extent more valuable than the amount of the second year as she can invest the money in the bank again and collect interest on it.

For the sake of clarity, we will present two investors: Charles Cash and Susie Sunny. Mr. Cash has 7000€ available that he invests with his bank at fixed interest p for 20 years. At the end of each year, he receives interest that is immediately invested again. With the compound interest rate equation, one can now calculate how much money K_n he will have at the end of the n years:

$$K_n = K_0 \cdot (1+p)^n = K_0 \cdot q^n \tag{10.8}$$

with q: interest factor: $q = 1 + p$.

Assume, for instance, an interest rate of 3%, then after 20 years Mr. Cash will have a sum of

$$K_{20} = K_0 \cdot (1+p)^{20} = 7000€ \cdot 1.03^{20} = 12\,643€. \tag{10.9}$$

Ms. Sunny also has 7000€ available and invests the money in a photovoltaic plant. Every year she receives an income from the power feed-in and invests this money, less the deductions of the operating costs, also at an interest rate of p at the bank. Here, too, we can work out how much money she has saved after 20 years:

$$K_{20} = K_{\text{Surplus}} \cdot (q^{19} + q^{18} + q^{17} + \cdots + q^1 + q^0). \tag{10.10}$$

The sum in Equation (10.10) can be simplified with the help of the mathematically well-known geometric series so that we finally obtain the so-called savings bank equation:

$$K_{20} = K_{\text{Surplus}} \cdot \frac{q^{20} - 1}{q - 1}. \tag{10.11}$$

Assume that for Ms. Sunny, there is again an annual surplus of 405€ so that her income after 20 years is

$$K_{20} = 405€ \cdot \frac{1.03^{20} - 1}{1.03 - 1} = 10\,883€. \tag{10.12}$$

In 20 years, Ms. Sunny thus has earned less money than Mr. Cash.

The property return is the interest that must be inserted in the Equations (10.8) and (10.11) so that in both cases, the same money amount is calculated. In other words:

We compare the investments in a photovoltaic plant with a money investment in a bank and define the appropriate rate of interest that applies to the same end amount as property return of the photovoltaic plant.

If we carry out this comparison for Ms. Sunny's plant, then we get a property return of 1.4%. In Figure 10.8, the capital development for this case of Charles Cash and Susie Sunny are compared. After 20 years, both have about 9300€.

The return given in the example of 1.4% is quite low.

In the EEG, the feed-in tariff originally depended on the actual prices of photovoltaic plants with a desired property return of 7.4%. Those high values, nowadays, only can be achieved with high self-consumption rates (Section 10.2.2.4).

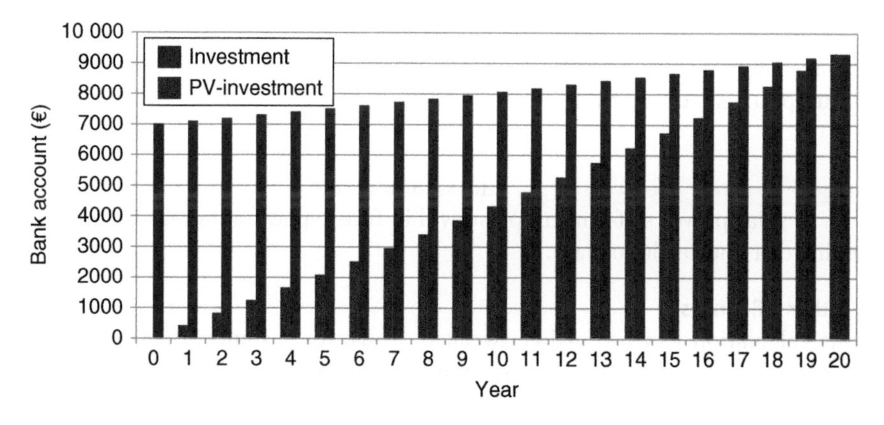

Figure 10.8 Explanation of property return: The investment in a photovoltaic plant is compared with an investment of money in which an amount is invested with compound interest for 20 years [119].

A PV installer has told me that one can get a return on equity of more than 10% with a photovoltaic plant. What does he mean by this? Isn't this in contradiction with the previously mentioned returns?

The term return on equity refers to the return of a plant with the capital used. Assume that Ms. Sunny has a top site in south Germany, and thus achieves a property return of 5%. In order to finance this she takes out a credit of 6000€ at a bank with a rate of interest of 3%. The remaining 1000€ is her own resources. With the borrowed 6000€, she achieves a profit of around 2%. The return on equity refers to the return on the 1000€ of own capital and achieves, for instance, a value of 12%.

The result of this calculation depends greatly on the assumptions. If Ms. Sunny were to take up the whole investment sum as credit, then the return on equity is infinite. This shows that this return on equity is no sensible measure on which an investment decision can be based.

10.2.2.4 Profit Increase Through Self-consumption of Solar Power

As already described in Section 8.3, the profit of a photovoltaic plant can be increased in that the largest possible portion of the energy generated is used by oneself. If Susie Sunny and her family have the typical energy demand of 4500 kWh per year, then she can assume a self-consumption rate a_{Self} of approximately 30% [104]. We will assume the price of the energy taken from the grid to be $k_{\text{Taken}} = 28$ cent/kWh. Thus, the calculation of Example 10.3 can now be modified:

Example 10.4 ***Amortization time of a 5 kW plant with 30% self-consumption***

We will first derive an "Average compensation" k_{Aver} from the two current parts:

$$k_{\text{Aver}} = a_{\text{Self}} \cdot k_{\text{Taken}} + (1 - a_{\text{Self}}) \cdot k_{\text{EEG}} = 0.3 \cdot 28 \ \text{ct/kWh} \\ + 0.7 \cdot 12 \ \text{ct/kWh} = 16.8 \ \text{cent/kWh}.$$

If we calculate Example 10.3 with this feed-in tariff, then the amortization time with the self-consumption is reduced from 17.3 to 11.5 years. The return increases from 1.4% to 6%. ■

With further decreasing feed-in tariffs and simultaneously increasing power consumption costs, solar power self-consumption will in future play an increasing role in the profitability of a photovoltaic plant.

10.2.2.5 Further Influences

The previous calculation can be further tweaked. Thus, for instance, one can integrate the general price increase on the assumption of an increase of almost 2% per annum.

Besides this, there is the question whether the modules will perform at full power also after 20 years. The producers of the modules generally give a power guarantee of 90%

within the first 10 years of operation and 80% on the nominal module power within 20 years. Thus, one can factor in an assumption of a conservative annual performance degradation of 1% in the expected feed-in tariff. On the other hand, experience with c-Si standard modules shows that the degradation is more likely to be 0.5% per annum so that this value should be sufficient.

Finally, it should not be forgotten that a photovoltaic plant is not valueless after 20 years. The components can fulfill their tasks for many more years. If the plant continues to be connected to the grid, it can reduce the own consumption costs. Further, it cannot be ignored that certain compensation will be paid also after 20 years, as macroeconomically it will still be more advantageous than, for instance, building new coal-fired power stations. An Excel program is available under *www.textbook-pv.org* with which simple profit calculations can be carried out in a traceable manner.

10.3 Surveillance, Monitoring, and Visualization

What do the three terms in this heading mean? Plant surveillance is an alarm system that gives early warning of failures or largely reduced yields. Plant monitoring goes beyond this aim: It is meant to compare the actual performance of a plant with other plants and in the best case give information on how the plant yield can be further increased. Plant visualization, on the other hand, is thought of as a descriptive information on the current status of the plant. In the following, we will look at the respective aspects of the three measures.

10.3.1 Methods of Plant Surveillance

The simplest type of plant surveillance is to occasionally look at the display of the inverter. This should show that the inverter is feeding the grid and that the plant is running in maximum power point (MPP) operation.

More accurate information is obtained when the generated energy is read off at the end of the month or the week and then compared with the yields of other plants. If the yield of the own plant is much lower, then the cause must be searched for. Nowadays, there are a series of online databases in which the yields of thousands of plants are available on a daily basis. Much more elegant than a manual comparison is the use of a data logger that compares the own daily yields with similarly oriented plants. If there is a large difference then, for instance, an SMS alarm message is output. The data logger's Internet connection has the additional advantage that the data is regularly saved on a central server and can even be called up years later. Examples of these types of databases are *www.sunnyportal.com* as well as *home.solarlog-web.net*.

An alternative for checking the plant based on reference plants is the immediate comparison of the current plant performance with the present radiation. For this purpose, a solar cell radiation sensor is mounted at the generator level as described in Section 8.1. Besides this, the measurement of the module temperature is helpful in this way to obtain a preferably good estimate of the expected plant power. Various commercial data loggers offer this special function and can also trigger an alarm per e-mail, SMS, or signal horn.

10.3.2 Monitoring PV Plants

10.3.2.1 Specific Yields

The yield of a photovoltaic plant depends primarily on the radiation energy H_G arriving at the generator. The observation time selected is usually a year, but sometimes a month or even a day. The reference yield Y_R is defined as (Figure 10.9):

$$Y_R = \frac{H_G}{E_{STC}}. \tag{10.13}$$

With this, one references the radiation energy to the full irradiance of the Sun under standard test conditions (STC). The radiation energy H_G is given in kWh/m^2, and if this is divided by 1000 W/m^2, the result is the unit hours. The reference yield Y_R thus gives the number of hours during which the Sun would have to shine with full force on the solar generator in order to generate the energy H_G. Similar to that described in Chapter 2, we call this the Sun full load hours:

The reference yield Y_R gives the number of Sun full load hours at the generator level per year.

Unfortunately, losses occur in a real photovoltaic plant so that the reference yield cannot be completely used for the production of electricity. This is shown clearly in Figure 10.10.

The result of the generator losses L_C (capture losses) means that only the generator yield Y_A (array yield) remains at the output of the solar generator.

$$Y_A = \frac{W_{DC}}{P_{STC}}. \tag{10.14}$$

In the following inverter including cables, there are also system losses L_S that finally lead to the end yield Y_F (final yield).

$$Y_F = \frac{W_{AC}}{P_{STC}}. \tag{10.15}$$

Both the generator yield and the final yield are referenced to the STC power of the plant so that here, too, the result is the unit h.

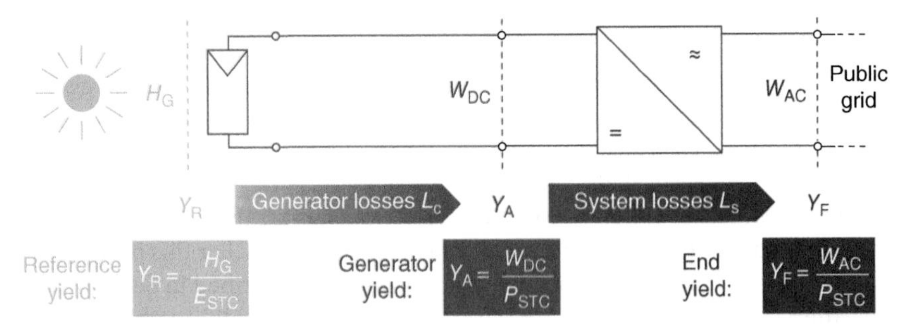

Figure 10.9 Specific yields of a photovoltaic plant: The generator yield Y_A is less than the generator loss and lower than the reference yield Y_R; the end yield Y_F again corresponds to the generator yield lessened by the system losses.

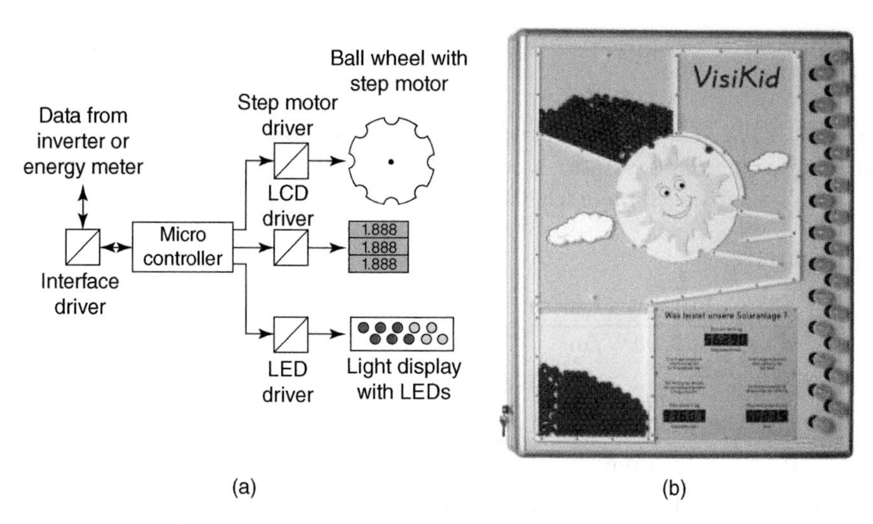

Figure 10.10 A child-centric visualization of a photovoltaic plant with VisiKid: Panel (a) is the internal structure of the prototype developed by the Münster UAS; panel (b) is a view of the commercial series device. Photo: IKS Photovoltaic GmbH.

The final yield Y_F gives the number of plant full load hours per year.

Naturally, this quantity is the most important size for the owner of the photovoltaic plant as this is used to determine how much energy is fed into the public grid per year. Up to now, we have also designated this as the specific annual yield w_{Year}.

10.3.2.2 Losses

How are these losses caused? There are many causes for generator losses L_C:

- The modules have less power than is stated in the data sheet.
- The module temperature is higher than 25 °C.
- The modules are soiled or partly shaded.
- There is a mismatch between modules of a string.
- The modules are not being operated in MPP.
- There are ohmic losses in the DC lines.

The system losses L_S are primarily caused by the inverter:

- The efficiency of the inverter is less than 100%.
- The inverter is underdimensioned and limits the output power during high input power.
- There are ohmic losses in the AC lines.

10.3.2.3 Performance Ratio

If the desire is to determine the efficiency with which the photovoltaic plant handles the available radiation energy, then the performance ratio (PR) can be used as a measure. This compares the final yield with the reference yield:

$$\text{PR} = \frac{Y_F}{Y_R}. \tag{10.16}$$

The performance ratio is typically between 75% and 85% and with very good plants, it can even reach higher values.

Example 10.5 ***Yield and performance ratio of a 5 kWp plant***
The annual radiation of a south-oriented roof with a pitch of 25° was 1100 kWh/m² in a year. The system installed there fed 4500 kWh into the grid in the same year. ■

The reference yield is

$$Y_R = \frac{H_G}{E_{STC}} = \frac{1100\,\text{kWh}/(\text{m}^2\,\text{a})}{1000\,\text{W}/\text{m}^2} = 1100\,\text{h}/\text{a}.$$

The final yield is

$$Y_F = \frac{W_{AC}}{P_{STC}} = \frac{4500\ \text{kWh}/\text{a}}{5\ \text{kWp}} = 900\ \text{h}/\text{a}.$$

The performance ratio is the quotient of the two results:

$$\text{PR} = \frac{Y_F}{Y_R} = \frac{900\,\text{h}/\text{a}}{1100\,\text{h}/\text{a}} = 81.8\% \approx 82\%.$$

10.3.2.4 Concrete Measures for Monitoring

The minimum demands of power monitoring are the acquisition of the radiation and the power fed into the grid. Here Y_R and Y_F are known, and the performance ratio as well as the total losses $L_{Tot} = L_C + L_S$ can be determined. However, proper information on improving the performance is obtained by means of a more accurate analysis. For this purpose, the data logger should also log the energy W_{DC} generated by the generator. The aim of this is to determine the generator and system losses, and the logger can provide information on whether there is improvement potential on the DC or the generator side. In addition, in the case of a malfunction, the error source can be quickly found. The acquisition of the module temperature also offers the possibility of dividing the generator losses into temperature-caused losses and other losses. In Section 10.4, we will look at simple monitoring results.

10.3.3 Visualization

A photovoltaic system installed on a roof is relatively unspectacular and can often not be seen, especially on office buildings. However, some undertakings, especially public bodies, want to show off their dedication to photovoltaics. Here through visualization, the data can be used. The standard depiction is a digital display with kilowatt hour generated as well as the current plant power. More informative is the use of a graphic display that shows the daily or monthly yield as a bar diagram.

In particular, schools and kindergartens like to show the power production in a way more appropriate for children. Unfortunately, the normal displays of power and energy are not very understandable for children (and parents…). This has led the Münster UAS to develop the visualization unit VisiKid [120]. Figure 10.10 shows the inner structure of the prototype as well as a view of the licensed series device. The power of the photovoltaic plant is shown for children by the number of illuminated bulbs, which are in fact highly efficient LEDs. At the same time, the display of the generated energy is shown by means of balls that are moved by a ball wheel. Each ball deposited into the bottom container depicts a fed-in kilowatt hour. At the end of the calendar month, the balls are poured back into the top container and the ball count starts at the beginning again. The number of devices sold so far shows that there is a considerable demand for this type of visualization.

10.4 Operating Results of Actual Installations

We will now look at the operating experiences of some photovoltaic plants.

10.4.1 Pitched Roof Installation from 1996

As a first example, we will consider a 2 kW system that was installed in Aachen in 1996. Aachen was one of the first cities in Germany that paid cost-covering feed-in tariffs for photovoltaic plants. The tariff at that time was 2 DM (about 1€) per kWh. Table 10.3 shows the most important data of the installation.

Figure 10.11 shows the final yield and the performance ratio of the installation in the course of the years. It is noticeable that in the first years, there were relatively strong fluctuations in the yield. These were caused mainly by considerable differences of the sun hours of these years. A specialty was the year 2003 in which Aachen yielded an annual global radiation of 1200 kWh/(m^2 a). For the orientation of the modules, this resulted in a reference yield Y_R of 1350 Sun full load hours. Also noticeable is the year 2001: In the summer of this year, a fuse in the inverter line of an inverter burned through. The damage went unnoticed for 2 months and caused a drastic reduction of the yield.

Much more informative than the final yield is the course of the performance ratio that in most years was between 70% and 75%. Apparently, we are dealing here with only a mediocre performing installation. The reasons here are two special aspects. On the one hand, the type of inverter used, the Sunny Boy 700 is one that is superseded by today's standards. Its peak efficiency is only 93.4% and the European efficiency is a poor 92%. Moreover, the PV generator – as was normal at the time – was clearly overdimensioned. Eighteen modules of 55 Wp were connected to an inverter with a maximum DC input power of 800 W. This corresponds to an overdimensioning factor k_{Over} of 1.24 (see Section 7.5).

Table 10.3 Data of the 2 kW plant in Aachen.

Site, year of construction	Aachen, 1996	Power (kWp)	1.98
Orientation	South	Solar modules	36 × Siemens SM 55
Module inclination (°)	45	Inverters	2 × SMA Sunny Boy 700
Number of strings	4	Overdimensioning factor k_{Over}	1.24
Length of string cable (m)	12	Cross section of string cable (mm^2)	2.50
Costs (DM/kWp)	15 000	Feed-in tariff (DM/kWp)	2

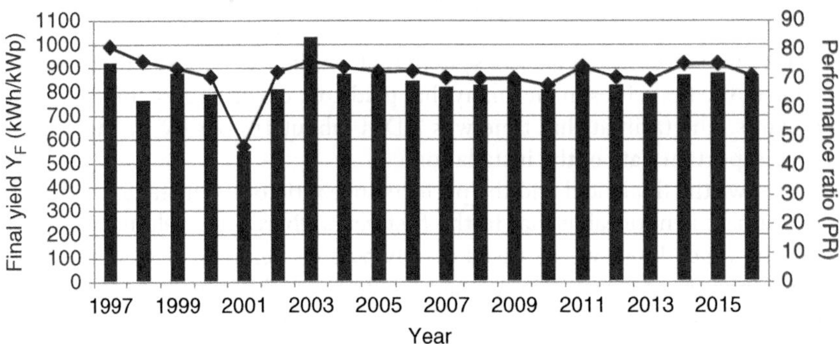

Figure 10.11 View and operating results of a 2 kW installation in Aachen: The poor yield in 2001 can be clearly seen and was caused by a burned-through fuse. Photo: M. Pankert.

In Figure 10.11, is seems as though the performance of the installation will continue to degrade. Is the reason for this a degradation of the modules?

Degradation after so many years of operation cannot be excluded. However, module soiling can also be a reason, as the installation was never cleaned.

10.4.2 Pitched Roof Installation from 2002

As a comparison, we will consider a 3.2 kWp installation that was installed in Steinfurt/Münsterland in 2002 (Upper modules in Figure 6.33). Table 10.4 shows the data for this plant.

Figure 10.12 shows the final yield and the performance ratio of the installation in the course of the years. In comparison with the installation in Aachen, the yield is somewhat better. In the first years, the performance ratio was an average of 4% above that of the other installation, and in the last years, this difference increased to about 8%. The reason was that in 2007, the installation was reduced by two modules, and the overdimensioning factor was reduced from 1.18 to 1.06.

Table 10.4 Data of the 3.2 kW plant in Steinfurt.

Site, year of construction	Steinfurt, 2002	Power (kWp)	3.2
Orientation	South	Solar modules	20 × Isofoton I-159
Module inclination (°)	48°	Inverters	1 × SMA Sunny Boy 2500
Number of strings	1	Overdimensioning factor k_{Over}	First 1.18, since 2007: 1.06
Length of string cable (m)	14	Cross section of string cable (mm^2)	6
Costs (€/kWp)	5200	Feed-in tariff (cents/kWh)	52.1

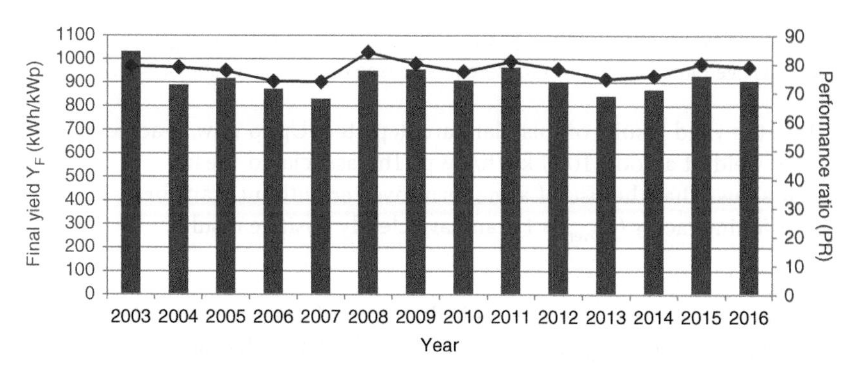

Figure 10.12 Operating results of the pitched roof plant in Steinfurt: In total, the PR values were somewhat higher in comparison to Figure 10.11. The installation was optimized slightly in 2007 so that the performance improved further.

10.4.3 Flat Roof from 2008

Finally, we will consider a modern 25 kWp plant that was erected in 2008 on the flat roof of the Münster University of Applied Sciences (see Figure 6.30). Table 10.5 shows the data for this plant.

Table 10.5 Data of the 25 kWp flat roof plant on the Münster UAS.

Site, year of construction	Steinfurt, 2008	Power (kWp)	24.84
Orientation	South	Solar modules	138 × Schüco S180-SP4
Module inclination (°)	25	Inverters	2 × SMA Sunny Boy 9000 TL 1 × SMA Sunny Boy 6000 TL
Number of strings	8	Overdimensioning factor k_{Over}	0.99 and 1.05
Length of string cable (m)	6–28	Cross section of string cable (mm^2)	6
Costs (€/kWp)	4496	Feed-in tariff (cents/kWh)	46.75

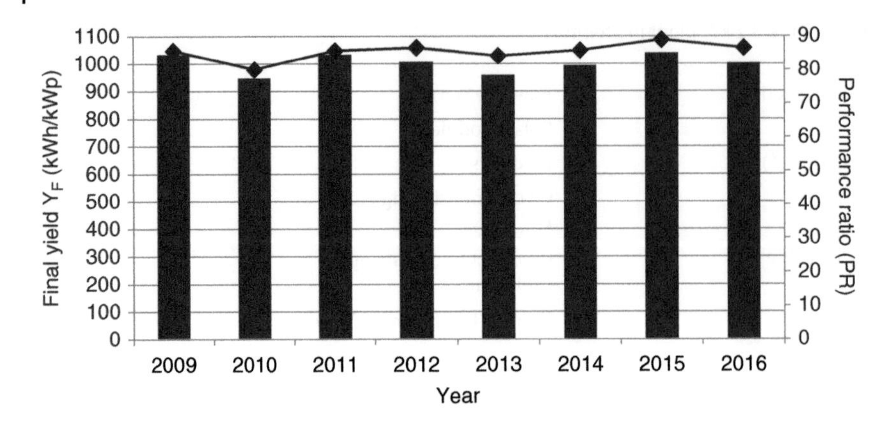

Figure 10.13 Results of a modern flat roof plant in Steinfurt: With the use of inverters without transformers and a low overdimensioning factor, there was a mean performance ratio of 84%.

In Figure 10.13, the yield results of this plant are depicted. Up to now, it achieved on average a specific yield of a good 1000 kWh/kWp. The associated mean performance ratio was 85%. It shows that the use of a modern inverter without transformer and a lower overdimensioning factor ($k_{Over} \approx 1$) can have clearly positive results.

In all three installations, it is noticeable that the performance ratio sinks from 2009 to 2010 and rises again in 2011. How can this be? In the calculation of the PR, the influence of the annual radiation is calculated out.

In addition to the annual radiation, there are also other influences to be taken into account. The temperature on the sunny days also plays a large role. Possibly, this was higher in 2010 than in 2009. The year 2011, again, had a sunny but relatively cool spring.

11

Future Development

This chapter deals with the possible future development of photovoltaics. First, we will look at the technical potentials of photovoltaics and consider the previous market and price development. Following, we learn to know today's power supply structure and the interplay of the different power plants.

Finally, some thoughts are given to how the future power supply with substantial participation of photovoltaics will look like. For this purpose, we consider various future scenarios and discuss the technical solutions for a complete conversion of the power supply system to renewable energies.

11.1 Potential of Photovoltaics

In Chapter 2, we estimated the area necessary for covering the whole of the primary energy needs of humanity by means of photovoltaics. The result was the Sahara miracle: A surface area of only 800 km × 800 km was required! As already mentioned, it makes no sense to concentrate the whole of the energy production at one site. Thus, we will limit the estimate of the potential to Germany.

11.1.1 Theoretical Potential

The theoretical potential is understood to be the whole of the radiation energy that reaches Germany in a year. In Chapter 2, we assumed an annual radiation of approximately 1000 kW h m^{-2}. With a surface area of the Federal Republic of Germany of 357 000 km^2, we obtain a theoretical potential of 375 10^{12} kW h. This corresponds to about 100 times the overall primary energy requirement of the country.

11.1.2 Technically Useful Radiation Energy

Naturally, only a small part of this large amount is usable. On the one hand, the whole of Germany should not disappear under a sea of solar modules, and on the other hand, solar modules only convert a part of the irradiance into electrical power. Possible sites for solar systems are especially roofs, facades, and free areas whose potential we will consider one after the other.

Photovoltaics – Fundamentals, Technology, and Practice, Second Edition. Konrad Mertens.

11.1.2.1 Roofs

According to [121], the whole roof surface area in Germany amounts to approximately 4345 Mio m^2 of which around 70% is pitched roofs and 30% is flat roofs. Of these, approximately 40% can be taken as being unsuitable for solar use [121] due to building restrictions (windows, chimneys, and shading).

In the case of pitched roofs, all those that have a maximum of 90° orientation turn to the south, meaning all roofs from east through south to west should be considered. Thus, only half the roof areas remain available. Also we want to reserve a third of the remaining surface area for solar heat so that there is an available pitched roof surface area A_{Pitch} for solar modules:

$$A_{\text{Pitch}} = 4345 \text{ Mio m}^2 \cdot 0.7 \cdot 0.6 \cdot 0.5 \cdot 2/3 = 608 \text{ Mio m}^2. \tag{11.1}$$

In the case of flat roofs, shading of the individual module rows must be prevented. We will, therefore, set the area utilization factor to about 45% in accordance with Example 10.2. With the reduction for solar heat described earlier, we will finally have available as usable module area A_{Flat} on flat roofs:

$$A_{\text{Flat}} = 4345 \text{ Mio m}^2 \cdot 0.3 \cdot 0.6 \cdot 0.45 \cdot 2/3 = 235 \text{ Mio m}^2. \tag{11.2}$$

In Chapter 2, we defined the annual radiation on a horizontal area in Germany has an average of 1000 kW h $(m^2\ a)^{-1}$. In the case of optimum alignment (azimuth angle $\alpha = 0°$, inclination angle $\beta = 35°$), the global radiation is approximately 1200 kW h $(m^2\ a)^{-1}$. The pitched roofs under consideration are angled from $\alpha = -90°$ to $+90°$, and we will assume a maximum inclination angle of $\beta = 60°$. According to Table 2.4, there are inclination losses of 0% to almost 30% so that we can set a mean of 15%. Finally, with Table 11.1, we obtain an overall usable radiation energy of 809 TW h a^{-1}.

11.1.2.2 Facades

The facade areas of 6660 Mio m^2 available in Germany are much greater than the roof areas. However, relatively many areas fall away due to structural restrictions (extensions, doors, windows, shading, etc.). Further, we are only interested in facades with southeast to southwest orientations. If we reserve half the remaining area for solar heat usage, then we are left with only approximately 200 Mio m^2 (3% of all façade area) for photovoltaics. The vertical facades receive an average 850 kW h $(m^2\ a)^{-1}$ of radiation energy so the total energy is 170 TW h a^{-1}.

Table 11.1 Usable radiation energy on suitable roof areas.

Type of area	Usable area (Mio m^2)	Mean inclination losses (%)	Mean global radiation (reference yield) (kW h $(m^2\ a)^{-1}$)	Overall usable radiation energy (TW h a^{-1})
Pitched roof	608	15	1020	527
Flat roof	235	0	1200	282
Total	842	—	—	809

11.1.2.3 Traffic Routes

In Germany, there are approximately 275 000 km of regional roads and rail routes. For this reason, it is also feasible to use them for photovoltaics. In reference [121], it is proposed to erect photovoltaic plants on both sides of almost 4% of these traffic routes. The idea is to use 2-m high glass–glass modules that permit radiation on both sides. Spread over all the cardinal points, one can assume around 1250 kW h $(m^2\ a)^{-1}$. With average losses (shading and soiling) of 15%, this provides a usable radiation energy of approximately 42 TW h a^{-1}.

11.1.2.4 Free Areas

The largest potential is in the use of free areas. In Ref. [1], for instance, it is assumed that use is made of agricultural set-aside areas. However, since 2008, these have largely been used for planting biomass and are no longer available for photovoltaics. Currently, in Germany, approximately 20% of all arable areas are in use for renewable raw materials, which corresponds to an area of 21 000 Mio m^2 [122]. We will assume that 1% of all arable areas will be made available for photovoltaics. This would be an area of approximately 1200 Mio m^2. With an area utilization ratio of 45%, this gives a module area of 540 Mio m^2 corresponding to a usable radiation energy of 648 TW h a^{-1}.

11.1.3 Technical Electrical Energy Generation Potential

How much electrical energy can be obtained from the module areas in Section 11.1.2 depends largely on the efficiency of the modules used. As we have seen from Chapter 5, standard solar modules are now available with an efficiency of almost 20%. In the next 10 years, this efficiency will continue to move upward. Thus, we will easily achieve a system efficiency of 18% (including the losses from inverter, cabling, and deviations of module temperatures from STC conditions). The results are shown in Table 11.2.

We have an installable power of 325 GWp and 302 TW h a^{-1} of electrical energy from photovoltaics. This corresponds to half the current German electrical energy consumption of approximately 600 TW h a^{-1}. Even without taking the free areas into account,

Table 11.2 Photovoltaic energy generation potential in Germany with a module efficiency of 20% and a system efficiency of 18%.

Type of area	Usable module area (Mio m^2)	Usable radiation energy (TW h a^{-1})	Installed power (η_{Mod} = 20%) (GWp)	Electrical energy (η_{Sys} = 18%) (TW h a^{-1})	Portion of energy requirement of 600 TW h a^{-1} (%)
Pitched roof	608	527	122	95	16
Flat roof	235	282	47	51	9
Façades	200	170	40	31	5
Traffic routes	38	42	8	8	1
Subtotal without free areas	1081	1021	217	183	31
Free areas	540	648	108	117	19
Total	1621	1669	325	302	50

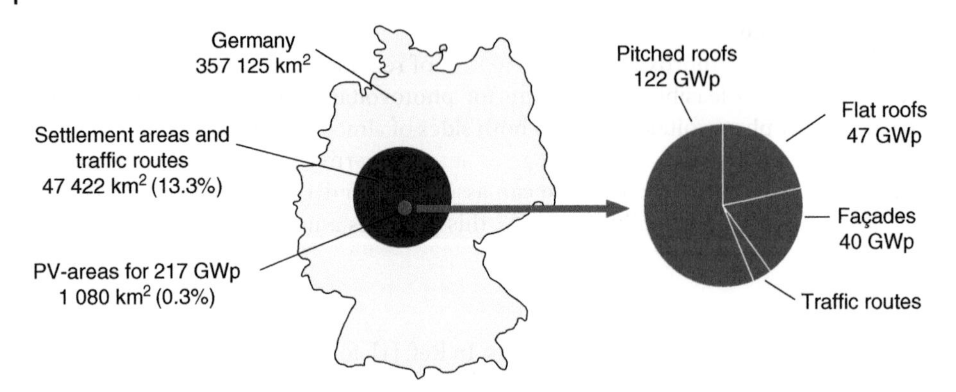

Figure 11.1 Surface area requirements for photovoltaics in Germany: A power of more than 200 GWp is available even without using free areas. The area required by this is not in competition with other types of utilization (e.g. biomass) [123].

we will still manage with a portion of 31%. Figure 11.1 shows the required module area as an amount of 0.3% of the total area of Germany. The areas taken into use are mainly on already available roofs and facades so that there is no competition for use.

11.1.4 Photovoltaics versus Biomass

If it is desired to generate much more electrical energy in the future (e.g. for electric cars or geothermal heat pumps for domestic heating), then free areas should also be used for photovoltaics. This represents an area competition to biomass. Therefore, a comparison of the two technologies should be made.

A hectare of energy corn annually yields approximately 45 t of biomass (fresh mash), which again results in a biogas yield of 8000 m^3. With a methane portion of 52%, this corresponds to an energy of 41 600 kW h. If the biogas is used in a block power station ($\eta_{Elec} = 40\%$), then this provides electrical energy of almost 17 000 kW h a^{-1} [124]. Per square meter of arable area we obtain an energy yield of $w_{Elec-Bio} = 1.7\,kW\,h\,(m^2\,a)^{-1}$. Compared with the solar radiation on a square meter of 1000 kW h (m^2 a)$^{-1}$, there is a yield degree of efficiency of only 0.17%.

If we consider the photovoltaics, then we can use as a reference, e.g. the relative modern flat-roof plant from Figure 10.13. With a module slope of 25°, it achieves a specific yield of n an open-air plant ($\eta_{Mod} = 20\%$, area utilization factor $f_{Util} = 0.45$), we get an area-referenced annual yield of 1000 kW h (kWp a)$^{-1}$. With this slope, we get analogue to Example 10.2, an area utilization factor of 0.4.

For an open-air plant with same slope and orientation and a module efficiency of 20%, for the area-referenced annual yield follows:

$$\begin{aligned} w_{Elec\_PV} = f_{Util} \cdot \eta_{Mod} \cdot E_{STC} \cdot Y_F &= 0.4 \cdot 0.2 \cdot 1\,kW\,m^{-2} \cdot 1000\,kW\,h\,(kWp\,a)^{-1} \\ &= 80\,kW\,h\,(m^2\,a)^{-1}. \end{aligned} \tag{11.3}$$

The yield efficiency of photovoltaics is therefore 8%, and this gives a relationship to biomass use of 47 : 1. However, both heat and electricity are produced in a biomass block power station and that can also be used. On the other hand, the planting of corn requires energy for working the soil, transport, fertilizer, and so on, that is not taken into account here.

In comparison to photovoltaics, biomass requires about 50 times the area to generate the same electrical energy.

Wouldn't the above comparison not even shift in favor of photovoltaics if the free areas were covered with east/west PV plants?

This we really should calculate. Let us assume that we build up the modules to east and west, respectively, all with an inclination of 15°, for example. In this case, we can position the module rows running from north to south directly beside each other, without causing a self-shading. Thus, the area utilization factor reaches more than 100%. However, according to Table 2.4, we get inclination losses of about 16%. If additionally 20% of the whole area is subtracted for access roads, etc., we end up at a yield efficiency of almost 14%. A comparison with the yield efficiency of biomass shows that photovoltaics is even more area efficient by a factor of 80!

Up to now, we have only considered the technical potential of photovoltaics. Whether this potential is actually realized in the erection of PV plants depends strongly on the legal and financial framework conditions. We will look at this more closely now.

11.2 Efficient Promotion Instruments

Germany already has a lot of experience with promotion programs for photovoltaics. At the start of the 1990s, the 1000 Roofs Program under the guidance of the Federal Ministry of Research commenced. This provided an investment subsidy for erection of on-grid plants amounting to 70% of the investment sum. The declared aim was the "Evaluation of the current state of the technology" and "Still needed development requirements" were to be determined [125]. Between 1991 and 1995, almost 2000 PV systems were erected within the framework of the program. Valuable operating experience was gathered with the use of a parallel measuring and evaluation program. The interaction of inverters with the power grid was investigated in particular, as there were originally great doubts from public utilities regarding safety. After the money pot was emptied, the short-term market that had developed, collapsed.

The availability of the high investment subsidy of 70% led to a flood of applications. Apparently, the prospect of state money, "cash on the barrelhead," is a high motivator. At the same time, the plant operators hardly took notice of the high installation costs, the plant prices were mostly very near the prescribed upper limit of 28 000 DM kWp^{-1} (14 000€ kWp^{-1}). The evaluation of the program also showed that many plants became defective (especially the inverters) after one or two years. Often, however, they were not repaired, as there was no financial support for repairs. In the following years, some of the federal states had further promotion programs based on investment subsidies. However, no permanent market developed, as these were quickly oversubscribed, depending on the sums available.

Only after the development of cost-covering feed-in tariffs by various city departments (Aachen Model) and finally the start of the Renewable Energy Law (EEG) in

2000 the situation did change. From the start, the EEG was conceived as a market introduction program and was meant to lead to a price reduction caused by mass production of PV plants.

What are the important advantages of this model in comparison to the promotion over investment subsidies?

1. As the money for feed-in compensation is raised by the electricity customers by paying the EEG levy, there are no cyclically empty funding pots as are typical with public investment programs. In this way, the market is stabilized, and investment willingness of the manufacturers is increased.
2. The plant operator only obtains a profit on his invested capital when he works economically. Thus, he is obliged to purchase technically mature and, also, economic technology. Besides this, he will repair the plant in his own interest and operate it as long as possible.
3. With the determination of a reduction of the feed-in tariff depending on the year of installation of the plant, the price and also the market volume can be controlled to a certain extent.

The unplanned lowering of the tariff in Germany in recent years had the object of preventing the market from overheating and at the same time making further growth possible. Unfortunately, other countries such as Spain and Italy were not so consistent in this. In these countries, too high feed-in tariffs were paid, which led to a cap of the promotable PV power and to a collapse of the respective markets. Yet, the aim of the reduction of costs due to mass production especially by the EEG (and the introduction of the concept in other countries) has been impressively reached, as we will see in the next section.

11.3 Price and Feed-in Tariff Development

11.3.1 Price Development of Solar Modules

Figure 11.2 shows the price development for solar modules since 1980. During that time, the prices have fallen drastically from 28€ to below 50 cents; the 60th of the original value. The curve shows some waves that were mainly caused by the temporary shortage of solar silicon.

A deeper understanding of the price development is given by the learning curve theory. It assumes that the production costs of a product (e.g. computers or solar modules) are continuously reduced with mass production. This is because the pressure of the competition causes production to become ever more efficient and that new technologies are introduced. The strength of the cost reduction is given by the learning rate:

> The learning rate is the percentage by which the costs of a product are reduced with a doubling of the quantity produced (accumulated).

A learning rate of 10%, for instance, means that the cost of the product is reduced by 10% with a doubling of the accumulated produced quantity. Figure 11.3 shows the price development from Figure 11.2 in a doubled logarithmic form. The *x*-axis describes

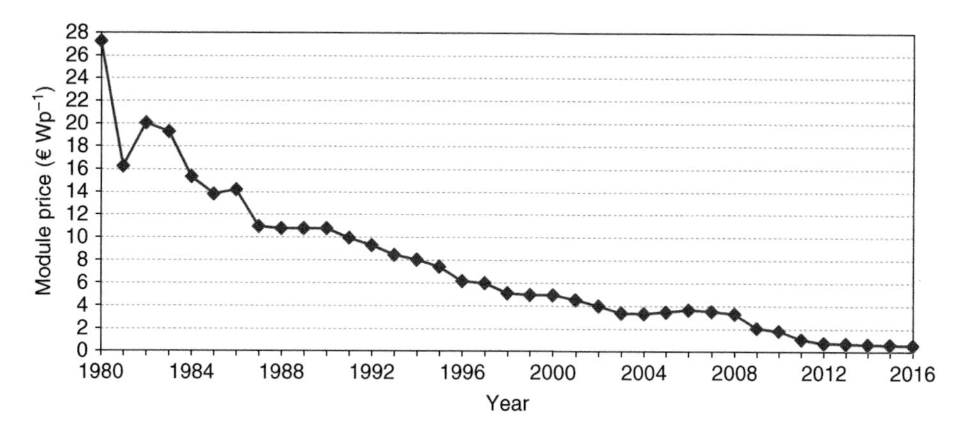

Figure 11.2 Price development of photovoltaic modules (inflation-weighted wholesale prices): Since 1980, the price has fallen to a 60th [126, 127].

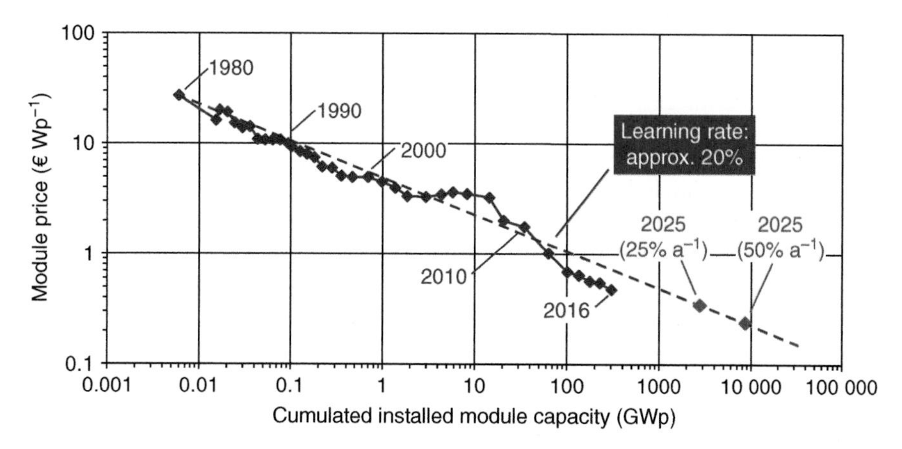

Figure 11.3 Module price development in doubled logarithmic depiction: For a doubling of the accumulated installed PV capacity, there is a price reduction of approximately 20% [126, 127].

the accumulated installed solar module power and the y-axis the price per Watt peak. Obviously, one can average the price development in this form of depiction as a straight line. The slope of the straight line shows that the learning rate of the solar modules is approximately 20%.

What will the price look like in the year 2025? This depends on what the future growth will look like. From Figure 1.17, we can see that the PV world market has grown by more than 50% annually since 2000. If we extrapolate this growth to 2025, then we will have an installed PV capacity of about 11 500 GWp. The price can then be expected to be approximately 20 cents. If the growth is only 25% then by 2025, we will land at a price of almost 40 cents (Figure 11.3). It is important to note that the x-axis in Figure 11.3 is not a time axis. The learning curve will only continue when the market (e.g. due to suitable framework conditions) continues to grow.

Figure 11.3 shows a clear downwards dip of the price curve for the last years. This was triggered by the extreme price pressure by Asiatic producers. In the long run, though, it is to be expected that the price curve will again approximate the dashed straight line.

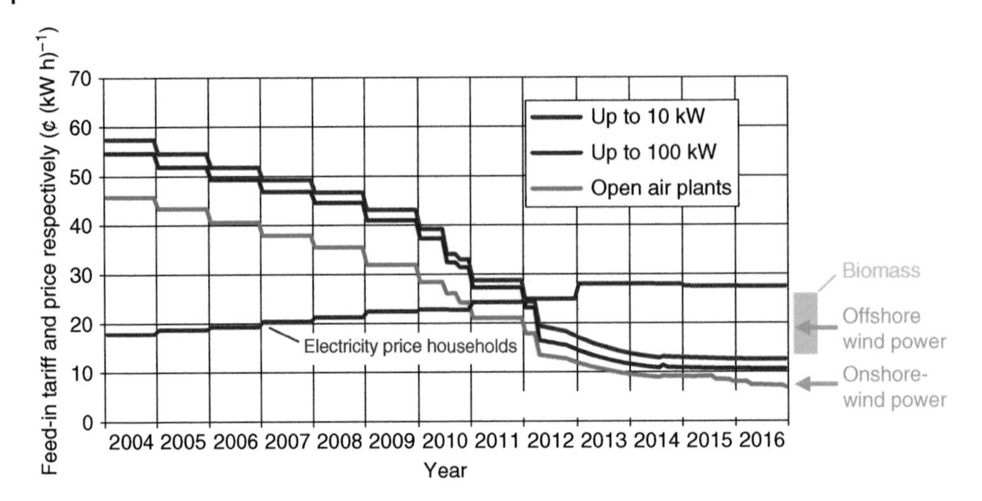

Figure 11.4 Development of the feed-in tariff for PV plants as well as domestic customer tariffs: Meanwhile, the PV feed-in tariff lies clearly below that one of offshore wind power and biomass.

11.3.2 Development of Feed-in Tariffs

The cost reduction of the solar modules was accompanied by a similar strong reduction of the total system costs (including inverter, mounting, etc.). Therefore, the feed-in tariff was accordingly reduced within the frame of the EEG. In Figure 11.4, the development since 2004 is depicted. Starting from 58 cent per kilowatt-hour, the feed-in tariff for small roof plants was reduced to 12 cent per kilowatt-hour in 2016, a reduction to less than a fourth of the original value. Large roof plants get 11 cent and open-air plants even only 6.6 cent per kilowatt-hour. With this, photovoltaics lies below the values of biomass and offshore wind power! Open-air plants, meanwhile, even lie below the values of onshore wind power!

Figure 11.4 additionally shows the development of the domestic energy consumer tariffs. About the year 2012, *grid parity* was reached. This denotes the point in time at which the electric energy from a PV plant is cheaper than the normal price of electricity for the end user. Occasionally, the opinion is expressed that no more feed-in compensation will be necessary from this point. But this does not take into account that the installation of a plant is only worthwhile for the customers when they can use the whole of the PV plant energy for themselves. However, this is seldom the case (especially for larger plants). For this reason, a certain form of feed-in compensation will continue to be necessary. Only in this way, a stable market will be achieved that will lead to further cost reductions.

11.4 Renewable Energies in Today's Power Supply System

We could already see the impressing development of renewable energies in the German power supply system in Figure 1.8. Now, we want to examine how the structure of electricity generation looks like in total. Moreover, we consider the interplay between renewable and conventional energies.

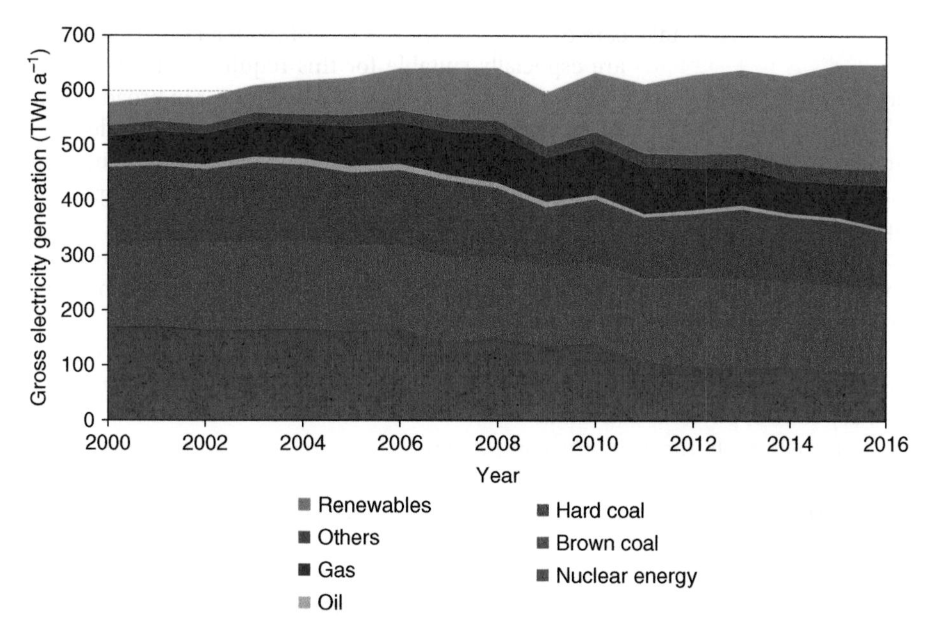

Figure 11.5 Development of the power generation structure: The renewable energies have augmented their contribution from 10% in 2000 up to 30% in 2016 [128].

11.4.1 Structure of Electricity Generation

In Figure 11.5, the electricity generation including the used energy carrier since the year 2000 is shown. The gross electricity generation (total electricity generation including the self-consumption of the power plants) was about 600 TW h per year over the whole time period. In 2000, the share of renewable energies was only 10%; this increased then relatively continuously up to the current value of 30%. They are already the most important energy carrier. Due to the phase-out decision, nuclear energy lies below 13%. Large fossil energy carrier is brown coal; it delivers a share of 23% followed by the hard coal with 17%. In sum, the conventional energies still dominate the electricity generation with a share of about 70%.

11.4.2 Types of Power Plants and Control Energy

In a classical electricity supply grid, so much energy must be generated continuously as just in that time is demanded by the consumers. To meet this requirement, different power plant types are available.

First, there are base load plants, which provide a certain base of energy. For this, they run continuously almost all the time with approximately the same power. Here, typically plants are used, which are particularly inflexible or have low fuel costs. In Germany, these are nuclear and brown coal plants as well as run-of-river plants.

Medium-load power plants cover the normal periodic fluctuations (day/night, foreseeable load changes). For this, they must be controllable in their power. Typically, hard coal and gas power plants (combined cycle gas turbines) come to employment.

Peak load power plants take care of short-term variations of the load curve. Pumped storage and gas power plants are especially suitable for this requirement. They have power-up times of only a few minutes and can vary the supplied power in broad areas.

The transmission grid operator is responsible for the balancing between offer and demand within his control zone. The control energy necessary for that is bought in advance from the power plant operators. The three different control energy types are distinguished as primary control, secondary control, and minutes reserve.

In case of primary control, the necessary control power has to be fully provided within 30 s and then has to be maintained for at least 15 min. To realize this, the respective power plants withhold some percentage of their generation power free as reserve. The power provision then automatically starts by continuously measuring the grid frequency. As soon as it differs more than ±10 mHz from the nominal value 50 Hz, the primary control power is activated.

Simultaneously with the primary control, the secondary control is started. This one has to provide its full power within 5 min and then at the latest, after 15 min has to take in turns with the primary control. For the secondary control, typically pumped storage gas turbine plants come to use.

Especially after power station outages, the minutes reserve (tertiary control) is used to take over from the secondary control, which then is freed for new control operations. It is classically ordered by phone; however, the requests are more and more sent by special servers. The minutes reserve has to be totally activated within 15 min.

The required control energy definitely can also be negative. In this case, the power provision it too large. Besides turning down power plants, controllable loads are used, e.g. electric arc furnaces or storage heaters.

The feed-in of renewable energies leads to the fact that conventional power plants have to react as well on varying demand as on uneven supply. This brings us to the question, how the today's electricity supply system is affected by that challenge.

11.4.3 Interplay Between Sun and Wind

How well photovoltaics and wind power actually fit together? To answer this question, Figure 11.6 shows the weekly electricity production of the year 2014 for solar, wind, and

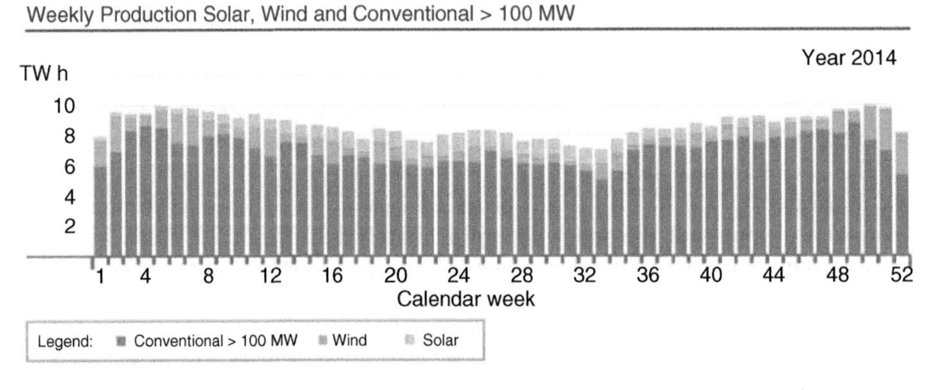

Figure 11.6 Weekly power production in Germany for the year 2014: PV and wind complement each other quite well [129].

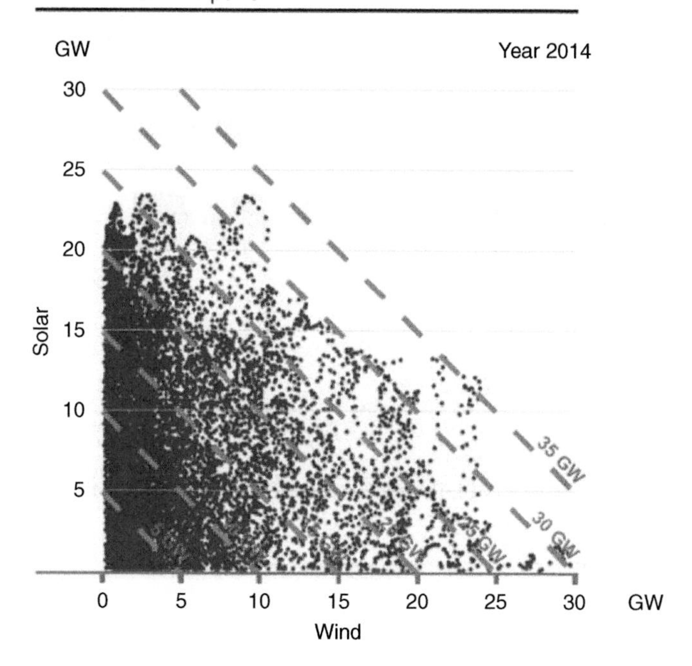

Figure 11.7 Occurred powers of wind power and photovoltaics in the year 2014: Although the maximum powers of both energy types came to 30 and 26 GW, respectively, the maximum total power was only 36 GW [129].

conventional energies. The two upper bars represent the energy amount that is fed-in by solar and wind power plants. It can be observed that they both complement each other quite well. In summer, the wind blows only sparsely, and photovoltaics offers high yields. In winter, the situation is quite the opposite.

To reveal the interplay between Sun and wind, in Figure 11.7, the actual powers of wind power and photovoltaic plants (15 min mean values) are depicted. The maximum powers of wind plants lie at about 30 GW, and those of photovoltaics by 23 GW. It is clearly visible that both peak powers never happen at the same time. Instead, both generation types in sum provide only a maximum power of 36 GW. This shows the large balancing potential between Sun and wind.

11.4.4 Exemplary Electricity Generation Courses

In Figure 11.8, the monthly electricity production for the year 2014 is shown. On the one hand, we see that in winter, the electricity demand is higher than in summer. On the other hand, it becomes clear that fossil and nuclear energy carrier still dominate the electricity generation with brown coal as the largest energy supplier.

Although wind and Sun complement each other well, both energy types occur with strong diurnal and seasonal fluctuations. As an example, Figure 11.9 shows the electricity production in the Holy Week 2014. At the beginning of the week, there are significant variations of wind power and photovoltaics. These variations can mostly be balanced with pumped storage, gas and hard coal power plants. At Easter Sunday, a

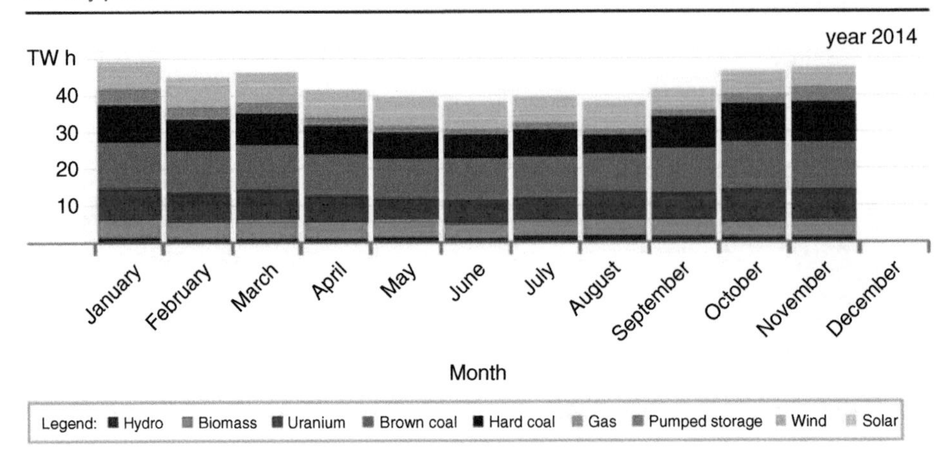

Figure 11.8 Monthly electricity production in year 2014: The fossil and nuclear energy carriers still dominate the energy production [129].

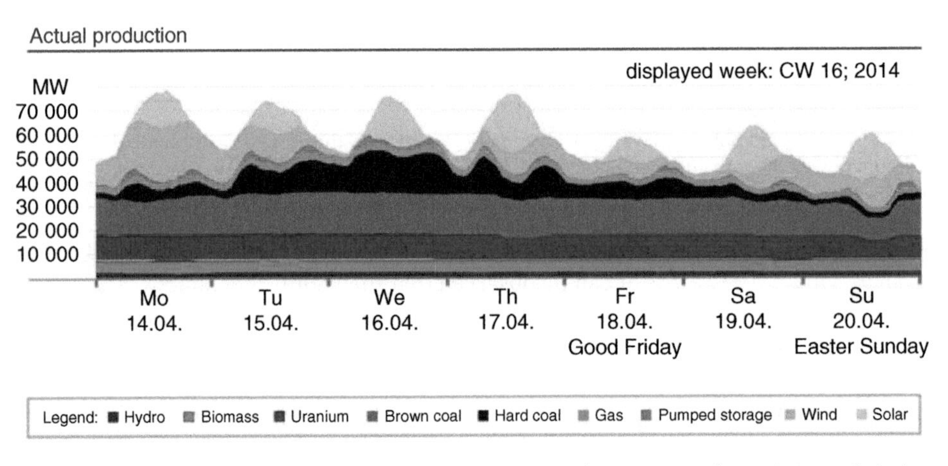

Figure 11.9 Electricity production in the Holy Week 2014: For the most part, the variations of wind and Sun can be compensated by throttling of gas and hard coal power plants. Only at Easter Sunday with sunshine and moderate demand, the reduction of brown coal and even nuclear power is necessary [129].

low energy demand is combined with good weather. Due to this, also, the brown coal and even the relative inflexible nuclear power have to be reduced in power.

Figure 11.9 shows the electricity production, which normally in total complies with the demand. However, there is the possibility to exchange electrical energy with other countries in the framework of the trans-European grid. For this, Figure 11.10 additionally shows the export of electricity in the considered Holy Week (depicted negatively). Only now it becomes clear that the demand at Easter Sunday was only at a maximum 47 GW. Simultaneously, electrical energy with a power of temporarily up to 12 GW was exported in other countries.

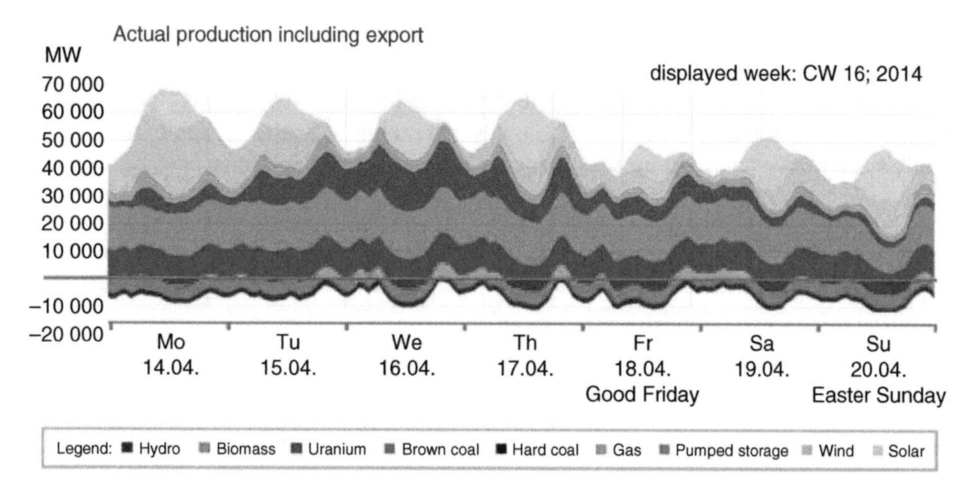

Figure 11.10 Electricity production including export in the Holy Week 2014: Besides the throttling of conventional power plants, also the export (depicted negatively) in other countries is used to balance the feed-in variations of the renewable energies [129].

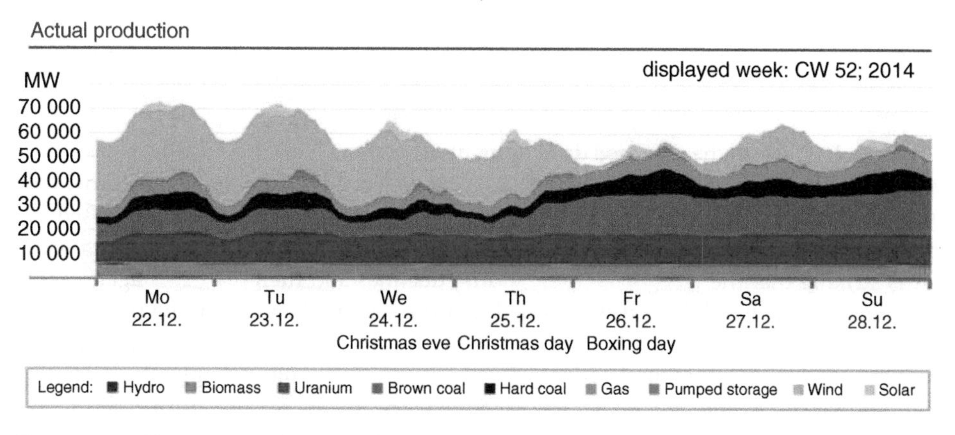

Figure 11.11 Electricity production in the Christmas Week 2014: Initially, the wind power dominates so that all conventional producers have to be throttled. However, at Boxing Day, wind and solar power only provide minimum contributions to the electricity generation [129].

An example with strong fluctuations of wind power can be seen in Figure 11.11 during the Christmas week in 2014. At the beginning of the week, wind power provided shortly 30 GW of power. This led to the fact that especially at night, the conventional plants had to be throttled below 25 GW. Totally different was the situation on Boxing Day: The wind had let up so that temporarily less than 1 GW from wind power and photovoltaics was generated.

The exemplary electricity production courses show that the further extension of renewable energies has to be accompanied by measures to balance the strongly fluctuating feed-in powers.

11.5 Thoughts on Future Energy Supply

What will the renewable energies contribute to the future energy supply? Which part will photovoltaics play? Which technologies guarantee a further reliable electricity supply system? We will light up these aspects in the following sections.

11.5.1 Consideration of Different Future Scenarios

"Forecasts are difficult especially when they concern the future" – this fine sentence attributed to Mark Twain also applies to an accurate forecast of the energy mix of the future. There are many prognoses that concern the conversion of the current energy supply. Depending on the assumptions, the political views, and one's own economic relationship to the subject of the energy economy, one obtains very different results.

We will first look at a study, which was carried out on behalf of the German Federal Ministry of the Environment (FME) [130]. It shall show concrete ways how the decided energy transition can be technically realized. As boundary conditions, the two main aims of the energy transition have to be kept in mind:

1. Transformation of the electricity supply system to minimum 80% up to the year 2050.
2. Reduction of the greenhouse emissions of Germany to minimum 80% up to the year 2050.

Figure 11.12 shows the assumed development of scenario "2011 A": The fossil energies will be mostly displaced, and only about 15% of the electricity demand will be provided by gas or coal. Here it is assumed, that these energy carriers will be mainly used in combined heat and power plants (CHP) to produce electricity and heat simultaneously.

It is striking that the electricity demand first declines and then increases again from 2030. On the one hand, here it is assumed that the energy demand can be reduced by efficiency measures to below 400 TW h a^{-1}. However, new consumers will be added with electric mobility and heat applications (Power to Heat). So, for example, the scenario assumes that half of the road traffic will be done with electric vehicles until 2050.

In 2050, largest part of the electricity generation will be provided by wind power plants with a share of 45%. According to this scenario, photovoltaics will only contribute a share of 11% and biomass of 10%.

In Figure 11.13, the installed power of the different generation plants is shown up to 2050. Wind power reaches 83 GW, which distributes to 51 GW onshore and 32 GW offshore according to the scenario. For PV, an installed power of 67 GWp is assumed. With this, it would only produce a fourth of the energy production of wind power, as can be seen in Figure 11.12. The reason lies in the small yearly full load hours of 1000 h, which for wind power can attain 2000 (onshore) up to 4500 h (offshore).

Figure 11.14 comprises a survey of different studies, which forecast possible situations in the year 2050. Variant A shows the already presented scenario "2011 A" of the FME study.

Variant B presents the result of the scenario "THG95" out of the same study. Here, the declared aim is to reduce the greenhouse emissions by 95% until the year 2050. The high electricity demand of 823 TW h a^{-1} is the result of the assumption that all cars

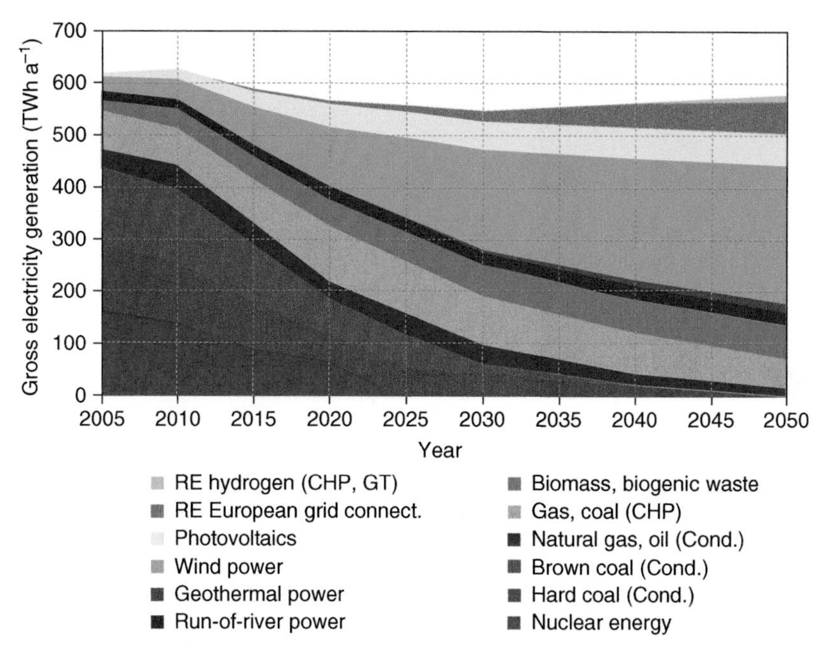

Figure 11.12 Far-reaching transition of the electricity supply to renewable energies up to the year 2050: Fossil energy carriers will only have a share of just 15% (Scenario "2011 A") [130].

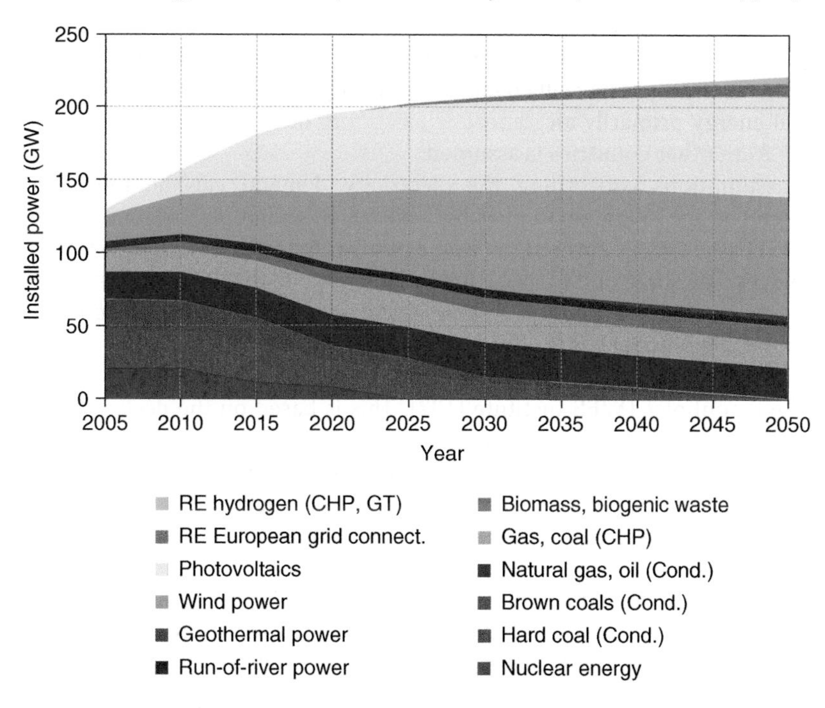

Figure 11.13 Installed capacities of the different power plants until 2050: Wind power and photovoltaics show by far the largest installed powers (Scenario "2011 A") [130].

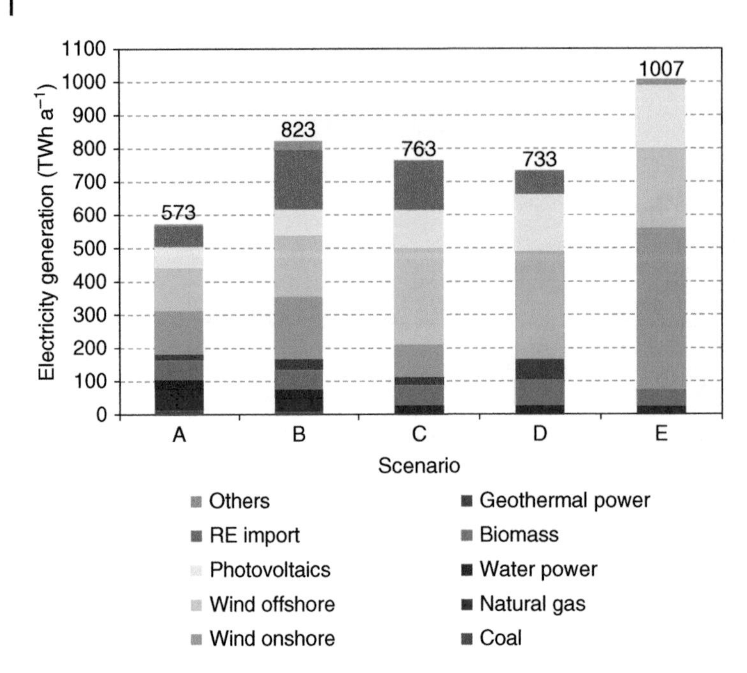

Figure 11.14 Results of different studies regarding the energy mix in 2050: The share of photovoltaics is estimated very differently [102, 130–132].

will be converted into electric vehicles in 2050. Moreover, large shares of the heat supply will be done by electricity (especially by means of heat pumps). For the additionally needed electrical energy primarily an import of electricity out of renewable energies ("RE electricity") from other countries is assumed.

Comparable assumptions concerning the electricity demand can be found in variant C, a study of the German Renewable Energy Research Association. Here however, a higher share (15%) is covered by "home photovoltaics."

The highest percentage share of PV can be found in variant D, the climate protection scenario of Quaschning [102]. Here, photovoltaics take over a fourth of the electricity demand with an installed capacity of 175 GWp.

To some extent, the maximum scenario comprises the study "business model energy transition" of the Fraunhofer IWES Institute [132]. This is based on the declared aim to completely cover the demands of the energy sectors electricity, heat, and traffic with the help of increased energy efficiency and use of renewable energies. For this, the electricity sector shall play the major role. Consequently, a very high electricity demand of about 1000 TW h a^{-1} is assumed. Photovoltaics shall contribute one-fifth with an installed power of about 200 GWp.

It is important to emphasize that all studies consider the total transition of the electricity supply system to renewable energies as realizable. Different is only the assumed extent of the inclusion of traffic and heat supply into the electricity system.

Which of the studies would be most likely to foresee the future development correctly?

Of course, this cannot be answered generally. The future development will strongly be dependent on the political and economic framework conditions. With regard to photovoltaics, it can be said that it will be the second important renewable energy (RE) after wind power. This is assured by the fact that PV is still getting cheaper and cheaper by going through the learning curve. At the same time, it has the lowest negative environment impact (high area efficiency, use of anyway sealed surfaces, no noise generation, no shadow flicker, etc.).

In fact, all the studies of the past years have underestimated the actual development of photovoltaics.

The import of renewable power assumed in some studies is based on solar power stations (solar heat and photovoltaic) as well as wind turbines being erected in North Africa. The power generated is then transported by means of low-loss HVDCT (high-voltage direct current transmission) to Central Europe. This is basically a feasible option, but up to now, it is unclear whether it would be cheaper than generating on one's own soil. For reasons of supply security, this imported power should be limited to a portion of 15%.

11.5.2 Options to Store Electrical Energy

As we have seen above, especially wind power and PV show a strongly varying energy supply. One possibility to balance these variations is the storage of electrical energy. Therefore, we will look, which technologies are suitable for this.

11.5.2.1 Pumped Storage Power Plants

Pumped storage power plants use surplus energy to pump water in a high-altitude water reservoir. The stored energy can later be used to cover demand gaps. The storage efficiency of about 80% is relatively high. In Germany, there are about 40 pumped storage plants with a total storage power of 7 GW and a capacity of about 40 GW h. The largest is the Goldisthal power plant in Thuringia with a power of 1 GW. If the storage is filled, this power can be provided over 8 h [99].

Up to the year 2020, the capacity of the German pumped storage plants shall be increased clearly by new-build and extension. After realization of all projects, a total capacity of 80 GW h will be available [133]. A further expansion of pumped storage plants in Germany will only be possible in a very limited way as almost no suitable sites are left.

Pumped storage power plants are very helpful in balancing the load and feed-in fluctuations. However, even the future capacity of 80 GW h is still very limited. The small storage effect shows the following thought experiment: If the whole electricity demand should be covered (mean load: 70 GW), this would be only possible for the time of 1 h.

Besides the extension of pumped storage plants in Germany, another option also is to reconstruct the existing hydroelectric power plants in Scandinavia into pumped storage power plants.

11.5.2.2 Compressed Air Storage

In compressed air energy storage (CAES) plants, electrically driven pumps are used to pump air into underground caverns. This compressed air can then be used to drive gas turbines for generating electricity. Concretely, the air replaces the compression stage during which normally two-thirds of the losses of a gas turbine occur. However, there are substantial losses in the storage of air: When compressing in the underground cavern, the air heats strongly and must be cooled. For this reason, the overall efficiency is a maximum of 55%. At present, new methods are being tested (AA-CAES – Advanced Adiabatic Compressed Air Energy Storage), in which the heat generated during compression is stored temporarily in a special heat storage. When discharging, this heats the air again so that an overall efficiency of 70% is possible [134].

In Germany, up to now, only the CAES plant Huntorf in Lower Saxony exists with a capacity of 560 MW h. In the whole country, about 200 Mio m^3 of salt caverns are available, which principally could be used for CAES. This would offer a storage capacity of 600–900 GW h depending on the plant efficiency [99].

11.5.2.3 Battery Storage

In Chapter 8, we have already dealt in detail with battery storages. Besides domestic home storages, frequently large-scale battery storages are deployed. They are mainly used for increasing the self-consumption of commercial enterprises and to ensure grid services (e.g. primary control).

One example is the 5 MW lithium ion storage of the Wemag AG in Northern Germany already mentioned in Chapter 8. The storage takes part in the primary control power market and stabilizes the grid frequency with its capacity of 5 MW h in the windy region of Mecklenburg. To achieve this, the batteries are only filled up to 50%. In case of a dropping grid frequency (see Section 11.4.2), the plant feeds power into the grid. Vice versa, the batteries are charged above a grid frequency of 50.1 Hz.

The worldwide largest battery storage was built by the Chinese company BYD (build your dream) near Hong Kong, to stabilize the grid. It consists of about 60 000 lithium iron phosphate cells and provides a power of 20 MW with a capacity of 40 MW h. The company wants to fulfill another of its dreams and plans – a larger storage plant with 1 GW and 200 MW h.

11.5.2.4 Electric Mobility

A storage that will be additionally available in the future is the batteries of electric vehicles. As the cars stand still most of the time, the batteries can be used to some extent as decentralized storages. A study of the FME assumes that in 2050, half of the cars would be operated electrically. In total, these will have a storage capacity of 550 GW h. It is estimated that about one-third of that capacity (180 GW) can be used as external storage for the electricity system. The available charge power will be in the order of 100 GW [135].

11.5.2.5 Hydrogen as Storage

Hydrogen has long been mentioned as a possible storage carrier for the renewable energies sector. In fact, it can easily be produced by means of electrolysis and then converted again into electricity by means of a fuel cell. The overall efficiency is 44% [131]. However, there is no hydrogen infrastructure (lines, storage, fuel cells, etc.) and these would have to be completely built up.

For this reason, it can be assumed that in the next years, the produced "renewable hydrogen" will be only additively fed into the natural gas grid. Experiences show that gas power plants can get by without any problems with a hydrogen share of up to 1–2% in the natural gas. In this case, the total efficiency from electricity to hydrogen and back to electricity lies in the range of 33–48% [99].

11.5.2.6 Power to Gas: Methanation

A relatively new technology is the methanation of electrical energy. For this purpose, use is, e.g. made of surplus wind or photovoltaic power to generate hydrogen by means of electrolysis. In a second step, the hydrogen is converted to methane in the so-called *Sabatier process* by the addition of carbon dioxide (see Figure 11.15). The generated RE methane can then be fed into the natural gas grid. In order to re-convert the methane to electric power, it is fed into a gas turbine (block power station, gas power station, or gas and steam power station). A decisive advantage is the existing gas network that can be seen as a giant storage. The storage capacity of the electrical energy is estimated to be approximately 120 TW h so that with the typical load of the German power grid of 70 GW even seasonal storage could be realized.

The overall efficiency from RE electricity over hydrogen and back to electricity lies in the range of 30–38%, depending on the used electrolysis and reconversion technology [99]. The main reason for the poor efficiency in comparison to pure hydrogen technology is the strongly exothermic Sabatier process.

Meanwhile, there exist several power to gas pilot plants with a total power of about 20 MW. The largest plant to convert electricity into methane with a power of 6 MW is operated by the car manufacturer Audi AG in Lower Saxony. It produces about 1000 t of renewable gas, which is used for gas-operated cars. A survey over the realized and planned power to gas projects can be found under *www.powertogas.info*.

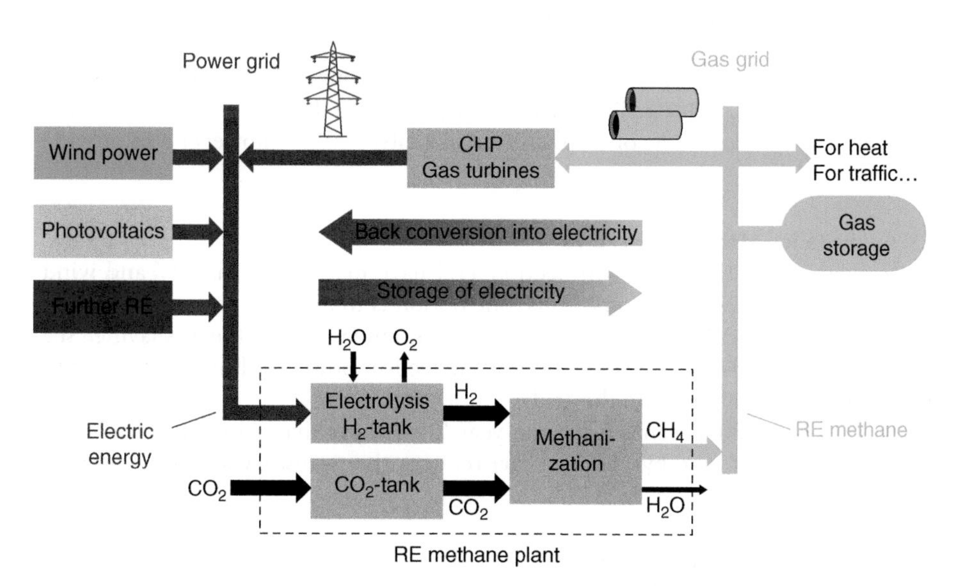

Figure 11.15 Principle depiction of electricity storage with power to gas: Surplus wind and PV power are methanized and fed into the gas grid. Then the RE methane can be reelectrified in gas turbines or used for heating and traffic purposes [99, 104].

11.5.3 Alternatives to Storage

Storages will become more and more important in the future electricity grid. However, they are relatively expensive. Therefore, before a massive storage expansion, it is wise to use other options (so-called flexibility options) to balance the strong feed-in and demand fluctuations.

11.5.3.1 Active Load Management by Smart Grids

If it is successful to consume a kilowatt-hour suitably postponed, this serves almost the same effect as an energy storage. Due to this reason, the flexibility option load management (Demand Side Management, DSM) is an attractive solution.

Exactly, this is the idea of smart grids, which influence the load curve. So some loads (e.g. cooling systems, electric storage heater, heat pumps) can be taken from the grid for some time without having a negative impact on the user. Other approaches use smart meters with different electricity tariffs, to run the washing machine or the dishwasher during the time of an electricity surplus. Thus, smart grids help to adapt the load curve to the energy offer and to reduce the necessary storage capacity.

Quickest, the load management is realizable in the range of industry. There, by remote control of only some large consumers already a high potential can be utilized. Typical applications are, e.g. the steel, copper, and aluminum production and the raw materials industry. The capacity that can be shifted by DSM is estimated in sum to be about 900 GW h [99].

11.5.3.2 Expansion of the Electricity Grids

Up to now, the bottlenecks in the grid mainly arise in the distribution grid due to local production peaks of PV and wind power. In addition, there will be bottlenecks in the transmission grid caused by the planned expansion of offshore wind power. Therefore, a further build-out of the grid will be indispensable. However, the question has to be posed, if the offshore energy production far from the locations of energy use makes sense in the planned extent. Additionally, it has to be kept in mind that an electricity grid only can provide a spatial balancing of surpluses and deficits. Storages however also provide a temporal balancing.

11.5.3.3 Limitation of the Feed-in Power

The electricity generation curves in Section 11.4 have made clear that PV and wind power show large power variations. Thus, one option is to limit the maximum feed-in power, e.g. in case of PV to 50% or 70% of the nominal power (peak shaving, see Section 8.4). This leads to a consolidation of the feed-in power and makes it easier to integrate high PV capacities into the grid.

Simulations of a 100% scenario for the year 2050 show that a minor curtailment of renewable energies (e.g. by 1% per year) reduces the necessary storage power by 30–40% [99].

11.5.3.4 Use of Flexible Power Plants

Instead of only poorly controllable conventional coal power plants, flexible plants have to be built out. Particularly suitable are gas power plants, which in the ideal case are powered with biogas or even better RE methane. To use the gas efficiently, the parallel utilization of heat in CHP plants is advisable. Especially in winter, there exists a high heat demand with at the same time low PV yields. In order to be, nevertheless, able to react flexibly on the electricity demand, the CHP plants should be equipped with heat storages.

11.6 Conclusion

To conclude, we can state that the complete transformation of the electricity power supply in Germany is technically and economically feasible. The necessary technologies are known and mostly already available.

Photovoltaics has sufficient potential in Germany for covering the overall electric energy requirements. However, Germany is a relatively densely populated country with limited free areas and only a fair amount of solar offering. Thus, in most other countries, there is even greater potential for solar power generation!

At the same time, the mass production of photovoltaic components reduces costs further so that it becomes together with wind onshore power by far the cheapest energy source. In this way, photovoltaics will become an important pillar for the future sustainable energy supply.

The building up of this new energy system is a huge challenge that we should meet with knowledge, imagination, and perseverance. If this book makes a contribution to this, it will have fulfilled its purpose.

12

Exercises

Exercises for Chapter 1

Exercise 1.1 Energy Content

You have a kilogram of coal available.

a. How much energy content does it contain in J and kW h?
b. How high could you theoretically lift 1 l of water with this energy?
c. To what speed (in km h^{-1}) could you theoretically accelerate a car (mass: 1 t) with this energy?

Exercise 1.2 Environmental Effects of the Present Energy Supply

a. What are the main problems with the present energy supply?
b. What is the mean temperature on Earth today and what could it be without the greenhouse effect?
c. Explain the greenhouse effect in terms of keywords and a sketch.

Exercise 1.3 Finiteness of Resources

a. Assume every person has the present primary energy requirements of a person in Germany. For how many people would the present world primary energy consumption last?
b. On the basis of Table 1.3, the scope of oil, natural gas, and coal with varying annual growths p are to be estimated starting from 2008. Derive a general equation *scope = f (annual extraction, reserves, p)*. Use the equation of the geometrical series (see Equations (10.10) and (10.11)).
c. Calculate the scope of oil, natural gas, and coal for an annual growth p of 2.2%.
d. Calculate the scope of oil, natural gas, and coal for an annual growth p of 4.4%.

Exercise 1.4 Properties of Renewable Energies

a. What are the three primary sources of renewable energies?
b. Name three advantages and three disadvantages of renewable energies.

Photovoltaics – Fundamentals, Technology, and Practice, Second Edition. Konrad Mertens.

Exercise 1.5 Yields of a Photovoltaic Plant

a. What is meant by STC, and what boundary conditions are associated with it?
b. Family Meyer consumes 3500 kW h of electricity in a year. What should be the minimum power of a photovoltaic plant in order to generate this amount of energy?
c. What module area is required when the modules have an efficiency of 15%?

Exercises for Chapter 2

Exercise 2.1 Solar Constants

a. Earth does not circle the Sun in an ideal circle, but rather along an elliptical path $r_{SE\_Max} = 152$ Mio km, $r_{SE\_Min} = 147$ Mio km). Between what values does the solar constant fluctuate in the course of a year?
b. What solar constant do you expect from Mercury that is approximately 58 Mio km from the Sun?

Exercise 2.2 Solar Spectrum

a. What does AM 0 stand for?
b. What does AM 1.5 mean, and what solar elevation angle is associated with it?
c. Why is the sky blue?
d. What causes sunset glow?

Exercise 2.3 Global Radiation

a. What effects lead to the existence of diffuse radiation?
b. What approximate portion of the global radiation does the diffuse radiation in Germany have?
c. What is meant by "Sun full load hours" and what rule of thumb could one use for Germany?
d. How does the value of the previous subpoint change for a surface facing south and tilted at an angle of 35°?

Exercise 2.4 Radiation on Tilted Surfaces

On a cloudless summer's day, the measured global radiation is $E_{G\_Hor} = 850\ \text{W m}^{-2}$. The Sun is at an angle of 50° to the horizon. Assume that there is no diffuse radiation.

a. What is the optimum inclination angle of the solar module and what is the irradiance E_{Gen} in this case?
b. What is the irradiance for an inclination angle of the solar module of 15° to the horizon? Now take a winter's day: $E_{Direct\_H} = E_{Diffus\_H} = 300\ \text{W m}^{-2}$. The Sun is at an angle of 25° to the horizon.
c. What is the optimum inclination angle of the solar module and what is the irradiance in this case? Determine the best angle by trial and error.
d. Solve the subpoint (c) not by trial and error, but by means of an extreme value consideration of the equation $E_{Gen} = f(\beta)$. Make use of the following addition theorem:

$$\cos(x_1 + x_2) = \cos(x_1) \cdot \cos(x_2) - \sin(x_1) \cdot \sin(x_2).$$

Exercises for Chapter 3

Exercise 3.1 Charge Carriers in Semiconductors

a. Calculate the intrinsic carrier concentration of silicon at 100 °C.
b. Explain the difference between field current and diffusion current.

Exercise 3.2 p–n Junction

a. Describe the origin of the space charge region in the p–n junction.
b. Sketch the band diagram of a p–n junction in a qualitative manner.
c. Calculate the diffusion voltage of a p–n junction of silicon at the following doping concentrations: $N_A = 5 \times 10^{16}\ cm^{-3}$, $N_D = 10^{18}\ cm^{-3}$.

Exercise 3.3 Light Absorption in Semiconductors

a. Verify the interconnection between the absorption coefficient and the penetration depth of light in a semiconductor (Equation (3.21)). For this use suitable values in Equation (3.20).
b. What is the photon energy of light at the wavelength $\lambda = 560$ nm?
c. What is the penetration depth of light at the wavelength $\lambda = 560$ nm in c-Si and a-Si?

Exercise 3.4 Antireflection Films

Light with a wavelength of 600 nm and an irradiance of $E_0 = 500\ W\ m^{-2}$ impinges vertically on to a semiconductor of amorphous silicon. At this wavelength, the material has a refractive index of 4.6.

a. What portion of the light is reflected at the semiconductor surface?
b. What thickness and what refractive index should an antireflective film possess in an ideal case?
c. In an actual case, silicon nitride (Si_3N_4) is used as an antireflective film. What should the thickness of the film be in this case, and how large is the remaining reflection factor?

Exercises for Chapter 4

Exercise 4.1 Recombination in the c-Si Solar Cell

a. What types of recombination do you know?
b. What is understood by the term *dead layer* in a cell?
c. A light photon is absorbed at a depth of 140 μm and generates an electron–hole pair. The average carrier life span is 7 μs. Will the generated electron probably contribute to the photocurrent?

Exercise 4.2 Absorption Efficiency of a c-Si Cell

Given is a c-Si cell of thickness $d = 140$ μm that is illuminated by light with a strength of $E_0 = 1000\ W\ m^{-2}$, ($\alpha = 100\ cm^{-1}$; $n = 3.3$; $\lambda = 1000$ nm).

a. How large is the penetration depth of the light?
b. How much light is reflected?

c. How much light is absorbed (assumption: rear side nonreflecting)?
d. How much light is absorbed when the rear side is mirrored and the front side is nonreflecting?
e. How big in the previous subpoint are the absorption efficiency and the spectral sensitivity if one assumes that every electron–hole pair contributes to the photocurrent?

Exercise 4.3 Single Diode Equivalent Circuit Diagram

a. Sketch the single-diode equivalent circuit diagram and derive the characteristic curve equation.
b. What losses are incurred by the series resistance R_S and what by the parallel resistance R_{Sh}?
c. Sketch the characteristic curve of a solar cell for a rising series resistance R_S and explain the sequence for V_{OC} and I_{SC}.
d. Sketch the characteristic curve of a solar cell for a falling shunt resistance R_{Sh} and explain the sequence for V_{OC} and I_{SC}.

Exercise 4.4 Spectral and Theoretical Efficiency

a. What does the term *spectral efficiency* mean and what effects lead to the fact that it is not 100%?
b. What further losses does the "theoretical efficiency" take into account?
c. How large is the theoretic efficiency of c-Si cells and how near have we come to this optimum?

Exercise 4.5 Spectral Efficiency of Monochromatic Light

Assume that we wish to convert monochromatic laser light into electrical energy as efficiently as possible. We select $\lambda = 1000$ nm as laser wavelength and the further data are $E = 1000\ \text{W m}^{-2}$, $A_{Cell} = 10\ \text{cm}^{-2}$, $m = 1$).

a. How large is the number N_{Ph} of photons per second that impinge on the cell and the maximum current density j_{Max}?
b. Determine the open-circuit voltage of the cell.
c. Give the idealized fill factor of the cell.
d. How great is the theoretical efficiency of the cell?
e. Answer the previous question for a concentration factor of $X = 1000$.

Exercises for Chapter 5

Exercise 5.1 Production of c-Si Solar Cells

a. What do the following abbreviations mean: SG-Si, MG-Si, UMG-Si, CZ-Si, FZ-Si, EFG?
b. List the seven main steps for producing a c-Si standard cell starting from the p-doped wafer.

Exercise 5.2 a-Si Thin-Film Cells

a. What is the basic difference between drift cells and diffusion cells?
b. Sketch the structure of an a-Si tandem cell with information on the materials and describe its function.
c. What is the Staebler–Wronski effect, and how can it be moderated?

Exercise 5.3 CIS Cells

a. Sketch the structure of a CIS cell
b. What two functions does the CdS layer perform?
c. Explain the difference between superstrate configuration and substrate configuration in a thin film cell.

Exercise 5.4 Concentrator Systems

a. Sketch the two most important principles of concentrator systems.
b. Explain why an increased irradiance leads to an increase in the efficiency of a solar cell.
c. Describe a solar cell with the following data: $V_{OC} = 600$ mV; $m = 1.5$; $\eta = 18\%$. What is the efficiency with the following concentration factors:
   i. $X_1 = 100$
   ii. $X_2 = 400$
d. Why does the efficiency not continue to increase with an increase of factor X?

Exercise 5.5 Ecology Questions

a. How do you judge the availability for cells of c-Si, CdTe, and CIS?
b. An energy demand of 5500 kWh kWp^{-1} was required for the production of a complete PV plant. Calculate the energy amortization time T_A and the energy returned on the energy invested EROEI:
   i. On a south-facing roof (35° pitch) in Germany.
   ii. On a west-facing roof (35° pitch) in Germany.

Exercises for Chapter 6

Exercise 6.1 Short-Circuit Current and Open-Circuit Voltage for a Variation of the Irradiance

A solar module with the data $V_{OC} = 43.2$ V, $I_{SC} = 10$ A is given. The installed cells have an ideality factor of $m = 1.5$.

a. How many cells are probably built into the module?
b. What short-circuit current will adjust itself with an irradiance E' of 500 W m^{-2}?
c. What open-circuit voltage will adjust itself with an irradiance E' of 500 W m^{-2}?

Exercise 6.2 Series Connection of Modules

a. Give two reasons for the use of bypass diodes.
b. Now two modules of the type from Exercise 6.1 are connected in series. Module A is radiated with 1000 W m^{-2} and module B with 500 W m^{-2}. Sketch the individual curves and the combined curve for the cases:
   i. The modules have no bypass diodes.
   ii. Both modules have at least one bypass diode.

Exercise 6.3 NOCT

a. What is the meaning of NOCT, and what boundary conditions are associated with it?
b. What will be the temperature and module power of the FS-102A module from First Solar (see Table 6.1) at 900 W m^{-2} and an ambient temperature of 30 °C?

Exercise 6.4 Mismatching

a. What is understood by mismatching?
b. On the basis of the generator I/V curves, explain: Why in the case of shading of a module, it is particularly disadvantageous to connect the modules into two instead of one string?

Exercises for Chapter 7

Exercise 7.1 Buck Converter

a. Sketch the circuit of a buck converter, and explain the functions of the individual components.
b. What are the advantages of a high-switching frequency?
c. What could be a disadvantage of a high-switching frequency?

Exercise 7.2 Feed-in Variations

Sketch the two feed-in variations *full feed-in* and *excess feed-in* of a photovoltaic plant.

Exercise 7.3 Inverter Variations

a. Give the advantages and disadvantages of the plant variations with central inverter, string inverter, and module-integrated inverter.
b. What is the advantage of a PWM bridge compared to a classic 50 Hz bridge?
c. In what cases should inverters without transformers not be installed without investigation?
d. Why inverters with high-frequency transformers are used?
e. Give three advantages of the three-phase inverter.

Exercise 7.4 Inverter Dimensioning

Assume that you have an inverter SMC 8000 TL from SMA available and wish to drive as many modules of type c-Si SW-280 from SolarWorld with it.

a. Determine the maximum possible number of modules per string.
b. Determine the minimum possible number of modules per string.
c. Determine the maximum possible number of strings.
d. Determine the optimum plant configuration.

Exercises for Chapter 8

Exercise 8.1 Battery Systems

a. Why does deep discharging reduce the life of lead batteries?
b. Why is a car battery not suitable for an off-grid solar plant?
c. In the data sheet of a battery, the capacity of $C_{10} = 150$ A h is given. What does this mean, and how long do you estimate the battery can be discharged with a current of 20 A?
d. What is understood by the I/V charging method?
e. Give two advantages of the shunt charge controller compared to the series controller.

Exercise 8.2 Lithium Ion Batteries

a. Name two advantages and two disadvantages of lithium ion batteries with respect to lead acid batteries.
b. What can happen inside a lithium ion battery in case of overcharging?
c. Name two advantages of lithium iron phosphate as a cathode material.

Exercise 8.3 Sodium Sulfur Batteries

a. Why sodium sulfur batteries are not suited as long-time storage?
b. What does the anode container serve for?
c. Why do the sodium sulfur batteries attain higher cycle numbers with respect to lead batteries?

Exercise 8.4 Redox Flow Batteries

a. Sketch the build-up of a redox flow battery with a designation of all relevant components.
b. Whereby the power of the battery is determined, whereby the capacity?
c. Name two advantages of this battery type with respect to its charging behavior.

Exercise 8.5 Home Storage Systems

a. Determine roughly the costs to store a kilowatt-hour for System 2 from Table 8.3.
b. What is meant by “peak-shaving”?
c. Which two technical conditions have to be observed for the solar storage funding program?

Exercise 8.6 Off-grid Systems

a. What does a typical solar home system look like?
b. What is understood by a hybrid system, and what are the advantages over pure solar off-grid systems?

Exercises for Chapter 9

Exercise 9.1 Radiation Sensors

a. What are the best class of pyranometers called and what is their accuracy?
b. What two possibilities do you know in order to measure only diffuse radiation?
c. Can a pyranometer be used as a reference sensor in a module flasher?

Exercise 9.2 Peak Power Measurement at Site

Sketch the structure of a characteristic curve measuring device for peak power determination with all relevant components.

Exercise 9.3 Thermographic Measuring Technology

a. In solar module checking what is the suitability of bright thermography and what of dark thermography?
b. In a temperature measurement of a solar module, the thermographic camera shows a value of 51 °C. However, in error, you have set the emission factor of the camera at 0.9 instead of 0.8. What is the actual temperature of the module?

Exercise 9.4 Electroluminescence Measuring Technology

a. Why is a normal CCD camera unsuitable for EL measurements of c-Si modules?
b. What types of cell errors can hardly be recognized in the thermography measurement but readily in the EL measurement?
c. Name two advantages and two disadvantages of the outdoor EL technique with respect to thermography.

Exercise 9.5 PID Effect

a. What is the main reason for the drastic power decay of PID affected solar modules?
b. How solar modules can be examined for PID vulnerability in the lab?
c. How can one identify with outdoor EL if a module string is affected by PID?

Exercises for Chapter 10

Exercise 10.1 Shading

An aerial tube of 5 cm in diameter is situated at a distance of 2 m from a photovoltaic plant.

a. Does the tube throw a deep shade on the plant?
b. How broad is the deep shade?
c. Assume the deep shading is unavoidable. Would you install the module rather vertically or horizontally?

Exercise 10.2 Yield Estimation

Farmer Jones would like to build a 30 kWp installation on a new barn in Attendorn, Germany. The roof has an orientation of 45° and a pitch of 12°. Determine the expected specific annual yield (final yield Y_F) in accordance with the following methods:

a. Use of the suitable fixed value for Germany of 900 kWh $(kWp\,a)^{-1}$ for optimum alignment and subsequent inclination loss reduction according to Table 2.4.
b. The fixed value in (a) is valid for a global radiation total H of 1000 kWh $(m^2\ a)^{-1}$. Take into account the approximate value for H' at Attendorn according to Figure 2.7 (about midway between Dortmund and Siegen).
c. Determine H″ at Attendorn using the PVGIS Internet database (see Table 10.2) and calculate analogously to the previous subpoint. Method: use the Internet address *re.jrc.ec.europa.eu/pvgis/apps4/pvest.php*, *Interactive Maps, Europe,* enter the site, type in a location and press *Search*, Select menu *Monthly radiation*, Radiation database: *Climate-SAF PVGIS*, tick *Horizontal radiation*, press *Calculate*.
d. Determine Y_F''' at site Attendorn via the PVGIS Internet database in that you press Menu *PV estimation*, then *Climate-SAF PVGIS* Radiation database, enter slope and azimuth, leave all other values and then press *Calculate*.

Exercise 10.3 Return Calculation

Farmer Jones of Exercise 10.2 purchases his plant for 42 000€ net. The feed-in tariff is 12 cents per kilowatt-hour; his customer tariff is 22 cents per kilowatt-hour. He achieves a self-consumption rate of 40%. As an expected annual yield, we use the results of Exercise 10.2d.

a. Calculate the amortization time.
b. Calculate the object return.

Exercise 10.4 Plant Monitoring

a. What is the difference between reference yield and Sun full load hours?
b. Does a southwestern aligned plant have a poorer performance ratio than a southern aligned one?
c. What values for the performance ratio can one expect for actual plants?

Exercises for Chapter 11

Exercise 11.1 Potential Estimation for Pitched Roofs

The Saarland (a small state in Germany) has an area of 2570 km^2.

a. What theoretical potential does this area possess?
b. Assuming approximately 0.3% of this area is available for the photovoltaics of suitable roof surfaces (east via south to west). What is the radiation energy on this surface?
c. What PV power can be installed on this area ($\eta_{Module} = 20\%$), and what is the electrical energy generation potential ($\eta_{System} = 18\%$)?

Exercise 11.2 Potential Estimations for Free Areas

Assume you have a hectare of free surface available in the Saarland.

a. What is the area utilization factor, when there is to be no self-shading (degree of latitude $\phi = 49°$, module inclination $\beta = 20°$)?
b. What PV capacity can be erected on this surface, and what is the electricity generation potential?
c. What electrical energy could be generated on this hectare if energy corn is planted instead of using it for photovoltaics?

Exercise 11.3 Market and Price Development

a. What is meant by the "learning rate"?
b. What is meant by "grid parity"?

Exercise 11.4 Nowadays Electricity Supply System

a. What is meant by "control energy?"
b. Which conditions apply for primary control?
c. How the primary control is made possible technically?
d. In what respect photovoltaics and wind power complement each other well?

Exercise 11.5 Future Energy Supply

a. Why most of the future scenarios assume that the electricity consumption will rise?
b. How many hours the existing electric vehicle batteries in 2050 could take over nowadays grid load in principle?
c. Which technology facilitates a "season storage for electricity," and which efficiency has to be expected there?

A

Solar Radiation Diagrams

Influence of Orientation and Inclination on the Yearly Radiations Sums for Different Locations

Location Bern, Switzerland

Azimuth α		Angle of slope β: 0°	5°	10°	15°	20°	25°	30°	35°	40°	45°	50°	55°	60°	65°	70°	75°	80°	85°	90°
		−180°	88.5	84.9	80.9	76.5	71.9	67.5	63.2	59.0	54.9	51.0	47.2	43.6	40.3	37.5	35.4	33.8	32.4	31.1	29.9
		−175°	88.5	84.9	80.9	76.6	72.0	67.6	63.2	59.1	55.0	51.1	47.3	43.7	40.4	37.6	35.5	33.9	32.6	31.2	30.0
	North	−170°	88.5	85.0	81.0	76.7	72.2	67.8	63.5	59.4	55.3	51.3	47.5	44.0	40.8	38.1	36.1	34.5	33.0	31.7	30.4
		−165°	88.5	85.1	81.1	76.9	72.5	68.1	63.9	59.7	55.7	51.8	48.0	44.6	41.4	38.9	37.0	35.3	33.8	32.3	31.1
		−160°	88.5	85.2	81.4	77.2	72.9	68.6	64.5	60.4	56.4	52.5	48.9	45.4	42.4	40.1	38.2	36.5	34.9	33.4	32.0
		−155°	88.5	85.2	81.6	77.5	73.4	69.2	65.1	61.1	57.2	53.4	49.8	46.6	43.8	41.6	39.6	37.8	36.1	34.6	33.1
		−150°	88.5	85.4	81.9	78.0	74.0	69.9	65.9	62.0	58.2	54.6	51.2	48.2	45.6	43.4	41.3	39.4	37.7	35.9	34.3
		−145°	88.5	85.5	82.2	78.5	74.7	70.8	66.9	63.1	59.4	56.0	52.9	50.1	47.6	45.3	43.2	41.2	39.3	37.5	35.7
		−140°	88.5	85.8	82.6	79.1	75.4	71.8	68.0	64.5	61.0	57.8	54.8	52.1	49.6	47.3	45.1	43.1	41.0	39.1	37.3
		−135°	88.5	86.0	83.0	79.7	76.3	72.8	69.3	65.9	62.7	59.7	56.8	54.2	51.8	49.4	47.1	45.0	42.9	40.9	38.9
	North east	−130°	88.5	86.1	83.4	80.4	77.2	74.0	70.7	67.6	64.5	61.6	59.0	56.4	54.0	51.5	49.2	47.0	44.8	42.7	40.6
		−125°	88.5	86.5	83.9	81.1	78.2	75.2	72.2	69.3	66.4	63.7	61.2	58.6	56.2	53.8	51.4	49.2	46.9	44.6	42.4
		−120°	88.5	86.7	84.4	81.9	79.2	76.5	73.7	71.1	68.4	65.9	63.4	60.9	58.5	56.1	53.7	51.4	48.9	46.6	44.2
		−115°	88.5	86.9	84.9	82.7	80.3	77.9	75.3	72.9	70.5	68.0	65.6	63.2	60.9	58.4	56.0	53.6	51.1	48.6	46.1
		−110°	88.5	87.2	85.5	83.6	81.4	79.2	76.9	74.7	72.5	70.2	67.9	65.5	63.2	60.8	58.3	55.8	53.2	50.7	48.0
		−105°	88.5	87.5	86.1	84.4	82.6	80.6	78.6	76.6	74.5	72.3	70.1	67.8	65.5	63.1	60.6	58.0	55.3	52.7	50.0
		−100°	88.5	87.7	86.6	85.2	83.7	82.0	80.3	78.4	76.5	74.4	72.3	70.1	67.8	65.4	62.8	60.2	57.5	54.6	51.9
		−95°	88.5	88.1	87.2	86.1	84.9	83.4	81.9	80.2	78.5	76.5	74.5	72.3	70.0	67.6	65.0	62.3	59.6	56.5	53.7
		−90°	88.5	88.4	87.8	87.0	86.0	84.8	83.5	82.0	80.4	78.5	76.6	74.4	72.1	69.8	67.1	64.4	61.6	58.5	55.4
	East	−85°	88.5	88.7	88.4	87.8	87.1	86.1	85.0	83.7	82.2	80.5	78.6	76.6	74.3	71.8	69.3	66.3	63.5	60.4	57.2
		−80°	88.5	89.0	89.0	88.7	88.2	87.5	86.5	85.4	84.0	82.4	80.6	78.6	76.3	73.8	71.2	68.3	65.2	62.1	58.8
		−75°	88.5	89.3	89.6	89.5	89.3	88.8	88.1	87.0	85.8	84.2	82.5	80.5	78.2	75.6	73.1	70.2	66.9	63.7	60.4
		−70°	88.5	89.5	90.1	90.3	90.3	90.0	89.5	88.6	87.4	86.1	84.2	82.3	80.1	77.5	74.7	71.8	68.6	65.2	61.7
		−65°	88.5	89.8	90.6	91.1	91.3	91.2	90.8	90.1	89.0	87.7	86.0	84.0	81.8	79.3	76.3	73.4	70.2	66.6	63.0
		−60°	88.5	90.0	91.2	91.9	92.3	92.4	92.1	91.5	90.5	89.3	87.7	85.6	83.4	80.9	78.0	74.8	71.5	68.0	64.2
		−55°	88.5	90.3	91.6	92.5	93.2	93.5	93.3	92.8	91.9	90.7	89.2	87.2	84.9	82.3	79.5	76.2	72.8	69.1	65.2
		−50°	88.5	90.5	92.1	93.2	94.1	94.4	94.4	94.1	93.2	92.1	90.6	88.7	86.3	83.6	80.7	77.5	73.8	70.1	66.1
		−45°	88.5	90.7	92.5	93.8	94.8	95.4	95.4	95.2	94.5	93.3	91.8	90.0	87.6	84.9	81.8	78.5	74.9	71.0	67.0
	South east	−40°	88.5	90.9	92.9	94.4	95.5	96.2	96.4	96.2	95.6	94.5	92.9	91.0	88.7	86.0	82.9	79.5	75.8	71.8	67.6
		−35°	88.5	91.1	93.2	94.9	96.1	97.0	97.3	97.1	96.6	95.5	94.0	92.0	89.6	87.0	83.9	80.3	76.5	72.5	68.1
		−30°	88.5	91.2	93.5	95.4	96.7	97.6	98.0	97.9	97.3	96.4	94.9	92.9	90.5	87.7	84.6	81.0	77.1	72.9	68.5
		−25°	88.5	91.4	93.8	95.7	97.2	98.1	98.6	98.6	98.0	97.0	95.6	93.7	91.2	88.4	85.1	81.5	77.5	73.2	68.7
		−20°	88.5	91.5	94.0	96.0	97.6	98.6	99.1	99.2	98.6	97.6	96.1	94.2	91.8	88.9	85.5	81.9	77.9	73.4	68.8
		−15°	88.5	91.6	94.1	96.3	97.9	98.9	99.5	99.5	99.1	98.1	96.6	94.6	92.2	89.3	85.9	82.1	78.0	73.6	68.8
		−10°	88.5	91.6	94.3	96.4	98.0	99.2	99.8	99.8	99.4	98.5	97.0	94.9	92.4	89.4	86.1	82.3	78.1	73.6	68.8
		−5°	88.5	91.7	94.4	96.5	98.2	99.3	99.9	100.0	99.5	98.6	97.1	95.1	92.5	89.6	86.1	82.3	78.2	73.6	68.7
	South	0°	88.5	91.7	94.4	96.5	98.2	99.3	99.9	100.0	99.5	98.5	97.0	95.1	92.6	89.6	86.2	82.3	78.1	73.5	68.7
		5°	88.5	91.7	94.4	96.5	98.2	99.3	99.9	100.0	99.5	98.6	97.1	95.1	92.5	89.6	86.1	82.3	78.2	73.6	68.7
		10°	88.5	91.6	94.3	96.4	98.0	99.2	99.8	99.8	99.4	98.5	97.0	94.9	92.4	89.4	86.1	82.3	78.1	73.6	68.8
		15°	88.5	91.6	94.1	96.3	97.9	98.9	99.5	99.5	99.1	98.1	96.6	94.6	92.2	89.3	85.9	82.1	78.0	73.6	68.8
		20°	88.5	91.5	94.0	96.0	97.6	98.6	99.1	99.2	98.6	97.6	96.1	94.2	91.8	88.9	85.5	81.9	77.9	73.4	68.8
		25°	88.5	91.4	93.8	95.7	97.2	98.1	98.6	98.6	98.0	97.0	95.6	93.7	91.2	88.4	85.1	81.5	77.5	73.2	68.7
		30°	88.5	91.2	93.5	95.4	96.7	97.6	98.0	97.9	97.3	96.4	94.9	92.9	90.5	87.7	84.6	81.0	77.1	72.9	68.5
		35°	88.5	91.1	93.2	94.9	96.1	97.0	97.3	97.1	96.6	95.5	94.0	92.0	89.6	87.0	83.9	80.3	76.5	72.5	68.1
		40°	88.5	90.9	92.9	94.4	95.5	96.2	96.4	96.2	95.6	94.5	92.9	91.0	88.7	86.0	82.9	79.5	75.8	71.8	67.6
		45°	88.5	90.7	92.5	93.8	94.8	95.4	95.4	95.2	94.5	93.3	91.8	90.0	87.6	84.9	81.8	78.5	74.9	71.0	67.0
	South west	50°	88.5	90.5	92.1	93.2	94.1	94.4	94.4	94.1	93.2	92.1	90.6	88.7	86.3	83.6	80.7	77.5	73.8	70.1	66.1
		55°	88.5	90.3	91.6	92.5	93.2	93.5	93.3	92.8	91.9	90.7	89.2	87.2	84.9	82.3	79.5	76.2	72.8	69.1	65.2
		60°	88.5	90.0	91.2	91.9	92.3	92.4	92.1	91.5	90.5	89.3	87.7	85.6	83.4	80.9	78.0	74.8	71.5	68.0	64.2
		65°	88.5	89.8	90.6	91.1	91.3	91.2	90.8	90.1	89.0	87.7	86.0	84.0	81.8	79.3	76.3	73.4	70.2	66.6	63.0
		70°	88.5	89.5	90.1	90.3	90.3	90.0	89.5	88.6	87.4	86.1	84.2	82.3	80.1	77.5	74.7	71.8	68.6	65.2	61.7
		75°	88.5	89.3	89.6	89.5	89.3	88.8	88.1	87.0	85.8	84.2	82.5	80.5	78.2	75.6	73.1	70.2	66.9	63.7	60.4
		80°	88.5	89.0	89.0	88.7	88.2	87.5	86.5	85.4	84.0	82.4	80.6	78.6	76.3	73.8	71.2	68.3	65.2	62.1	58.8
		85°	88.5	88.7	88.4	87.8	87.1	86.1	85.0	83.7	82.2	80.5	78.6	76.6	74.3	71.8	69.3	66.3	63.5	60.4	57.2
	West	90°	88.5	88.4	87.8	87.0	86.0	84.8	83.5	82.0	80.4	78.5	76.6	74.4	72.1	69.8	67.1	64.4	61.6	58.5	55.4
		95°	88.5	88.1	87.2	86.1	84.9	83.4	81.9	80.2	78.5	76.5	74.5	72.3	70.0	67.6	65.0	62.3	59.6	56.5	53.7
		100°	88.5	87.7	86.6	85.2	83.7	82.0	80.3	78.4	76.5	74.4	72.3	70.1	67.8	65.4	62.8	60.2	57.5	54.6	51.9
		105°	88.5	87.5	86.1	84.4	82.6	80.6	78.6	76.6	74.5	72.3	70.1	67.8	65.5	63.1	60.6	58.0	55.3	52.7	50.0
		110°	88.5	87.2	85.5	83.6	81.4	79.2	76.9	74.7	72.5	70.2	67.9	65.5	63.2	60.8	58.3	55.8	53.2	50.7	48.0
		115°	88.5	86.9	84.9	82.7	80.3	77.9	75.3	72.9	70.5	68.0	65.6	63.2	60.9	58.4	56.0	53.6	51.1	48.6	46.1
		120°	88.5	86.7	84.4	81.9	79.2	76.5	73.7	71.1	68.4	65.9	63.4	60.9	58.5	56.1	53.7	51.4	48.9	46.6	44.2
		125°	88.5	86.5	83.9	81.1	78.2	75.2	72.2	69.3	66.4	63.7	61.2	58.6	56.2	53.8	51.4	49.2	46.9	44.6	42.4
		130°	88.5	86.1	83.4	80.4	77.2	74.0	70.7	67.6	64.5	61.6	59.0	56.4	54.0	51.5	49.2	47.0	44.8	42.7	40.6
		135°	88.5	86.0	83.0	79.7	76.3	72.8	69.3	65.9	62.7	59.7	56.8	54.2	51.8	49.4	47.1	45.0	42.9	40.9	38.9
		140°	88.5	85.8	82.6	79.1	75.4	71.8	68.0	64.5	61.0	57.8	54.8	52.1	49.6	47.3	45.1	43.1	41.0	39.1	37.3
	North west	145°	88.5	85.5	82.2	78.5	74.7	70.8	66.9	63.1	59.4	56.0	52.9	50.1	47.6	45.3	43.2	41.2	39.3	37.5	35.7
		150°	88.5	85.4	81.9	78.0	74.0	69.9	65.9	62.0	58.2	54.6	51.2	48.2	45.6	43.4	41.3	39.4	37.7	35.9	34.3
		155°	88.5	85.2	81.6	77.5	73.4	69.2	65.1	61.1	57.2	53.4	49.8	46.6	43.8	41.6	39.6	37.8	36.1	34.6	33.1
		160°	88.5	85.2	81.4	77.2	72.9	68.6	64.5	60.4	56.4	52.5	48.9	45.4	42.4	40.1	38.2	36.5	34.9	33.4	32.0
		165°	88.5	85.1	81.1	76.9	72.5	68.1	63.9	59.7	55.7	51.8	48.0	44.6	41.4	38.9	37.0	35.3	33.8	32.3	31.1
		170°	88.5	85.0	81.0	76.7	72.2	67.8	63.5	59.4	55.3	51.3	47.5	44.0	40.8	38.1	36.1	34.5	33.0	31.7	30.4
	North	175°	88.5	84.9	80.9	76.6	72.0	67.6	63.2	59.1	55.0	51.1	47.3	43.7	40.4	37.6	35.5	33.9	32.6	31.2	30.0
		180°	88.5	84.9	80.9	76.5	71.9	67.5	63.2	59.0	54.9	51.0	47.2	43.6	40.3	37.5	35.4	33.8	32.4	31.1	29.9

Location Buenos Aires, Argentina

	Angle of slope β																			
Azimuth α		0°	5°	10°	15°	20°	25°	30°	35°	40°	45°	50°	55°	60°	65°	70°	75°	80°	85°	90°
	−180°	92.1	88.8	85.2	81.0	76.5	71.7	66.6	61.7	57.2	52.9	48.7	44.7	40.8	37.2	34.0	31.1	28.9	27.5	26.3
North	−175°	92.1	88.9	85.2	81.1	76.6	71.8	66.8	61.9	57.4	53.1	48.9	44.9	41.1	37.6	34.3	31.5	29.3	27.8	26.7
	−170°	92.1	89.0	85.3	81.3	76.8	72.0	67.1	62.3	57.8	53.4	49.3	45.3	41.5	38.0	34.9	32.1	30.1	28.6	27.3
	−165°	92.1	89.0	85.5	81.5	77.1	72.4	67.6	62.8	58.3	54.0	49.9	46.0	42.3	38.8	35.7	33.1	31.1	29.5	28.2
	−160°	92.1	89.1	85.6	81.7	77.5	72.9	68.2	63.5	59.1	54.8	50.8	46.9	43.2	39.9	36.9	34.5	32.5	30.8	29.4
	−155°	92.1	89.2	85.8	82.1	77.9	73.5	69.0	64.4	60.0	55.8	51.8	48.0	44.5	41.2	38.5	36.1	34.1	32.3	30.7
	−150°	92.1	89.3	86.1	82.4	78.4	74.2	69.9	65.5	61.2	57.1	53.1	49.4	46.1	43.0	40.3	37.9	35.8	33.9	32.1
	−145°	92.1	89.5	86.4	82.9	79.1	75.1	70.9	66.6	62.5	58.5	54.7	51.2	48.0	44.9	42.2	39.8	37.6	35.6	33.7
	−140°	92.1	89.7	86.7	83.4	79.8	75.9	72.0	67.9	64.0	60.2	56.6	53.2	50.0	47.0	44.3	41.8	39.5	37.3	35.3
North east	−135°	92.1	89.8	87.1	83.9	80.5	76.9	73.2	69.4	65.6	62.0	58.6	55.2	52.2	49.2	46.5	43.9	41.5	39.2	37.1
	−130°	92.1	90.0	87.5	84.5	81.4	78.0	74.4	70.9	67.3	63.9	60.6	57.4	54.4	51.5	48.7	46.1	43.6	41.2	38.9
	−125°	92.1	90.2	87.9	85.2	82.2	79.0	75.8	72.4	69.1	65.8	62.7	59.6	56.6	53.7	50.9	48.2	45.7	43.2	40.7
	−120°	92.1	90.5	88.4	85.9	83.1	80.2	77.2	74.0	71.0	67.8	64.8	61.8	58.8	56.0	53.2	50.4	47.7	45.2	42.6
	−115°	92.1	90.7	88.8	86.6	84.1	81.4	78.6	75.6	72.7	69.8	66.9	64.0	61.1	58.2	55.4	52.6	49.8	47.1	44.4
	−110°	92.1	90.9	89.3	87.3	85.1	82.6	80.0	77.3	74.6	71.8	69.0	66.1	63.3	60.4	57.5	54.7	51.8	49.1	46.2
	−105°	92.1	91.2	89.8	88.0	86.0	83.8	81.5	79.0	76.3	73.7	71.0	68.2	65.4	62.6	59.7	56.7	53.8	50.9	48.0
	−100°	92.1	91.4	90.3	88.8	87.0	85.1	82.9	80.6	78.2	75.6	73.0	70.3	67.5	64.6	61.7	58.7	55.7	52.7	49.6
	−95°	92.1	91.6	90.8	89.5	88.0	86.3	84.3	82.2	79.9	77.5	74.9	72.3	69.5	66.6	63.6	60.6	57.5	54.5	51.2
East	−90°	92.1	91.9	91.3	90.3	89.0	87.5	85.7	83.8	81.6	79.3	76.8	74.1	71.4	68.5	65.5	62.3	59.2	56.0	52.8
	−85°	92.1	92.2	91.8	91.1	90.0	88.7	87.1	85.3	83.2	81.0	78.6	76.0	73.2	70.3	67.3	64.2	60.8	57.4	54.2
	−80°	92.1	92.5	92.3	91.8	90.9	89.8	88.4	86.8	84.8	82.7	80.3	77.7	75.0	71.9	68.9	65.7	62.3	58.8	55.3
	−75°	92.1	92.7	92.8	92.5	91.9	90.9	89.7	88.2	86.4	84.3	81.9	79.4	76.7	73.7	70.4	67.1	63.7	60.1	56.5
	−70°	92.1	92.9	93.3	93.2	92.8	92.0	90.9	89.5	87.8	85.8	83.5	80.9	78.1	75.2	71.9	68.4	64.9	61.3	57.5
	−65°	92.1	93.2	93.8	93.9	93.7	93.1	92.1	90.8	89.2	87.2	85.0	82.3	79.5	76.5	73.2	69.7	66.0	62.2	58.4
	−60°	92.1	93.4	94.2	94.6	94.5	94.1	93.2	92.0	90.4	88.5	86.3	83.8	80.8	77.6	74.3	70.7	67.0	63.1	59.1
	−55°	92.1	93.6	94.6	95.2	95.3	95.0	94.2	93.2	91.6	89.7	87.5	85.0	82.1	78.9	75.4	71.7	67.8	63.8	59.7
	−50°	92.1	93.8	95.0	95.7	96.0	95.9	95.2	94.2	92.8	90.9	88.6	86.0	83.1	79.9	76.4	72.6	68.6	64.4	60.1
South east	−45°	92.1	94.0	95.4	96.2	96.7	96.6	96.1	95.2	93.8	92.0	89.7	87.1	84.1	80.8	77.2	73.3	69.2	64.9	60.5
	−40°	92.1	94.1	95.7	96.7	97.3	97.4	96.9	96.0	94.7	92.9	90.7	88.0	85.0	81.6	77.9	74.0	69.7	65.3	60.8
	−35°	92.1	94.3	96.0	97.1	97.8	98.0	97.6	96.8	95.5	93.8	91.5	88.8	85.8	82.4	78.6	74.5	70.2	65.6	60.9
	−30°	92.1	94.4	96.2	97.5	98.3	98.5	98.3	97.5	96.2	94.4	92.2	89.5	86.4	82.9	79.1	74.9	70.4	65.8	60.9
	−25°	92.1	94.5	96.4	97.8	98.7	99.0	98.8	98.1	96.8	95.0	92.8	90.1	86.9	83.4	79.5	75.2	70.6	65.8	60.8
	−20°	92.1	94.6	96.6	98.1	99.0	99.4	99.2	98.5	97.3	95.5	93.3	90.5	87.3	83.7	79.7	75.3	70.6	65.8	60.7
	−15°	92.1	94.7	96.8	98.3	99.2	99.7	99.5	98.9	97.6	95.9	93.6	90.8	87.6	83.9	79.8	75.4	70.7	65.7	60.5
	−10°	92.1	94.7	96.8	98.4	99.4	99.9	99.7	99.1	97.9	96.1	93.9	91.1	87.8	84.1	80.0	75.4	70.6	65.6	60.2
	−5°	92.1	94.7	96.9	98.5	99.5	100.0	99.9	99.2	98.0	96.2	94.0	91.2	87.9	84.1	80.0	75.4	70.6	65.4	60.1
South	0°	92.1	94.7	96.9	98.5	99.5	100.0	99.9	99.2	98.0	96.2	93.9	91.1	87.8	84.1	80.0	75.4	70.5	65.4	60.0
	5°	92.1	94.7	96.9	98.5	99.5	100.0	99.9	99.2	98.0	96.2	94.0	91.2	87.9	84.1	80.0	75.4	70.6	65.4	60.1
	10°	92.1	94.7	96.8	98.4	99.4	99.9	99.7	99.1	97.9	96.1	93.9	91.1	87.8	84.1	80.0	75.4	70.6	65.6	60.2
	15°	92.1	94.7	96.8	98.3	99.2	99.7	99.5	98.9	97.6	95.9	93.6	90.8	87.6	83.9	79.8	75.4	70.7	65.7	60.5
	20°	92.1	94.6	96.6	98.1	99.0	99.4	99.2	98.5	97.3	95.5	93.3	90.5	87.3	83.7	79.7	75.3	70.6	65.8	60.7
	25°	92.1	94.5	96.4	97.8	98.7	99.0	98.8	98.1	96.8	95.0	92.8	90.1	86.9	83.4	79.5	75.2	70.6	65.8	60.8
	30°	92.1	94.4	96.2	97.5	98.3	98.5	98.3	97.5	96.2	94.4	92.2	89.5	86.4	82.9	79.1	74.9	70.4	65.8	60.9
	35°	92.1	94.3	96.0	97.1	97.8	98.0	97.6	96.8	95.5	93.8	91.5	88.8	85.8	82.4	78.6	74.5	70.2	65.6	60.9
	40°	92.1	94.1	95.7	96.7	97.3	97.4	96.9	96.0	94.7	92.9	90.7	88.0	85.0	81.6	77.9	74.0	69.7	65.3	60.8
South west	45°	92.1	94.0	95.4	96.2	96.7	96.6	96.1	95.2	93.8	92.0	89.7	87.1	84.1	80.8	77.2	73.3	69.2	64.9	60.5
	50°	92.1	93.8	95.0	95.7	96.0	95.9	95.2	94.2	92.8	90.9	88.6	86.0	83.1	79.9	76.4	72.6	68.6	64.4	60.1
	55°	92.1	93.6	94.6	95.2	95.3	95.0	94.2	93.2	91.6	89.7	87.5	85.0	82.1	78.9	75.4	71.7	67.8	63.8	59.7
	60°	92.1	93.4	94.2	94.6	94.5	94.1	93.2	92.0	90.4	88.5	86.3	83.8	80.8	77.6	74.3	70.7	67.0	63.1	59.1
	65°	92.1	93.2	93.8	93.9	93.7	93.1	92.1	90.8	89.2	87.2	85.0	82.3	79.5	76.5	73.2	69.7	66.0	62.2	58.4
	70°	92.1	92.9	93.3	93.2	92.8	92.0	90.9	89.5	87.8	85.8	83.5	80.9	78.1	75.2	71.9	68.4	64.9	61.3	57.5
	75°	92.1	92.7	92.8	92.5	91.9	90.9	89.7	88.2	86.4	84.3	81.9	79.4	76.7	73.7	70.4	67.1	63.7	60.1	56.5
	80°	92.1	92.5	92.3	91.8	90.9	89.8	88.4	86.8	84.8	82.7	80.3	77.7	75.0	71.9	68.9	65.7	62.3	58.8	55.3
	85°	92.1	92.2	91.8	91.1	90.0	88.7	87.1	85.3	83.2	81.0	78.6	76.0	73.2	70.3	67.3	64.2	60.8	57.4	54.2
West	90°	92.1	91.9	91.3	90.3	89.0	87.5	85.7	83.8	81.6	79.3	76.8	74.1	71.4	68.5	65.5	62.3	59.2	56.0	52.8
	95°	92.1	91.6	90.8	89.5	88.0	86.3	84.3	82.2	79.9	77.5	74.9	72.3	69.5	66.6	63.6	60.6	57.5	54.5	51.2
	100°	92.1	91.4	90.3	88.8	87.0	85.1	82.9	80.6	78.2	75.6	73.0	70.3	67.5	64.6	61.7	58.7	55.7	52.7	49.6
	105°	92.1	91.2	89.8	88.0	86.0	83.8	81.5	79.0	76.3	73.7	71.0	68.2	65.4	62.6	59.7	56.7	53.8	50.9	48.0
	110°	92.1	90.9	89.3	87.3	85.1	82.6	80.0	77.3	74.6	71.8	69.0	66.1	63.3	60.4	57.5	54.7	51.8	49.1	46.2
	115°	92.1	90.7	88.8	86.6	84.1	81.4	78.6	75.6	72.7	69.8	66.9	64.0	61.1	58.2	55.4	52.6	49.8	47.1	44.4
	120°	92.1	90.5	88.4	85.9	83.1	80.2	77.2	74.0	71.0	67.8	64.8	61.8	58.8	56.0	53.2	50.4	47.7	45.2	42.6
	125°	92.1	90.2	87.9	85.2	82.2	79.0	75.8	72.4	69.1	65.8	62.7	59.6	56.6	53.7	50.9	48.2	45.7	43.2	40.7
	130°	92.1	90.0	87.5	84.5	81.4	78.0	74.4	70.9	67.3	63.9	60.6	57.4	54.4	51.5	48.7	46.1	43.6	41.2	38.9
North west	135°	92.1	89.8	87.1	83.9	80.5	76.9	73.2	69.4	65.6	62.0	58.6	55.2	52.2	49.2	46.5	43.9	41.5	39.2	37.1
	140°	92.1	89.7	86.7	83.4	79.8	75.9	72.0	67.9	64.0	60.2	56.6	53.2	50.0	47.0	44.3	41.8	39.5	37.3	35.3
	145°	92.1	89.5	86.4	82.9	79.1	75.1	70.9	66.6	62.5	58.5	54.7	51.2	48.0	44.9	42.2	39.8	37.6	35.6	33.7
	150°	92.1	89.3	86.1	82.4	78.4	74.2	69.9	65.5	61.2	57.1	53.1	49.4	46.1	43.0	40.3	37.9	35.8	33.9	32.1
	155°	92.1	89.2	85.8	82.1	77.9	73.5	69.0	64.4	60.0	55.8	51.8	48.0	44.5	41.2	38.5	36.1	34.1	32.3	30.7
	160°	92.1	89.1	85.6	81.7	77.5	72.9	68.2	63.5	59.1	54.8	50.8	46.9	43.2	39.9	36.9	34.5	32.5	30.8	29.4
	165°	92.1	89.0	85.5	81.5	77.1	72.4	67.6	62.8	58.3	54.0	49.9	46.0	42.3	38.8	35.7	33.1	31.1	29.5	28.2
North	170°	92.1	89.0	85.3	81.3	76.8	72.0	67.1	62.3	57.8	53.4	49.3	45.3	41.5	38.0	34.9	32.1	30.1	28.6	27.3
	175°	92.1	88.9	85.2	81.1	76.6	71.8	66.8	61.9	57.4	53.1	48.9	44.9	41.1	37.6	34.3	31.5	29.3	27.8	26.7
	180°	92.1	88.8	85.2	81.0	76.5	71.7	66.6	61.7	57.2	52.9	48.7	44.7	40.8	37.2	34.0	31.1	28.9	27.5	26.3

Location Cairo, Egypt

Azimuth α		Angle of slope β																		
		0°	5°	10°	15°	20°	25°	30°	35°	40°	45°	50°	55°	60°	65°	70°	75°	80°	85°	90°
	−180°	93.7	90.8	87.3	83.4	79.1	74.3	69.3	64.1	59.1	54.4	49.9	45.6	41.5	37.7	34.2	31.0	28.3	26.3	25.1
North	−175°	93.7	90.8	87.4	83.4	79.1	74.4	69.4	64.2	59.3	54.6	50.0	45.7	41.6	37.8	34.3	31.1	28.4	26.5	25.2
	−170°	93.7	90.8	87.4	83.6	79.3	74.6	69.7	64.6	59.6	54.9	50.4	46.1	42.0	38.2	34.6	31.5	28.9	27.0	25.8
	−165°	93.7	90.9	87.5	83.7	79.6	75.0	70.1	65.1	60.1	55.4	50.9	46.7	42.6	38.8	35.3	32.3	29.8	28.0	26.6
	−160°	93.7	91.0	87.7	84.0	79.9	75.4	70.7	65.8	60.9	56.2	51.7	47.5	43.4	39.7	36.3	33.3	31.1	29.2	27.7
	−155°	93.7	91.1	87.9	84.3	80.3	76.0	71.4	66.7	61.9	57.2	52.7	48.5	44.6	40.9	37.6	34.9	32.6	30.7	29.0
	−150°	93.7	91.2	88.1	84.6	80.8	76.6	72.2	67.6	63.0	58.4	54.0	49.9	46.0	42.5	39.4	36.6	34.3	32.3	30.5
	−145°	93.7	91.3	88.4	85.0	81.4	77.3	73.1	68.7	64.3	59.9	55.6	51.6	47.8	44.4	41.3	38.5	36.1	33.9	32.0
	−140°	93.7	91.4	88.7	85.5	82.0	78.1	74.1	69.9	65.7	61.5	57.4	53.5	49.9	46.5	43.4	40.6	38.0	35.7	33.6
North east	−135°	93.7	91.6	89.0	86.0	82.7	79.0	75.2	71.2	67.1	63.2	59.3	55.5	52.0	48.7	45.6	42.7	40.0	37.6	35.4
	−130°	93.7	91.8	89.4	86.5	83.4	80.0	76.3	72.5	68.7	65.0	61.2	57.6	54.1	50.9	47.8	44.9	42.1	39.6	37.2
	−125°	93.7	92.0	89.7	87.1	84.1	81.0	77.5	74.0	70.4	66.8	63.2	59.7	56.3	53.1	50.0	47.1	44.3	41.6	39.1
	−120°	93.7	92.2	90.1	87.7	85.0	82.0	78.8	75.5	72.1	68.6	65.2	61.9	58.6	55.3	52.2	49.3	46.4	43.6	41.0
	−115°	93.7	92.4	90.6	88.4	85.8	83.1	80.1	77.0	73.8	70.5	67.2	64.0	60.8	57.6	54.4	51.4	48.5	45.6	42.8
	−110°	93.7	92.6	91.0	89.0	86.7	84.1	81.4	78.5	75.5	72.3	69.2	66.1	62.9	59.7	56.6	53.5	50.5	47.6	44.7
	−105°	93.7	92.8	91.5	89.7	87.6	84.8	82.7	80.0	77.2	74.2	71.2	68.1	65.0	61.8	58.7	55.6	52.4	49.5	46.4
	−100°	93.7	93.0	91.9	90.4	88.5	86.4	84.0	81.5	78.8	76.0	73.0	70.0	67.0	63.8	60.7	57.5	54.3	51.2	48.1
	−95°	93.7	93.3	92.4	91.1	89.4	87.5	84.8	83.0	80.4	77.7	74.9	71.9	68.9	65.8	62.5	59.4	56.1	52.9	49.8
East	−90°	93.7	93.5	92.8	91.7	90.3	88.6	86.6	84.4	82.0	79.4	76.6	73.7	70.7	67.5	64.3	61.1	57.8	54.3	51.2
	−85°	93.7	93.7	93.3	92.4	91.2	89.7	87.8	85.8	83.4	81.0	78.3	75.4	72.4	69.3	65.9	62.8	59.3	55.7	52.4
	−80°	93.7	94.0	93.7	93.1	92.1	90.7	89.0	87.1	84.9	82.5	79.9	77.0	74.0	70.9	67.4	64.2	60.8	57.1	53.6
	−75°	93.7	94.2	94.2	93.7	92.9	91.7	90.2	88.4	86.3	83.9	81.4	78.5	75.5	72.4	68.9	65.5	62.0	58.3	54.6
	−70°	93.7	94.4	94.6	94.4	93.7	92.7	91.3	89.7	87.6	85.2	82.8	79.9	76.8	73.7	70.3	66.6	63.1	59.3	55.5
	−65°	93.7	94.6	95.0	95.0	94.5	93.7	92.4	90.8	88.9	86.5	84.1	81.3	78.1	74.9	71.4	67.7	64.0	60.2	56.2
	−60°	93.7	94.8	95.4	95.6	95.3	94.6	93.4	91.9	90.0	87.7	85.2	82.5	79.3	75.9	72.4	68.6	64.8	60.9	56.8
	−55°	93.7	95.0	95.8	96.1	96.0	95.4	94.3	92.9	91.2	88.9	86.3	83.5	80.4	76.9	73.3	69.4	65.5	61.4	57.2
	−50°	93.7	95.2	96.2	96.6	96.6	96.2	95.2	93.8	92.1	90.0	87.3	84.5	81.3	77.8	74.0	70.1	66.0	61.8	57.5
South east	−45°	93.7	95.4	96.5	97.1	97.2	96.9	96.0	94.7	93.0	90.9	88.3	85.3	82.2	78.6	74.7	70.7	66.4	62.1	57.7
	−40°	93.7	95.5	96.8	97.5	97.8	97.5	96.8	95.5	93.8	91.7	89.1	86.1	82.9	79.2	75.3	71.1	66.7	62.3	57.7
	−35°	93.7	95.7	97.1	97.9	98.3	98.1	97.5	96.2	94.5	92.4	89.9	86.8	83.4	79.8	75.7	71.4	66.9	62.3	57.6
	−30°	93.7	95.8	97.3	98.3	98.7	98.6	98.0	96.9	95.2	93.0	90.5	87.5	84.0	80.2	76.1	71.7	67.1	62.3	57.5
	−25°	93.7	95.9	97.5	98.6	99.1	99.0	98.5	97.4	95.8	93.6	91.0	87.9	84.4	80.6	76.4	71.9	67.1	62.2	57.2
	−20°	93.7	96.0	97.7	98.8	99.4	99.4	98.8	97.8	96.2	94.1	91.4	88.3	84.8	80.9	76.5	71.9	67.0	62.0	56.8
	−15°	93.7	96.0	97.8	99.0	99.6	99.7	99.2	98.1	96.5	94.4	91.7	88.6	85.0	81.0	76.6	71.9	66.9	61.7	56.4
	−10°	93.7	96.1	97.9	99.1	99.8	99.9	99.4	98.3	96.7	94.6	91.9	88.8	85.1	81.1	76.6	71.8	66.7	61.4	56.0
	−5°	93.7	96.1	97.9	99.2	99.9	100.0	99.5	98.5	96.8	94.7	92.0	88.8	85.1	81.1	76.6	71.7	66.6	61.2	55.7
South	0°	93.7	96.1	97.9	99.2	99.9	100.0	99.5	98.5	96.8	94.6	91.9	88.8	85.1	81.0	76.5	71.6	66.5	61.0	55.5
	5°	93.7	96.1	97.9	99.2	99.9	100.0	99.5	98.5	96.8	94.7	92.0	88.8	85.1	81.1	76.6	71.7	66.6	61.2	55.7
	10°	93.7	96.1	97.9	99.1	99.8	99.9	99.4	98.3	96.7	94.6	91.9	88.8	85.1	81.1	76.6	71.8	66.7	61.4	56.0
	15°	93.7	96.0	97.8	99.0	99.6	99.7	99.2	98.1	96.5	94.4	91.7	88.6	85.0	81.0	76.6	71.9	66.9	61.7	56.4
	20°	93.7	96.0	97.7	98.8	99.4	99.4	98.8	97.8	96.2	94.1	91.4	88.3	84.8	80.9	76.5	71.9	67.0	62.0	56.8
	25°	93.7	95.9	97.5	98.6	99.1	99.0	98.5	97.4	95.8	93.6	91.0	87.9	84.4	80.6	76.4	71.9	67.1	62.2	57.2
	30°	93.7	95.8	97.3	98.3	98.7	98.6	98.0	96.9	95.2	93.0	90.5	87.5	84.0	80.2	76.1	71.7	67.1	62.3	57.5
	35°	93.7	95.7	97.1	97.9	98.3	98.1	97.5	96.2	94.5	92.4	89.9	86.8	83.4	79.8	75.7	71.4	66.9	62.3	57.6
	40°	93.7	95.5	96.8	97.5	97.8	97.5	96.8	95.5	93.8	91.7	89.1	86.1	82.9	79.2	75.3	71.1	66.7	62.3	57.7
South west	45°	93.7	95.4	96.5	97.1	97.2	96.9	96.0	94.7	93.0	90.9	88.3	85.3	82.2	78.6	74.7	70.7	66.4	62.1	57.7
	50°	93.7	95.2	96.2	96.6	96.6	96.2	95.2	93.8	92.1	90.0	87.3	84.5	81.3	77.8	74.0	70.1	66.0	61.8	57.5
	55°	93.7	95.0	95.8	96.1	96.0	95.4	94.3	92.9	91.2	88.9	86.3	83.5	80.4	76.9	73.3	69.4	65.5	61.4	57.2
	60°	93.7	94.8	95.4	95.6	95.3	94.6	93.4	91.9	90.0	87.7	85.2	82.5	79.3	75.9	72.4	68.6	64.8	60.9	56.8
	65°	93.7	94.6	95.0	95.0	94.5	93.7	92.4	90.8	88.9	86.5	84.1	81.3	78.1	74.9	71.4	67.7	64.0	60.2	56.2
	70°	93.7	94.4	94.6	94.4	93.7	92.7	91.3	89.7	87.6	85.2	82.8	79.9	76.8	73.7	70.3	66.6	63.1	59.3	55.5
	75°	93.7	94.2	94.2	93.7	92.9	91.7	90.2	88.4	86.3	83.9	81.4	78.5	75.5	72.4	68.9	65.5	62.0	58.3	54.6
	80°	93.7	94.0	93.7	93.1	92.1	90.7	89.0	87.1	84.9	82.5	79.9	77.0	74.0	70.9	67.4	64.2	60.8	57.1	53.6
	85°	93.7	93.7	93.3	92.4	91.2	89.7	87.8	85.8	83.4	81.0	78.3	75.4	72.4	69.3	65.9	62.8	59.3	55.7	52.4
West	90°	93.7	93.5	92.8	91.7	90.3	88.6	86.6	84.4	82.0	79.4	76.6	73.7	70.7	67.5	64.3	61.1	57.8	54.3	51.2
	95°	93.7	93.3	92.4	91.1	89.4	87.5	84.8	83.0	80.4	77.7	74.9	71.9	68.9	65.8	62.5	59.4	56.1	52.9	49.8
	100°	93.7	93.0	91.9	90.4	88.5	86.4	84.0	81.5	78.8	76.0	73.0	70.0	67.0	63.8	60.7	57.5	54.3	51.2	48.1
	105°	93.7	92.8	91.5	89.7	87.6	84.8	82.7	80.0	77.2	74.2	71.2	68.1	65.0	61.8	58.7	55.6	52.4	49.5	46.4
	110°	93.7	92.6	91.0	89.0	86.7	84.1	81.4	78.5	75.5	72.3	69.2	66.1	62.9	59.7	56.6	53.5	50.5	47.6	44.7
	115°	93.7	92.4	90.6	88.4	85.8	83.1	80.1	77.0	73.8	70.5	67.2	64.0	60.8	57.6	54.4	51.4	48.5	45.6	42.8
	120°	93.7	92.2	90.1	87.7	85.0	82.0	78.8	75.5	72.1	68.6	65.2	61.9	58.6	55.3	52.2	49.3	46.4	43.6	41.0
	125°	93.7	92.0	89.7	87.1	84.1	81.0	77.5	74.0	70.4	66.8	63.2	59.7	56.3	53.1	50.0	47.1	44.3	41.6	39.1
	130°	93.7	91.8	89.4	86.5	83.4	80.0	76.3	72.5	68.7	65.0	61.2	57.6	54.1	50.9	47.8	44.9	42.1	39.6	37.2
North west	135°	93.7	91.6	89.0	86.0	82.7	79.0	75.2	71.2	67.1	63.2	59.3	55.5	52.0	48.7	45.6	42.7	40.0	37.6	35.4
	140°	93.7	91.4	88.7	85.5	82.0	78.1	74.1	69.9	65.7	61.5	57.4	53.5	49.9	46.5	43.4	40.6	38.0	35.7	33.6
	145°	93.7	91.3	88.4	85.0	81.4	77.3	73.1	68.7	64.3	59.9	55.6	51.6	47.8	44.4	41.3	38.5	36.1	33.9	32.0
	150°	93.7	91.2	88.1	84.6	80.8	76.6	72.2	67.6	63.0	58.4	54.0	49.9	46.0	42.5	39.4	36.6	34.3	32.3	30.5
	155°	93.7	91.1	87.9	84.3	80.3	76.0	71.4	66.7	61.9	57.2	52.7	48.5	44.6	40.9	37.6	34.9	32.6	30.7	29.0
	160°	93.7	91.0	87.7	84.0	79.9	75.4	70.7	65.8	60.9	56.2	51.7	47.5	43.4	39.7	36.3	33.3	31.1	29.2	27.7
	165°	93.7	90.9	87.5	83.7	79.6	75.0	70.1	65.1	60.1	55.4	50.9	46.7	42.6	38.8	35.3	32.3	29.8	28.0	26.6
	170°	93.7	90.8	87.4	83.6	79.3	74.6	69.7	64.6	59.6	54.9	50.4	46.1	42.0	38.2	34.6	31.5	28.9	27.0	25.8
North	175°	93.7	90.8	87.4	83.4	79.1	74.4	69.4	64.2	59.3	54.6	50.0	45.7	41.6	37.8	34.3	31.1	28.4	26.5	25.2
	180°	93.7	90.8	87.3	83.4	79.1	74.3	69.3	64.1	59.1	54.4	49.9	45.6	41.5	37.7	34.2	31.0	28.3	26.3	25.1

Location London, UK

Azimuth α		Angle of slope β																		
		0°	5°	10°	15°	20°	25°	30°	35°	40°	45°	50°	55°	60°	65°	70°	75°	80°	85°	90°
	−180°	86.4	82.6	78.2	73.8	69.5	65.2	61.0	57.0	53.1	49.4	46.0	42.7	39.8	37.6	36.0	34.5	33.1	31.8	30.6
North	−175°	86.4	82.6	78.3	73.9	69.5	65.3	61.1	57.0	53.1	49.4	46.0	42.7	39.8	37.7	36.0	34.6	33.2	31.9	30.6
	−170°	86.4	82.6	78.4	74.0	69.7	65.5	61.3	57.2	53.4	49.7	46.1	43.0	40.1	38.2	36.5	35.0	33.7	32.2	31.0
	−165°	86.4	82.6	78.6	74.2	70.0	65.8	61.6	57.7	53.8	50.1	46.6	43.4	40.8	38.9	37.3	35.7	34.3	32.9	31.5
	−160°	86.4	82.7	78.7	74.6	70.3	66.3	62.2	58.2	54.4	50.8	47.3	44.3	41.9	39.9	38.3	36.7	35.2	33.7	32.2
	−155°	86.4	82.9	79.0	74.9	70.9	66.8	62.8	58.9	55.2	51.6	48.3	45.5	43.3	41.4	39.6	37.9	36.3	34.7	33.3
	−150°	86.4	83.1	79.4	75.5	71.5	67.5	63.6	59.9	56.2	52.7	49.6	47.0	44.9	43.0	41.1	39.3	37.6	36.0	34.4
	−145°	86.4	83.3	79.7	76.0	72.2	68.3	64.6	60.9	57.4	54.2	51.3	48.9	46.7	44.6	42.7	40.9	39.1	37.4	35.6
	−140°	86.4	83.4	80.2	76.6	73.0	69.3	65.7	62.2	58.9	55.9	53.2	50.8	48.5	46.5	44.6	42.6	40.7	38.9	37.1
	−135°	86.4	83.6	80.6	77.2	73.9	70.4	67.0	63.7	60.6	57.8	55.2	52.8	50.6	48.4	46.4	44.5	42.4	40.5	38.6
North east	−130°	86.4	83.9	81.0	77.9	74.8	71.6	68.4	65.4	62.4	59.8	57.3	54.9	52.7	50.6	48.4	46.3	44.3	42.2	40.2
	−125°	86.4	84.1	81.6	78.7	75.8	72.9	70.0	67.1	64.5	61.9	59.5	57.2	55.0	52.8	50.6	48.4	46.2	44.1	41.9
	−120°	86.4	84.4	82.1	79.5	76.9	74.2	71.6	69.0	66.4	64.0	61.7	59.4	57.3	55.0	52.8	50.6	48.4	46.1	43.8
	−115°	86.4	84.7	82.6	80.4	78.0	75.6	73.3	70.9	68.6	66.3	64.0	61.8	59.6	57.4	55.1	52.9	50.5	48.1	45.7
	−110°	86.4	84.9	83.3	81.3	79.3	77.1	74.9	72.8	70.6	68.5	66.3	64.2	62.0	59.8	57.5	55.2	52.7	50.2	47.7
	−105°	86.4	85.3	83.9	82.2	80.4	78.6	76.6	74.7	72.7	70.7	68.6	66.6	64.4	62.2	59.8	57.4	54.9	52.3	49.7
	−100°	86.4	85.6	84.5	83.2	81.7	80.1	78.4	76.6	74.8	73.0	70.9	68.9	66.8	64.5	62.2	59.6	57.1	54.5	51.6
	−95°	86.4	85.9	85.1	84.1	82.9	81.6	80.1	78.6	76.9	75.2	73.3	71.2	69.1	66.8	64.5	61.9	59.3	56.6	53.8
East	−90°	86.4	86.3	85.7	85.0	84.1	83.0	81.8	80.4	78.9	77.3	75.5	73.5	71.4	69.1	66.7	64.1	61.3	58.6	55.7
	−85°	86.4	86.5	86.4	85.9	85.3	84.5	83.4	82.3	81.0	79.5	77.7	75.7	73.7	71.3	68.9	66.3	63.5	60.5	57.6
	−80°	86.4	86.9	87.0	86.9	86.4	85.9	85.1	84.1	82.9	81.5	79.8	77.9	75.8	73.5	70.9	68.4	65.5	62.4	59.3
	−75°	86.4	87.2	87.6	87.8	87.6	87.3	86.7	85.9	84.8	83.4	81.9	80.0	77.9	75.6	73.1	70.2	67.4	64.3	61.0
	−70°	86.4	87.5	88.2	88.7	88.8	88.7	88.2	87.6	86.6	85.4	83.9	82.0	79.9	77.7	75.1	72.2	69.2	66.1	62.6
	−65°	86.4	87.8	88.8	89.5	89.8	89.9	89.7	89.2	88.4	87.2	85.7	84.0	81.8	79.5	77.0	74.1	70.9	67.6	64.2
	−60°	86.4	88.0	89.3	90.3	90.9	91.1	91.1	90.7	90.0	88.9	87.5	85.8	83.8	81.3	78.7	75.8	72.5	69.1	65.5
	−55°	86.4	88.3	89.8	91.0	91.9	92.3	92.5	92.1	91.6	90.6	89.2	87.5	85.6	83.1	80.2	77.3	74.1	70.5	66.8
	−50°	86.4	88.6	90.3	91.7	92.7	93.4	93.6	93.5	93.0	92.1	90.8	89.0	87.1	84.7	81.9	78.7	75.5	71.8	68.0
South east	−45°	86.4	88.8	90.8	92.4	93.5	94.3	94.8	94.8	94.3	93.5	92.3	90.5	88.4	86.1	83.3	80.2	76.7	73.0	69.1
	−40°	86.4	89.0	91.1	92.9	94.3	95.3	95.8	95.9	95.5	94.8	93.6	91.9	89.8	87.2	84.5	81.4	77.8	74.0	70.0
	−35°	86.4	89.2	91.5	93.4	95.0	96.1	96.7	96.9	96.6	95.8	94.7	93.1	91.1	88.5	85.6	82.4	78.7	74.8	70.7
	−30°	86.4	89.4	91.9	94.0	95.6	96.8	97.5	97.8	97.6	96.8	95.6	94.0	92.0	89.5	86.5	83.2	79.5	75.6	71.3
	−25°	86.4	89.5	92.1	94.3	96.1	97.4	98.1	98.5	98.3	97.7	96.5	94.9	92.7	90.3	87.3	84.0	80.2	76.1	71.7
	−20°	86.4	89.5	92.4	94.7	96.5	97.9	98.8	99.1	98.9	98.3	97.3	95.6	93.4	90.9	87.9	84.5	80.7	76.4	72.0
	−15°	86.4	89.6	92.6	94.9	96.8	98.2	99.2	99.6	99.4	98.8	97.7	96.1	94.0	91.3	88.2	84.8	81.0	76.8	72.3
	−10°	86.4	89.7	92.6	95.0	97.0	98.5	99.5	99.9	99.7	99.1	98.0	96.3	94.2	91.7	88.6	85.0	81.1	77.0	72.4
	−5°	86.4	89.7	92.7	95.2	97.2	98.6	99.6	100.0	99.9	99.3	98.1	96.5	94.3	91.8	88.7	85.2	81.2	76.9	72.3
South	0°	86.4	89.8	92.7	95.2	97.2	98.7	99.6	100.0	100.0	99.4	98.2	96.5	94.4	91.7	88.6	85.0	81.1	76.9	72.3
	5°	86.4	89.7	92.7	95.2	97.2	98.6	99.6	100.0	99.9	99.3	98.1	96.5	94.3	91.8	88.7	85.2	81.2	76.9	72.3
	10°	86.4	89.7	92.6	95.0	97.0	98.5	99.5	99.9	99.7	99.1	98.0	96.3	94.2	91.7	88.6	85.0	81.1	77.0	72.4
	15°	86.4	89.6	92.6	94.9	96.8	98.2	99.2	99.6	99.4	98.8	97.7	96.1	94.0	91.3	88.2	84.8	81.0	76.8	72.3
	20°	86.4	89.5	92.4	94.7	96.5	97.9	98.8	99.1	98.9	98.3	97.3	95.6	93.4	90.9	87.9	84.5	80.7	76.4	72.0
	25°	86.4	89.5	92.1	94.3	96.1	97.4	98.1	98.5	98.3	97.7	96.5	94.9	92.7	90.3	87.3	84.0	80.2	76.1	71.7
	30°	86.4	89.4	91.9	94.0	95.6	96.8	97.5	97.8	97.6	96.8	95.6	94.0	92.0	89.5	86.5	83.2	79.5	75.6	71.3
	35°	86.4	89.2	91.5	93.4	95.0	96.1	96.7	96.9	96.6	95.8	94.7	93.1	91.1	88.5	85.6	82.4	78.7	74.8	70.7
	40°	86.4	89.0	91.1	92.9	94.3	95.3	95.8	95.9	95.5	94.8	93.6	91.9	89.8	87.2	84.5	81.4	77.8	74.0	70.0
South west	45°	86.4	88.8	90.8	92.4	93.5	94.3	94.8	94.8	94.3	93.5	92.3	90.5	88.4	86.1	83.3	80.2	76.7	73.0	69.1
	50°	86.4	88.6	90.3	91.7	92.7	93.4	93.6	93.5	93.0	92.1	90.8	89.0	87.1	84.7	81.9	78.7	75.5	71.8	68.0
	55°	86.4	88.3	89.8	91.0	91.9	92.3	92.5	92.1	91.6	90.6	89.2	87.5	85.6	83.1	80.2	77.3	74.1	70.5	66.8
	60°	86.4	88.0	89.3	90.3	90.9	91.1	91.1	90.7	90.0	88.9	87.5	85.8	83.8	81.3	78.7	75.8	72.5	69.1	65.5
	65°	86.4	87.8	88.8	89.5	89.8	89.9	89.7	89.2	88.4	87.2	85.7	84.0	81.8	79.5	77.0	74.1	70.9	67.6	64.2
	70°	86.4	87.5	88.2	88.7	88.8	88.7	88.2	87.6	86.6	85.4	83.9	82.0	79.9	77.7	75.1	72.2	69.2	66.1	62.6
	75°	86.4	87.2	87.6	87.8	87.6	87.3	86.7	85.9	84.8	83.4	81.9	80.0	77.9	75.6	73.1	70.2	67.4	64.3	61.0
	80°	86.4	86.9	87.0	86.9	86.4	85.9	85.1	84.1	82.9	81.5	79.8	77.9	75.8	73.5	70.9	68.4	65.5	62.4	59.3
	85°	86.4	86.5	86.4	85.9	85.3	84.5	83.4	82.3	81.0	79.5	77.7	75.7	73.7	71.3	68.9	66.3	63.5	60.5	57.6
West	90°	86.4	86.3	85.7	85.0	84.1	83.0	81.8	80.4	78.9	77.3	75.5	73.5	71.4	69.1	66.7	64.1	61.3	58.6	55.7
	95°	86.4	85.9	85.1	84.1	82.9	81.6	80.1	78.6	76.9	75.2	73.3	71.2	69.1	66.8	64.5	61.9	59.3	56.6	53.8
	100°	86.4	85.6	84.5	83.2	81.7	80.1	78.4	76.6	74.8	73.0	70.9	68.9	66.8	64.5	62.2	59.6	57.1	54.5	51.6
	105°	86.4	85.3	83.9	82.2	80.4	78.6	76.6	74.7	72.7	70.7	68.6	66.6	64.4	62.2	59.8	57.4	54.9	52.3	49.7
	110°	86.4	84.9	83.3	81.3	79.3	77.1	74.9	72.8	70.6	68.5	66.3	64.2	62.0	59.8	57.5	55.2	52.7	50.2	47.7
	115°	86.4	84.7	82.6	80.4	78.0	75.6	73.3	70.9	68.6	66.3	64.0	61.8	59.6	57.4	55.1	52.9	50.5	48.1	45.7
	120°	86.4	84.4	82.1	79.5	76.9	74.2	71.6	69.0	66.4	64.0	61.7	59.4	57.3	55.0	52.8	50.6	48.4	46.1	43.8
	125°	86.4	84.1	81.6	78.7	75.8	72.9	70.0	67.1	64.5	61.9	59.5	57.2	55.0	52.8	50.6	48.4	46.2	44.1	41.9
	130°	86.4	83.9	81.0	77.9	74.8	71.6	68.4	65.4	62.4	59.8	57.3	54.9	52.7	50.6	48.4	46.3	44.3	42.2	40.2
	135°	86.4	83.6	80.6	77.2	73.9	70.4	67.0	63.7	60.6	57.8	55.2	52.8	50.6	48.4	46.4	44.5	42.4	40.5	38.6
	140°	86.4	83.4	80.2	76.6	73.0	69.3	65.7	62.2	58.9	55.9	53.2	50.8	48.5	46.5	44.6	42.6	40.7	38.9	37.1
North west	145°	86.4	83.3	79.7	76.0	72.2	68.3	64.6	60.9	57.4	54.2	51.3	48.9	46.7	44.6	42.7	40.9	39.1	37.4	35.6
	150°	86.4	83.1	79.4	75.5	71.5	67.5	63.6	59.9	56.2	52.7	49.6	47.0	44.9	43.0	41.1	39.3	37.6	36.0	34.4
	155°	86.4	82.9	79.0	74.9	70.9	66.8	62.8	58.9	55.2	51.6	48.3	45.5	43.3	41.4	39.6	37.9	36.3	34.7	33.3
	160°	86.4	82.7	78.7	74.6	70.3	66.3	62.2	58.2	54.4	50.8	47.3	44.3	41.9	39.9	38.3	36.7	35.2	33.7	32.2
	165°	86.4	82.6	78.6	74.2	70.0	65.8	61.6	57.7	53.8	50.1	46.6	43.4	40.8	38.9	37.3	35.7	34.3	32.9	31.5
	170°	86.4	82.6	78.4	74.0	69.7	65.5	61.3	57.2	53.4	49.7	46.1	43.0	40.1	38.2	36.5	35.0	33.7	32.2	31.0
North	175°	86.4	82.6	78.3	73.9	69.5	65.3	61.1	57.0	53.1	49.4	46.0	42.7	39.8	37.7	36.0	34.6	33.2	31.9	30.6
	180°	86.4	82.6	78.2	73.8	69.5	65.2	61.0	57.0	53.1	49.4	46.0	42.7	39.8	37.6	36.0	34.5	33.1	31.8	30.6

Location Marseille, France

	Azimuth α	Angle of slope β: 0°	5°	10°	15°	20°	25°	30°	35°	40°	45°	50°	55°	60°	65°	70°	75°	80°	85°	90°
North	−180°	86.1	82.0	77.4	72.4	67.2	62.1	57.2	52.6	48.1	43.8	39.7	36.0	32.6	29.5	27.1	25.6	24.6	23.7	23.0
	−175°	86.1	82.0	77.4	72.4	67.3	62.1	57.3	52.6	48.2	43.9	39.8	36.0	32.6	29.6	27.2	25.8	24.7	23.9	23.1
	−170°	86.1	82.1	77.5	72.6	67.4	62.3	57.5	52.9	48.4	44.1	40.1	36.3	32.9	30.0	27.8	26.4	25.3	24.3	23.5
	−165°	86.1	82.1	77.7	72.8	67.8	62.7	57.9	53.3	48.9	44.6	40.7	37.0	33.6	30.8	28.8	27.3	26.1	25.2	24.3
North east	−160°	86.1	82.2	77.9	73.2	68.2	63.2	58.5	54.0	49.6	45.4	41.5	37.9	34.7	32.1	30.2	28.6	27.4	26.3	25.3
	−155°	86.1	82.3	78.1	73.6	68.8	64.0	59.2	54.8	50.5	46.4	42.6	39.1	36.1	33.8	31.8	30.2	28.9	27.6	26.5
	−150°	86.1	82.5	78.4	74.1	69.5	64.8	60.2	55.8	51.6	47.7	44.0	40.8	38.0	35.7	33.7	32.0	30.5	29.1	27.9
	−145°	86.1	82.7	78.8	74.6	70.2	65.8	61.4	57.1	53.0	49.2	45.8	42.8	40.1	37.8	35.7	33.9	32.3	30.8	29.3
	−140°	86.1	82.9	79.2	75.3	71.1	66.9	62.7	58.6	54.8	51.1	47.9	44.9	42.3	40.0	37.9	36.0	34.2	32.6	31.0
	−135°	86.1	83.1	79.7	76.0	72.1	68.1	64.1	60.3	56.7	53.3	50.1	47.3	44.7	42.3	40.2	38.2	36.3	34.5	32.8
	−130°	86.1	83.4	80.2	76.7	73.1	69.3	65.7	62.1	58.7	55.5	52.5	49.7	47.2	44.8	42.6	40.5	38.5	36.5	34.7
	−125°	86.1	83.6	80.7	77.6	74.2	70.8	67.3	64.0	60.9	57.8	55.0	52.3	49.8	47.3	45.1	42.9	40.8	38.7	36.7
	−120°	86.1	83.9	81.3	78.4	75.3	72.2	69.1	66.0	63.0	60.2	57.5	54.9	52.4	49.9	47.7	45.3	43.1	40.9	38.8
	−115°	86.1	84.2	81.9	79.3	76.5	73.7	70.9	68.0	65.3	62.6	60.1	57.5	55.1	52.6	50.2	47.9	45.5	43.3	40.9
East	−110°	86.1	84.5	82.5	80.2	77.8	75.3	72.7	70.1	67.6	65.1	62.6	60.1	57.7	55.3	52.8	50.4	47.9	45.5	43.1
	−105°	86.1	84.8	83.2	81.2	79.1	76.8	74.6	72.2	69.8	67.4	65.1	62.7	60.3	57.9	55.4	52.9	50.3	47.8	45.3
	−100°	86.1	85.2	83.8	82.2	80.4	78.4	76.4	74.2	72.1	69.8	67.6	65.3	62.9	60.4	57.9	55.4	52.7	50.1	47.4
	−95°	86.1	85.5	84.5	83.2	81.7	80.0	78.2	76.3	74.3	72.2	70.1	67.8	65.5	63.0	60.4	57.8	55.1	52.2	49.6
	−90°	86.1	85.9	85.2	84.1	82.9	81.6	80.0	78.3	76.6	74.6	72.5	70.3	67.9	65.5	62.8	60.2	57.4	54.4	51.5
	−85°	86.1	86.2	85.8	85.2	84.2	83.1	81.8	80.3	78.7	76.8	74.9	72.7	70.3	67.9	65.2	62.4	59.6	56.6	53.3
	−80°	86.1	86.5	86.5	86.1	85.5	84.6	83.6	82.3	80.8	79.0	77.1	75.0	72.6	70.2	67.5	64.5	61.6	58.6	55.3
	−75°	86.1	86.8	87.1	87.1	86.7	86.1	85.3	84.1	82.8	81.1	79.3	77.2	74.8	72.3	69.7	66.7	63.5	60.4	57.0
	−70°	86.1	87.2	87.8	88.0	87.9	87.6	87.0	86.0	84.8	83.2	81.4	79.3	77.0	74.3	71.7	68.7	65.4	62.1	58.6
South east	−65°	86.1	87.5	88.4	88.9	89.1	89.0	88.5	87.7	86.6	85.2	83.4	81.4	79.1	76.4	73.5	70.5	67.2	63.6	60.1
	−60°	86.1	87.8	89.0	89.8	90.2	90.3	90.1	89.3	88.4	87.1	85.3	83.2	81.0	78.4	75.3	72.2	68.8	65.1	61.4
	−55°	86.1	88.1	89.6	90.6	91.3	91.5	91.5	90.9	90.0	88.8	87.2	85.0	82.7	80.1	77.1	73.7	70.3	66.5	62.6
	−50°	86.1	88.4	90.1	91.4	92.3	92.7	92.8	92.4	91.6	90.4	88.9	86.8	84.3	81.7	78.6	75.2	71.6	67.7	63.6
	−45°	86.1	88.6	90.6	92.1	93.2	93.8	94.0	93.8	93.0	91.8	90.4	88.4	85.9	83.1	80.1	76.6	72.8	68.8	64.6
	−40°	86.1	88.8	91.0	92.8	94.0	94.8	95.2	95.1	94.5	93.3	91.7	89.8	87.4	84.5	81.3	77.8	73.9	69.8	65.4
	−35°	86.1	89.0	91.5	93.4	94.8	95.8	96.1	96.1	95.7	94.6	92.9	91.0	88.6	85.8	82.4	78.8	74.9	70.6	66.1
	−30°	86.1	89.2	91.8	93.9	95.5	96.5	97.1	97.1	96.7	95.7	94.2	92.1	89.7	86.8	83.5	79.7	75.7	71.3	66.6
	−25°	86.1	89.3	92.1	94.3	96.0	97.2	97.9	97.9	97.5	96.6	95.2	93.2	90.6	87.6	84.2	80.4	76.3	71.8	67.0
South	−20°	86.1	89.5	92.4	94.7	96.5	97.8	98.5	98.7	98.2	97.3	95.9	93.9	91.4	88.3	84.9	81.0	76.7	72.1	67.3
	−15°	86.1	89.6	92.6	95.0	96.9	98.3	99.0	99.2	98.9	97.9	96.4	94.5	92.0	89.0	85.4	81.5	77.1	72.4	67.4
	−10°	86.1	89.7	92.7	95.2	97.2	98.5	99.4	99.7	99.3	98.4	97.0	94.9	92.3	89.3	85.8	81.7	77.3	72.6	67.4
	−5°	86.1	89.7	92.8	95.4	97.3	98.8	99.6	99.9	99.6	98.8	97.3	95.2	92.6	89.5	86.0	82.0	77.5	72.6	67.5
	0°	86.1	89.7	92.8	95.4	97.4	98.9	99.7	100.0	99.7	98.8	97.3	95.3	92.8	89.7	86.0	82.0	77.5	72.7	67.5
	5°	86.1	89.7	92.8	95.4	97.3	98.8	99.6	99.9	99.6	98.8	97.3	95.2	92.6	89.5	86.0	82.0	77.5	72.6	67.5
	10°	86.1	89.7	92.7	95.2	97.2	98.5	99.4	99.7	99.3	98.4	97.0	94.9	92.3	89.3	85.8	81.7	77.3	72.6	67.4
	15°	86.1	89.6	92.6	95.0	96.9	98.3	99.0	99.2	98.9	97.9	96.4	94.5	92.0	89.0	85.4	81.5	77.1	72.4	67.4
	20°	86.1	89.5	92.4	94.7	96.5	97.8	98.5	98.7	98.2	97.3	95.9	93.9	91.4	88.3	84.9	81.0	76.7	72.1	67.3
South west	25°	86.1	89.3	92.1	94.3	96.0	97.2	97.9	97.9	97.5	96.6	95.2	93.2	90.6	87.6	84.2	80.4	76.3	71.8	67.0
	30°	86.1	89.2	91.8	93.9	95.5	96.5	97.1	97.1	96.7	95.7	94.2	92.1	89.7	86.8	83.5	79.7	75.7	71.3	66.6
	35°	86.1	89.0	91.5	93.4	94.8	95.8	96.1	96.1	95.7	94.6	92.9	91.0	88.6	85.8	82.4	78.8	74.9	70.6	66.1
	40°	86.1	88.8	91.0	92.8	94.0	94.8	95.2	95.1	94.5	93.3	91.7	89.8	87.4	84.5	81.3	77.8	73.9	69.8	65.4
	45°	86.1	88.6	90.6	92.1	93.2	93.8	94.0	93.8	93.0	91.8	90.4	88.4	85.9	83.1	80.1	76.6	72.8	68.8	64.6
	50°	86.1	88.4	90.1	91.4	92.3	92.7	92.8	92.4	91.6	90.4	88.9	86.8	84.3	81.7	78.6	75.2	71.6	67.7	63.6
	55°	86.1	88.1	89.6	90.6	91.3	91.5	91.5	90.9	90.0	88.8	87.2	85.0	82.7	80.1	77.1	73.7	70.3	66.5	62.6
	60°	86.1	87.8	89.0	89.8	90.2	90.3	90.1	89.3	88.4	87.1	85.3	83.2	81.0	78.4	75.3	72.2	68.8	65.1	61.4
	65°	86.1	87.5	88.4	88.9	89.1	89.0	88.5	87.7	86.6	85.2	83.4	81.4	79.1	76.4	73.5	70.5	67.2	63.6	60.1
West	70°	86.1	87.2	87.8	88.0	87.9	87.6	87.0	86.0	84.8	83.2	81.4	79.3	77.0	74.3	71.7	68.7	65.4	62.1	58.6
	75°	86.1	86.8	87.1	87.1	86.7	86.1	85.3	84.1	82.8	81.1	79.3	77.2	74.8	72.3	69.7	66.7	63.5	60.4	57.0
	80°	86.1	86.5	86.5	86.1	85.5	84.6	83.6	82.3	80.8	79.0	77.1	75.0	72.6	70.2	67.5	64.5	61.6	58.6	55.3
	85°	86.1	86.2	85.8	85.2	84.2	83.1	81.8	80.3	78.7	76.8	74.9	72.7	70.3	67.9	65.2	62.4	59.6	56.6	53.3
	90°	86.1	85.9	85.2	84.1	82.9	81.6	80.0	78.3	76.6	74.6	72.5	70.3	67.9	65.5	62.8	60.2	57.4	54.4	51.5
	95°	86.1	85.5	84.5	83.2	81.7	80.0	78.2	76.3	74.3	72.2	70.1	67.8	65.5	63.0	60.4	57.8	55.1	52.2	49.6
	100°	86.1	85.2	83.8	82.2	80.4	78.4	76.4	74.2	72.1	69.8	67.6	65.3	62.9	60.4	57.9	55.4	52.7	50.1	47.4
	105°	86.1	84.8	83.2	81.2	79.1	76.8	74.6	72.2	69.8	67.4	65.1	62.7	60.3	57.9	55.4	52.9	50.3	47.8	45.3
	110°	86.1	84.5	82.5	80.2	77.8	75.3	72.7	70.1	67.6	65.1	62.6	60.1	57.7	55.3	52.8	50.4	47.9	45.5	43.1
North west	115°	86.1	84.2	81.9	79.3	76.5	73.7	70.9	68.0	65.3	62.6	60.1	57.5	55.1	52.6	50.2	47.9	45.5	43.3	40.9
	120°	86.1	83.9	81.3	78.4	75.3	72.2	69.1	66.0	63.0	60.2	57.5	54.9	52.4	49.9	47.7	45.3	43.1	40.9	38.8
	125°	86.1	83.6	80.7	77.6	74.2	70.8	67.3	64.0	60.9	57.8	55.0	52.3	49.8	47.3	45.1	42.9	40.8	38.7	36.7
	130°	86.1	83.4	80.2	76.7	73.1	69.3	65.7	62.1	58.7	55.5	52.5	49.7	47.2	44.8	42.6	40.5	38.5	36.5	34.7
	135°	86.1	83.1	79.7	76.0	72.1	68.1	64.1	60.3	56.7	53.3	50.1	47.3	44.7	42.3	40.2	38.2	36.3	34.5	32.8
	140°	86.1	82.9	79.2	75.3	71.1	66.9	62.7	58.6	54.8	51.1	47.9	44.9	42.3	40.0	37.9	36.0	34.2	32.6	31.0
	145°	86.1	82.7	78.8	74.6	70.2	65.8	61.4	57.1	53.0	49.2	45.8	42.8	40.1	37.8	35.7	33.9	32.3	30.8	29.3
	150°	86.1	82.5	78.4	74.1	69.5	64.8	60.2	55.8	51.6	47.7	44.0	40.8	38.0	35.7	33.7	32.0	30.5	29.1	27.9
	155°	86.1	82.3	78.1	73.6	68.8	64.0	59.2	54.8	50.5	46.4	42.6	39.1	36.1	33.8	31.8	30.2	28.9	27.6	26.5
	160°	86.1	82.2	77.9	73.2	68.2	63.2	58.5	54.0	49.6	45.4	41.5	37.9	34.7	32.1	30.2	28.6	27.4	26.3	25.3
North	165°	86.1	82.1	77.7	72.8	67.8	62.7	57.9	53.3	48.9	44.6	40.7	37.0	33.6	30.8	28.8	27.3	26.1	25.2	24.3
	170°	86.1	82.1	77.5	72.6	67.4	62.3	57.5	52.9	48.4	44.1	40.1	36.3	32.9	30.0	27.8	26.4	25.3	24.3	23.5
	175°	86.1	82.0	77.4	72.4	67.3	62.1	57.3	52.6	48.2	43.9	39.8	36.0	32.6	29.6	27.2	25.8	24.7	23.9	23.1
	180°	86.1	82.0	77.4	72.4	67.2	62.1	57.2	52.6	48.1	43.8	39.7	36.0	32.6	29.5	27.1	25.6	24.6	23.7	23.0

Location Munich, Germany

Azimuth α		Angle of slope β: 0°	5°	10°	15°	20°	25°	30°	35°	40°	45°	50°	55°	60°	65°	70°	75°	80°	85°	90°
	−180°	86.5	82.7	78.4	73.8	69.2	64.8	60.5	56.3	52.1	48.1	44.4	41.0	37.9	35.5	33.7	32.3	31.0	29.8	28.7
North	−175°	86.5	82.7	78.4	73.9	69.3	64.9	60.6	56.4	52.2	48.3	44.5	41.1	38.1	35.6	33.9	32.4	31.2	29.9	28.8
	−170°	86.5	82.7	78.5	74.0	69.4	65.0	60.8	56.6	52.5	48.6	44.9	41.5	38.5	36.1	34.4	33.0	31.6	30.4	29.3
	−165°	86.5	82.8	78.7	74.3	69.7	65.4	61.2	57.0	53.0	49.1	45.5	42.1	39.3	37.0	35.3	33.9	32.4	31.2	29.9
	−160°	86.5	83.0	78.9	74.6	70.1	65.9	61.7	57.6	53.6	49.9	46.3	43.1	40.4	38.3	36.5	35.0	33.5	32.1	30.8
	−155°	86.5	83.0	79.2	75.0	70.7	66.4	62.4	58.4	54.5	50.8	47.4	44.4	41.9	39.9	38.0	36.4	34.8	33.3	31.9
	−150°	86.5	83.3	79.5	75.5	71.4	67.3	63.3	59.3	55.6	52.1	48.8	46.1	43.6	41.6	39.6	37.9	36.3	34.7	33.1
	−145°	86.5	83.4	79.9	76.1	72.1	68.1	64.2	60.5	56.9	53.6	50.6	47.9	45.6	43.4	41.5	39.6	37.9	36.1	34.5
	−140°	86.5	83.6	80.3	76.7	73.0	69.2	65.5	61.9	58.5	55.3	52.5	49.9	47.6	45.4	43.4	41.5	39.6	37.8	36.0
North east	−135°	86.5	83.9	80.7	77.4	73.9	70.4	66.9	63.5	60.3	57.3	54.6	52.1	49.8	47.6	45.5	43.4	41.4	39.5	37.6
	−130°	86.5	84.1	81.2	78.1	74.9	71.6	68.4	65.3	62.2	59.5	56.8	54.4	52.0	49.8	47.6	45.5	43.5	41.4	39.4
	−125°	86.5	84.4	81.8	79.0	76.0	73.0	70.0	67.0	64.3	61.6	59.1	56.7	54.4	52.1	49.9	47.7	45.5	43.3	41.3
	−120°	86.5	84.6	82.3	79.8	77.1	74.4	71.6	69.0	66.4	63.9	61.5	59.1	56.8	54.5	52.2	50.0	47.7	45.5	43.1
	−115°	86.5	84.9	82.9	80.7	78.3	75.9	73.3	71.0	68.5	66.2	63.9	61.6	59.3	57.0	54.6	52.3	49.9	47.6	45.2
	−110°	86.5	85.2	83.5	81.6	79.5	77.3	75.1	72.9	70.7	68.5	66.3	64.0	61.8	59.5	57.0	54.7	52.1	49.7	47.3
	−105°	86.5	85.5	84.1	82.4	80.7	78.8	76.9	74.9	72.8	70.8	68.7	66.5	64.2	61.9	59.5	57.0	54.5	51.9	49.3
	−100°	86.5	85.9	84.7	83.4	81.9	80.3	78.6	76.8	75.0	73.0	71.0	68.9	66.7	64.4	61.9	59.3	56.8	54.1	51.3
	−95°	86.5	86.1	85.3	84.4	83.1	81.9	80.4	78.8	77.1	75.3	73.3	71.3	69.0	66.7	64.3	61.6	59.0	56.2	53.3
East	−90°	86.5	86.4	86.0	85.3	84.4	83.3	82.1	80.7	79.2	77.5	75.6	73.6	71.4	69.0	66.6	63.9	61.2	58.4	55.3
	−85°	86.5	86.7	86.6	86.2	85.6	84.7	83.8	82.6	81.2	79.6	77.9	75.9	73.7	71.3	68.8	66.1	63.2	60.3	57.3
	−80°	86.5	87.1	87.3	87.1	86.7	86.2	85.4	84.4	83.1	81.7	79.9	78.1	75.9	73.5	71.0	68.2	65.3	62.1	59.0
	−75°	86.5	87.4	87.9	88.0	87.9	87.6	87.0	86.1	85.0	83.7	82.0	80.1	78.0	75.6	72.9	70.2	67.3	63.9	60.6
	−70°	86.5	87.7	88.4	88.9	89.0	88.9	88.4	87.9	86.8	85.6	84.0	82.1	80.0	77.6	74.9	72.0	69.0	65.7	62.1
	−65°	86.5	88.0	89.0	89.7	90.1	90.2	89.9	89.4	88.5	87.3	85.9	84.0	81.9	79.6	76.8	73.7	70.6	67.3	63.6
	−60°	86.5	88.3	89.6	90.5	91.1	91.4	91.3	91.0	90.1	89.0	87.6	85.9	83.6	81.2	78.5	75.5	72.1	68.7	65.0
	−55°	86.5	88.5	90.1	91.3	92.1	92.6	92.7	92.4	91.7	90.7	89.3	87.6	85.3	82.7	80.1	77.0	73.6	69.9	66.2
	−50°	86.5	88.7	90.5	92.0	93.0	93.6	93.9	93.7	93.2	92.1	90.7	89.0	87.0	84.4	81.4	78.4	74.9	71.2	67.3
South east	−45°	86.5	89.0	91.0	92.6	93.8	94.6	95.0	94.9	94.4	93.6	92.1	90.4	88.3	85.8	82.8	79.6	76.1	72.2	68.1
	−40°	86.5	89.2	91.4	93.2	94.5	95.5	96.0	96.0	95.6	94.7	93.5	91.6	89.4	87.0	84.0	80.7	77.0	73.1	69.0
	−35°	86.5	89.4	91.8	93.7	95.3	96.2	96.9	97.0	96.6	95.8	94.6	92.8	90.6	87.9	85.0	81.6	77.9	73.9	69.6
	−30°	86.5	89.6	92.1	94.1	95.8	97.0	97.6	97.9	97.5	96.7	95.5	93.8	91.6	88.9	85.8	82.4	78.6	74.4	70.1
	−25°	86.5	89.7	92.4	94.6	96.3	97.6	98.3	98.6	98.3	97.5	96.1	94.4	92.3	89.6	86.5	83.0	79.1	74.9	70.4
	−20°	86.5	89.8	92.6	94.9	96.7	98.0	98.8	99.1	98.9	98.1	96.9	95.0	92.8	90.1	87.0	83.5	79.5	75.2	70.6
	−15°	86.5	89.9	92.7	95.1	97.0	98.4	99.2	99.5	99.3	98.7	97.4	95.6	93.3	90.5	87.3	83.7	79.7	75.3	70.7
	−10°	86.5	89.9	92.8	95.3	97.2	98.6	99.5	99.8	99.6	98.9	97.6	95.9	93.6	90.8	87.6	83.9	79.9	75.5	70.7
	−5°	86.5	90.0	92.9	95.3	97.3	98.7	99.6	100.0	99.8	99.0	97.7	96.0	93.7	91.0	87.7	84.0	79.9	75.5	70.7
South	0°	86.5	90.0	92.9	95.3	97.3	98.7	99.6	100.0	99.8	99.0	97.8	96.0	93.7	90.9	87.6	83.9	79.9	75.3	70.6
	5°	86.5	90.0	92.9	95.3	97.3	98.7	99.6	100.0	99.8	99.0	97.7	96.0	93.7	91.0	87.7	84.0	79.9	75.5	70.7
	10°	86.5	89.9	92.8	95.3	97.2	98.6	99.5	99.8	99.6	98.9	97.6	95.9	93.6	90.8	87.6	83.9	79.9	75.5	70.7
	15°	86.5	89.9	92.7	95.1	97.0	98.4	99.2	99.5	99.3	98.7	97.4	95.6	93.3	90.5	87.3	83.7	79.7	75.3	70.7
	20°	86.5	89.8	92.6	94.9	96.7	98.0	98.8	99.1	98.9	98.1	96.9	95.0	92.8	90.1	87.0	83.5	79.5	75.2	70.6
	25°	86.5	89.7	92.4	94.6	96.3	97.6	98.3	98.6	98.3	97.5	96.1	94.4	92.3	89.6	86.5	83.0	79.1	74.9	70.4
	30°	86.5	89.6	92.1	94.1	95.8	97.0	97.6	97.9	97.5	96.7	95.5	93.8	91.6	88.9	85.8	82.4	78.6	74.4	70.1
	35°	86.5	89.4	91.8	93.7	95.3	96.2	96.9	97.0	96.6	95.8	94.6	92.8	90.6	87.9	85.0	81.6	77.9	73.9	69.6
	40°	86.5	89.2	91.4	93.2	94.5	95.5	96.0	96.0	95.6	94.7	93.5	91.6	89.4	87.0	84.0	80.7	77.0	73.1	69.0
South west	45°	86.5	89.0	91.0	92.6	93.8	94.6	95.0	94.9	94.4	93.6	92.1	90.4	88.3	85.8	82.8	79.6	76.1	72.2	68.1
	50°	86.5	88.7	90.5	92.0	93.0	93.6	93.9	93.7	93.2	92.1	90.7	89.0	87.0	84.4	81.4	78.4	74.9	71.2	67.3
	55°	86.5	88.5	90.1	91.3	92.1	92.6	92.7	92.4	91.7	90.7	89.3	87.6	85.3	82.7	80.1	77.0	73.6	69.9	66.2
	60°	86.5	88.3	89.6	90.5	91.1	91.4	91.3	91.0	90.1	89.0	87.6	85.9	83.6	81.2	78.5	75.5	72.1	68.7	65.0
	65°	86.5	88.0	89.0	89.7	90.1	90.2	89.9	89.4	88.5	87.3	85.9	84.0	81.9	79.6	76.8	73.7	70.6	67.3	63.6
	70°	86.5	87.7	88.4	88.9	89.0	88.9	88.4	87.9	86.8	85.6	84.0	82.1	80.0	77.6	74.9	72.0	69.0	65.7	62.1
	75°	86.5	87.4	87.9	88.0	87.9	87.6	87.0	86.1	85.0	83.7	82.0	80.1	78.0	75.6	72.9	70.2	67.3	63.9	60.6
	80°	86.5	87.1	87.3	87.1	86.7	86.2	85.4	84.4	83.1	81.7	79.9	78.1	75.9	73.5	71.0	68.2	65.3	62.1	59.0
	85°	86.5	86.7	86.6	86.2	85.6	84.7	83.8	82.6	81.2	79.6	77.9	75.9	73.7	71.3	68.8	66.1	63.2	60.3	57.3
West	90°	86.5	86.4	86.0	85.3	84.4	83.3	82.1	80.7	79.2	77.5	75.6	73.6	71.4	69.0	66.6	63.9	61.2	58.4	55.3
	95°	86.5	86.1	85.3	84.4	83.1	81.9	80.4	78.8	77.1	75.3	73.3	71.3	69.0	66.7	64.3	61.6	59.0	56.2	53.3
	100°	86.5	85.9	84.7	83.4	81.9	80.3	78.6	76.8	75.0	73.0	71.0	68.9	66.7	64.4	61.9	59.3	56.8	54.1	51.3
	105°	86.5	85.5	84.1	82.4	80.7	78.8	76.9	74.9	72.8	70.8	68.7	66.5	64.2	61.9	59.5	57.0	54.5	51.9	49.3
	110°	86.5	85.2	83.5	81.6	79.5	77.3	75.1	72.9	70.7	68.5	66.3	64.0	61.8	59.5	57.0	54.7	52.1	49.7	47.3
	115°	86.5	84.9	82.9	80.7	78.3	75.9	73.3	71.0	68.5	66.2	63.9	61.6	59.3	57.0	54.6	52.3	49.9	47.6	45.2
	120°	86.5	84.6	82.3	79.8	77.1	74.4	71.6	69.0	66.4	63.9	61.5	59.1	56.8	54.5	52.2	50.0	47.7	45.5	43.1
	125°	86.5	84.4	81.8	79.0	76.0	73.0	70.0	67.0	64.3	61.6	59.1	56.7	54.4	52.1	49.9	47.7	45.5	43.3	41.3
	130°	86.5	84.1	81.2	78.1	74.9	71.6	68.4	65.3	62.2	59.5	56.8	54.4	52.0	49.8	47.6	45.5	43.5	41.4	39.4
	135°	86.5	83.9	80.7	77.4	73.9	70.4	66.9	63.5	60.3	57.3	54.6	52.1	49.8	47.6	45.5	43.4	41.4	39.5	37.6
North west	140°	86.5	83.6	80.3	76.7	73.0	69.2	65.5	61.9	58.5	55.3	52.5	49.9	47.6	45.4	43.4	41.5	39.6	37.8	36.0
	145°	86.5	83.4	79.9	76.1	72.1	68.1	64.2	60.5	56.9	53.6	50.6	47.9	45.6	43.4	41.5	39.6	37.9	36.1	34.5
	150°	86.5	83.3	79.5	75.5	71.4	67.3	63.3	59.3	55.6	52.1	48.8	46.1	43.6	41.6	39.6	37.9	36.3	34.7	33.1
	155°	86.5	83.0	79.2	75.0	70.7	66.4	62.4	58.4	54.5	50.8	47.4	44.4	41.9	39.9	38.0	36.4	34.8	33.3	31.9
	160°	86.5	83.0	78.9	74.6	70.1	65.9	61.7	57.6	53.6	49.9	46.3	43.1	40.4	38.3	36.5	35.0	33.5	32.1	30.8
	165°	86.5	82.8	78.7	74.3	69.7	65.4	61.2	57.0	53.0	49.1	45.5	42.1	39.3	37.0	35.3	33.9	32.4	31.2	29.9
	170°	86.5	82.7	78.5	74.0	69.4	65.0	60.8	56.6	52.5	48.6	44.9	41.5	38.5	36.1	34.4	33.0	31.6	30.4	29.3
North	175°	86.5	82.7	78.4	73.9	69.3	64.9	60.6	56.4	52.2	48.3	44.5	41.1	38.1	35.6	33.9	32.4	31.2	29.9	28.8
	180°	86.5	82.7	78.4	73.8	69.2	64.8	60.5	56.3	52.1	48.1	44.4	41.0	37.9	35.5	33.7	32.3	31.0	29.8	28.7

Location San Francisco, USA

Azimuth α		Angle of slope β																		
		0°	5°	10°	15°	20°	25°	30°	35°	40°	45°	50°	55°	60°	65°	70°	75°	80°	85°	90°
North	−180°	89.0	85.2	81.0	76.3	71.2	65.8	60.5	55.6	50.9	46.3	41.9	37.8	34.1	30.7	27.7	25.3	23.9	23.0	22.3
	−175°	89.0	85.3	81.1	76.4	71.4	66.1	60.9	55.9	51.2	46.6	42.3	38.2	34.5	31.1	28.1	25.7	24.3	23.3	22.6
	−170°	89.0	85.3	81.2	76.7	71.7	66.6	61.3	56.5	51.7	47.2	42.9	38.8	35.1	31.7	28.8	26.5	25.0	24.0	23.2
	−165°	89.0	85.5	81.4	77.0	72.2	67.2	62.1	57.2	52.5	48.0	43.7	39.7	36.0	32.7	29.8	27.7	26.1	25.0	24.0
North east	−160°	89.0	85.6	81.8	77.4	72.8	67.9	62.9	58.1	53.4	49.0	44.8	40.8	37.2	33.9	31.2	29.1	27.5	26.2	25.1
	−155°	89.0	85.7	82.1	77.9	73.5	68.8	64.0	59.2	54.6	50.2	46.1	42.2	38.6	35.5	33.0	30.8	29.1	27.6	26.3
	−150°	89.0	86.0	82.4	78.5	74.2	69.7	65.1	60.5	56.0	51.7	47.6	43.9	40.4	37.5	34.9	32.7	30.8	29.2	27.7
	−145°	89.0	86.2	82.8	79.1	75.1	70.8	66.4	62.0	57.6	53.4	49.5	45.8	42.5	39.6	37.0	34.7	32.7	30.9	29.3
	−140°	89.0	86.4	83.3	79.8	75.9	72.0	67.8	63.6	59.4	55.4	51.6	48.1	44.8	41.8	39.2	36.9	34.7	32.8	30.9
	−135°	89.0	86.6	83.7	80.5	76.9	73.1	69.3	65.3	61.3	57.5	53.8	50.4	47.2	44.3	41.6	39.1	36.9	34.8	32.8
	−130°	89.0	86.8	84.2	81.3	78.0	74.5	70.8	67.1	63.3	59.7	56.2	52.9	49.7	46.8	44.1	41.5	39.1	36.9	34.8
	−125°	89.0	87.1	84.8	82.1	79.1	75.8	72.4	68.9	65.4	62.0	58.6	55.4	52.3	49.4	46.6	44.0	41.5	39.0	36.8
	−120°	89.0	87.4	85.4	82.9	80.2	77.2	74.1	70.9	67.5	64.3	61.1	58.0	55.0	52.0	49.1	46.4	43.8	41.3	38.9
	−115°	89.0	87.7	86.0	83.8	81.3	78.7	75.8	72.8	69.7	66.7	63.6	60.5	57.5	54.6	51.7	48.9	46.2	43.5	41.0
East	−110°	89.0	88.0	86.5	84.7	82.5	80.1	77.5	74.8	71.9	69.0	66.0	63.1	60.2	57.2	54.3	51.4	48.6	45.8	43.1
	−105°	89.0	88.3	87.1	85.6	83.7	81.6	79.3	76.7	74.1	71.3	68.5	65.6	62.7	59.8	56.9	53.9	51.0	48.1	45.2
	−100°	89.0	88.6	87.8	86.5	85.0	83.1	81.0	78.7	76.2	73.6	70.9	68.1	65.2	62.3	59.4	56.3	53.3	50.3	47.2
	−95°	89.0	88.9	88.4	87.5	86.2	84.6	82.7	80.6	78.3	75.9	73.3	70.6	67.7	64.7	61.8	58.6	55.5	52.5	49.2
	−90°	89.0	89.2	89.0	88.3	87.4	86.0	84.3	82.5	80.4	78.1	75.5	72.9	70.1	67.1	64.1	60.9	57.6	54.5	51.2
	−85°	89.0	89.5	89.6	89.2	88.5	87.4	86.0	84.3	82.4	80.2	77.8	75.2	72.4	69.3	66.4	63.1	59.7	56.4	53.1
	−80°	89.0	89.8	90.2	90.1	89.6	88.8	87.6	86.1	84.3	82.3	79.9	77.3	74.6	71.5	68.4	65.2	61.7	58.2	54.7
	−75°	89.0	90.1	90.8	90.9	90.7	90.1	89.2	87.8	86.1	84.2	81.9	79.4	76.7	73.7	70.3	67.1	63.6	59.9	56.2
	−70°	89.0	90.4	91.3	91.8	91.8	91.4	90.6	89.5	87.9	86.1	83.9	81.3	78.6	75.6	72.3	68.8	65.3	61.5	57.6
South east	−65°	89.0	90.7	91.9	92.5	92.8	92.6	92.0	91.0	89.5	87.8	85.7	83.2	80.4	77.4	74.1	70.5	66.8	62.9	58.9
	−60°	89.0	90.9	92.3	93.3	93.7	93.7	93.3	92.4	91.1	89.4	87.4	85.0	82.1	79.1	75.7	72.0	68.2	64.3	60.1
	−55°	89.0	91.1	92.8	93.9	94.6	94.8	94.5	93.8	92.6	90.9	88.9	86.6	83.7	80.5	77.1	73.4	69.5	65.4	61.1
	−50°	89.0	91.3	93.2	94.6	95.4	95.8	95.6	95.0	93.9	92.3	90.3	88.0	85.2	81.9	78.4	74.7	70.6	66.4	61.9
	−45°	89.0	91.6	93.6	95.1	96.2	96.7	96.6	96.1	95.1	93.6	91.6	89.2	86.4	83.2	79.6	75.7	71.6	67.2	62.7
	−40°	89.0	91.8	94.0	95.6	96.8	97.5	97.6	97.0	96.2	94.8	92.8	90.3	87.6	84.3	80.6	76.7	72.4	67.9	63.2
	−35°	89.0	91.9	94.2	96.1	97.4	98.1	98.3	98.0	97.0	95.7	93.8	91.3	88.5	85.2	81.5	77.4	73.1	68.4	63.6
	−30°	89.0	92.0	94.5	96.5	97.9	98.7	99.0	98.7	97.8	96.5	94.6	92.2	89.3	86.0	82.2	78.0	73.6	68.8	63.8
	−25°	89.0	92.1	94.7	96.7	98.2	99.1	99.5	99.3	98.5	97.1	95.2	92.8	89.9	86.5	82.7	78.5	73.9	69.0	63.9
South	−20°	89.0	92.2	94.9	97.0	98.5	99.5	99.9	99.7	99.0	97.6	95.6	93.3	90.3	86.9	82.9	78.7	74.0	69.1	63.8
	−15°	89.0	92.2	95.0	97.1	98.7	99.7	100.1	99.9	99.2	97.9	96.0	93.5	90.5	87.0	83.2	78.7	74.0	68.9	63.6
	−10°	89.0	92.3	95.0	97.2	98.8	99.8	100.2	100.1	99.3	98.0	96.1	93.6	90.6	87.1	83.1	78.7	73.8	68.7	63.3
	−5°	89.0	92.3	95.0	97.1	98.8	99.8	100.2	100.0	99.2	97.9	96.0	93.5	90.5	86.9	82.8	78.4	73.6	68.4	62.9
	0°	89.0	92.2	94.9	97.0	98.7	99.6	100.0	99.8	99.0	97.6	95.6	93.2	90.1	86.6	82.5	78.1	73.2	68.0	62.5
	5°	89.0	92.3	95.0	97.1	98.8	99.8	100.2	100.0	99.2	97.9	96.0	93.5	90.5	86.9	82.8	78.4	73.6	68.4	62.9
	10°	89.0	92.3	95.0	97.2	98.8	99.8	100.2	100.1	99.3	98.0	96.1	93.6	90.6	87.1	83.1	78.7	73.8	68.7	63.3
	15°	89.0	92.2	95.0	97.1	98.7	99.7	100.1	99.9	99.2	97.9	96.0	93.5	90.5	87.0	83.2	78.7	74.0	68.9	63.6
	20°	89.0	92.2	94.9	97.0	98.5	99.5	99.9	99.7	99.0	97.6	95.6	93.3	90.3	86.9	82.9	78.7	74.0	69.1	63.8
South west	25°	89.0	92.1	94.7	96.7	98.2	99.1	99.5	99.3	98.5	97.1	95.2	92.8	89.9	86.5	82.7	78.5	73.9	69.0	63.9
	30°	89.0	92.0	94.5	96.5	97.9	98.7	99.0	98.7	97.8	96.5	94.6	92.2	89.3	86.0	82.2	78.0	73.6	68.8	63.8
	35°	89.0	91.9	94.2	96.1	97.4	98.1	98.3	98.0	97.0	95.7	93.8	91.3	88.5	85.2	81.5	77.4	73.1	68.4	63.6
	40°	89.0	91.8	94.0	95.6	96.8	97.5	97.6	97.0	96.2	94.8	92.8	90.3	87.6	84.3	80.6	76.7	72.4	67.9	63.2
	45°	89.0	91.6	93.6	95.1	96.2	96.7	96.6	96.1	95.1	93.6	91.6	89.2	86.4	83.2	79.6	75.7	71.6	67.2	62.7
	50°	89.0	91.3	93.2	94.6	95.4	95.8	95.6	95.0	93.9	92.3	90.3	88.0	85.2	81.9	78.4	74.7	70.6	66.4	61.9
	55°	89.0	91.1	92.8	93.9	94.6	94.8	94.5	93.8	92.6	90.9	88.9	86.6	83.7	80.5	77.1	73.4	69.5	65.4	61.1
	60°	89.0	90.9	92.3	93.3	93.7	93.7	93.3	92.4	91.1	89.4	87.4	85.0	82.1	79.1	75.7	72.0	68.2	64.3	60.1
	65°	89.0	90.7	91.9	92.5	92.8	92.6	92.0	91.0	89.5	87.8	85.7	83.2	80.4	77.4	74.1	70.5	66.8	62.9	58.9
West	70°	89.0	90.4	91.3	91.8	91.8	91.4	90.6	89.5	87.9	86.1	83.9	81.3	78.6	75.6	72.3	68.8	65.3	61.5	57.6
	75°	89.0	90.1	90.8	90.9	90.7	90.1	89.2	87.8	86.1	84.2	81.9	79.4	76.7	73.7	70.3	67.1	63.6	59.9	56.2
	80°	89.0	89.8	90.2	90.1	89.6	88.8	87.6	86.1	84.3	82.3	79.9	77.3	74.6	71.5	68.4	65.2	61.7	58.2	54.7
	85°	89.0	89.5	89.6	89.2	88.5	87.4	86.0	84.3	82.4	80.2	77.8	75.2	72.4	69.3	66.4	63.1	59.7	56.4	53.1
	90°	89.0	89.2	89.0	88.3	87.4	86.0	84.3	82.5	80.4	78.1	75.5	72.9	70.1	67.1	64.1	60.9	57.6	54.5	51.2
	95°	89.0	88.9	88.4	87.5	86.2	84.6	82.7	80.6	78.3	75.9	73.3	70.6	67.7	64.7	61.8	58.6	55.5	52.5	49.2
	100°	89.0	88.6	87.8	86.5	85.0	83.1	81.0	78.7	76.2	73.6	70.9	68.1	65.2	62.3	59.4	56.3	53.3	50.3	47.2
	105°	89.0	88.3	87.1	85.6	83.7	81.6	79.3	76.7	74.1	71.3	68.5	65.6	62.7	59.8	56.9	53.9	51.0	48.1	45.2
	110°	89.0	88.0	86.5	84.7	82.5	80.1	77.5	74.8	71.9	69.0	66.0	63.1	60.2	57.2	54.3	51.4	48.6	45.8	43.1
North west	115°	89.0	87.7	86.0	83.8	81.3	78.7	75.8	72.8	69.7	66.7	63.6	60.5	57.5	54.6	51.7	48.9	46.2	43.5	41.0
	120°	89.0	87.4	85.4	82.9	80.2	77.2	74.1	70.9	67.5	64.3	61.1	58.0	55.0	52.0	49.1	46.4	43.8	41.3	38.9
	125°	89.0	87.1	84.8	82.1	79.1	75.8	72.4	68.9	65.4	62.0	58.6	55.4	52.3	49.4	46.6	44.0	41.5	39.0	36.8
	130°	89.0	86.8	84.2	81.3	78.0	74.5	70.8	67.1	63.3	59.7	56.2	52.9	49.7	46.8	44.1	41.5	39.1	36.9	34.8
	135°	89.0	86.6	83.7	80.5	76.9	73.1	69.3	65.3	61.3	57.5	53.8	50.4	47.2	44.3	41.6	39.1	36.9	34.8	32.8
	140°	89.0	86.4	83.3	79.8	75.9	72.0	67.8	63.6	59.4	55.4	51.6	48.1	44.8	41.8	39.2	36.9	34.7	32.8	30.9
	145°	89.0	86.2	82.8	79.1	75.1	70.8	66.4	62.0	57.6	53.4	49.5	45.8	42.5	39.6	37.0	34.7	32.7	30.9	29.3
	150°	89.0	86.0	82.4	78.5	74.2	69.7	65.1	60.5	56.0	51.7	47.6	43.9	40.4	37.5	34.9	32.7	30.8	29.2	27.7
	155°	89.0	85.7	82.1	77.9	73.5	68.8	64.0	59.2	54.6	50.2	46.1	42.2	38.6	35.5	33.0	30.8	29.1	27.6	26.3
	160°	89.0	85.6	81.8	77.4	72.8	67.9	62.9	58.1	53.4	49.0	44.8	40.8	37.2	33.9	31.2	29.1	27.5	26.2	25.1
North	165°	89.0	85.5	81.4	77.0	72.2	67.2	62.1	57.2	52.5	48.0	43.7	39.7	36.0	32.7	29.8	27.7	26.1	25.0	24.0
	170°	89.0	85.3	81.2	76.7	71.7	66.6	61.3	56.5	51.7	47.2	42.9	38.8	35.1	31.7	28.8	26.5	25.0	24.0	23.2
	175°	89.0	85.3	81.1	76.4	71.4	66.1	60.9	55.9	51.2	46.6	42.3	38.2	34.5	31.1	28.1	25.7	24.3	23.3	22.6
	180°	89.0	85.2	81.0	76.3	71.2	65.8	60.5	55.6	50.9	46.3	41.9	37.8	34.1	30.7	27.7	25.3	23.9	23.0	22.3

Location Vienna, Austria

		Angle of slope β																		
		0°	5°	10°	15°	20°	25°	30°	35°	40°	45°	50°	55°	60°	65°	70°	75°	80°	85°	90°
Azimuth α																				
North	−180°	93.7	90.8	87.3	83.4	79.1	74.3	69.3	64.1	59.1	54.4	49.9	45.6	41.5	37.7	34.2	31.0	28.3	26.3	25.1
	−175°	93.7	90.8	87.4	83.4	79.1	74.4	69.4	64.2	59.3	54.6	50.0	45.7	41.6	37.8	34.3	31.1	28.4	26.5	25.2
	−170°	93.7	90.8	87.4	83.6	79.3	74.6	69.7	64.6	59.6	54.9	50.4	46.1	42.0	38.2	34.6	31.5	28.9	27.0	25.8
	−165°	93.7	90.9	87.5	83.7	79.6	75.0	70.1	65.1	60.1	55.4	50.9	46.7	42.6	38.8	35.3	32.3	29.8	28.0	26.6
	−160°	93.7	91.0	87.7	84.0	79.9	75.4	70.7	65.8	60.9	56.2	51.7	47.5	43.4	39.7	36.3	33.3	31.1	29.2	27.7
	−155°	93.7	91.1	87.9	84.3	80.3	76.0	71.4	66.7	61.9	57.2	52.7	48.5	44.6	40.9	37.6	34.9	32.6	30.7	29.0
	−150°	93.7	91.2	88.1	84.6	80.8	76.6	72.2	67.6	63.0	58.4	54.0	49.9	46.0	42.5	39.4	36.6	34.3	32.3	30.5
	−145°	93.7	91.3	88.4	85.0	81.4	77.3	73.1	68.7	64.3	59.9	55.6	51.6	47.8	44.4	41.3	38.5	36.1	33.9	32.0
	−140°	93.7	91.4	88.7	85.5	82.0	78.1	74.1	69.9	65.7	61.5	57.4	53.5	49.9	46.5	43.4	40.6	38.0	35.7	33.6
North east	−135°	93.7	91.6	89.0	86.0	82.7	79.0	75.2	71.2	67.1	63.2	59.3	55.5	52.0	48.7	45.6	42.7	40.0	37.6	35.4
	−130°	93.7	91.8	89.4	86.5	83.4	80.0	76.3	72.5	68.7	65.0	61.2	57.6	54.1	50.9	47.8	44.9	42.1	39.6	37.2
	−125°	93.7	92.0	89.7	87.1	84.1	81.0	77.5	74.0	70.4	66.8	63.2	59.7	56.3	53.1	50.0	47.1	44.3	41.6	39.1
	−120°	93.7	92.2	90.1	87.7	85.0	82.0	78.8	75.5	72.1	68.6	65.2	61.9	58.6	55.3	52.2	49.3	46.4	43.6	41.0
	−115°	93.7	92.4	90.6	88.4	85.8	83.1	80.1	77.0	73.8	70.5	67.2	64.0	60.8	57.6	54.4	51.4	48.5	45.6	42.8
	−110°	93.7	92.6	91.0	89.0	86.7	84.1	81.4	78.5	75.5	72.3	69.2	66.1	62.9	59.7	56.6	53.5	50.5	47.6	44.7
	−105°	93.7	92.8	91.5	89.7	87.6	84.8	82.7	80.0	77.2	74.2	71.2	68.1	65.0	61.8	58.7	55.6	52.4	49.5	46.4
	−100°	93.7	93.0	91.9	90.4	88.5	86.4	84.0	81.5	78.8	76.0	73.0	70.0	67.0	63.8	60.7	57.5	54.3	51.2	48.1
	−95°	93.7	93.3	92.4	91.1	89.4	87.5	84.8	83.0	80.4	77.7	74.9	71.9	68.9	65.8	62.5	59.4	56.1	52.9	49.8
East	−90°	93.7	93.5	92.8	91.7	90.3	88.6	86.6	84.4	82.0	79.4	76.6	73.7	70.7	67.5	64.3	61.1	57.8	54.3	51.2
	−85°	93.7	93.7	93.3	92.4	91.2	89.7	87.8	85.8	83.4	81.0	78.3	75.4	72.4	69.3	65.9	62.8	59.3	55.7	52.4
	−80°	93.7	94.0	93.7	93.1	92.1	90.7	89.0	87.1	84.9	82.5	79.9	77.0	74.0	70.9	67.4	64.2	60.8	57.1	53.6
	−75°	93.7	94.2	94.2	93.7	92.9	91.7	90.2	88.4	86.3	83.9	81.4	78.5	75.5	72.4	68.9	65.5	62.0	58.3	54.6
	−70°	93.7	94.4	94.6	94.4	93.7	92.7	91.3	89.7	87.6	85.2	82.8	79.9	76.8	73.7	70.3	66.6	63.1	59.3	55.5
	−65°	93.7	94.6	95.0	95.0	94.5	93.7	92.4	90.8	88.9	86.5	84.1	81.3	78.1	74.9	71.4	67.7	64.0	60.2	56.2
	−60°	93.7	94.8	95.4	95.6	95.3	94.6	93.4	91.9	90.0	87.7	85.2	82.5	79.3	75.9	72.4	68.6	64.8	60.9	56.8
	−55°	93.7	95.0	95.8	96.1	96.0	95.4	94.3	92.9	91.2	88.9	86.3	83.5	80.4	76.9	73.3	69.4	65.5	61.4	57.2
	−50°	93.7	95.2	96.2	96.6	96.6	96.2	95.2	93.8	92.1	90.0	87.3	84.5	81.3	77.8	74.0	70.1	66.0	61.8	57.5
South east	−45°	93.7	95.4	96.5	97.1	97.2	96.9	96.0	94.7	93.0	90.9	88.3	85.3	82.2	78.6	74.7	70.7	66.4	62.1	57.7
	−40°	93.7	95.5	96.8	97.5	97.8	97.5	96.8	95.5	93.8	91.7	89.1	86.1	82.9	79.2	75.3	71.1	66.7	62.3	57.7
	−35°	93.7	95.7	97.1	97.9	98.3	98.1	97.5	96.2	94.5	92.4	89.9	86.8	83.4	79.8	75.7	71.4	66.9	62.3	57.6
	−30°	93.7	95.8	97.3	98.3	98.7	98.6	98.0	96.9	95.2	93.0	90.5	87.5	84.0	80.2	76.1	71.7	67.1	62.3	57.5
	−25°	93.7	95.9	97.5	98.6	99.1	99.0	98.5	97.4	95.8	93.6	91.0	87.9	84.4	80.6	76.4	71.9	67.1	62.2	57.2
	−20°	93.7	96.0	97.7	98.8	99.4	99.4	98.8	97.8	96.2	94.1	91.4	88.3	84.8	80.9	76.5	71.9	67.0	62.0	56.8
	−15°	93.7	96.0	97.8	99.0	99.6	99.7	99.2	98.1	96.5	94.4	91.7	88.6	85.0	81.0	76.6	71.9	66.9	61.7	56.4
	−10°	93.7	96.1	97.9	99.1	99.8	99.9	99.4	98.3	96.7	94.6	91.9	88.8	85.1	81.1	76.6	71.8	66.7	61.4	56.0
	−5°	93.7	96.1	97.9	99.2	99.9	100.0	99.5	98.5	96.8	94.7	92.0	88.8	85.1	81.1	76.6	71.7	66.6	61.2	55.7
South	0°	93.7	96.1	97.9	99.2	99.9	100.0	99.5	98.5	96.8	94.6	91.9	88.8	85.1	81.0	76.5	71.6	66.5	61.0	55.5
	5°	93.7	96.1	97.9	99.2	99.9	100.0	99.5	98.5	96.8	94.7	92.0	88.8	85.1	81.1	76.6	71.7	66.6	61.2	55.7
	10°	93.7	96.1	97.9	99.1	99.8	99.9	99.4	98.3	96.7	94.6	91.9	88.8	85.1	81.1	76.6	71.8	66.7	61.4	56.0
	15°	93.7	96.0	97.8	99.0	99.6	99.7	99.2	98.1	96.5	94.4	91.7	88.6	85.0	81.0	76.6	71.9	66.9	61.7	56.4
	20°	93.7	96.0	97.7	98.8	99.4	99.4	98.8	97.8	96.2	94.1	91.4	88.3	84.8	80.9	76.5	71.9	67.0	62.0	56.8
	25°	93.7	95.9	97.5	98.6	99.1	99.0	98.5	97.4	95.8	93.6	91.0	87.9	84.4	80.6	76.4	71.9	67.1	62.2	57.2
	30°	93.7	95.8	97.3	98.3	98.7	98.6	98.0	96.9	95.2	93.0	90.5	87.5	84.0	80.2	76.1	71.7	67.1	62.3	57.5
	35°	93.7	95.7	97.1	97.9	98.3	98.1	97.5	96.2	94.5	92.4	89.9	86.8	83.4	79.8	75.7	71.4	66.9	62.3	57.6
	40°	93.7	95.5	96.8	97.5	97.8	97.5	96.8	95.5	93.8	91.7	89.1	86.1	82.9	79.2	75.3	71.1	66.7	62.3	57.7
South west	45°	93.7	95.4	96.5	97.1	97.2	96.9	96.0	94.7	93.0	90.9	88.3	85.3	82.2	78.6	74.7	70.7	66.4	62.1	57.7
	50°	93.7	95.2	96.2	96.6	96.6	96.2	95.2	93.8	92.1	90.0	87.3	84.5	81.3	77.8	74.0	70.1	66.0	61.8	57.5
	55°	93.7	95.0	95.8	96.1	96.0	95.4	94.3	92.9	91.2	88.9	86.3	83.5	80.4	76.9	73.3	69.4	65.5	61.4	57.2
	60°	93.7	94.8	95.4	95.6	95.3	94.6	93.4	91.9	90.0	87.7	85.2	82.5	79.3	75.9	72.4	68.6	64.8	60.9	56.8
	65°	93.7	94.6	95.0	95.0	94.5	93.7	92.4	90.8	88.9	86.5	84.1	81.3	78.1	74.9	71.4	67.7	64.0	60.2	56.2
	70°	93.7	94.4	94.6	94.4	93.7	92.7	91.3	89.7	87.6	85.2	82.8	79.9	76.8	73.7	70.3	66.6	63.1	59.3	55.5
	75°	93.7	94.2	94.2	93.7	92.9	91.7	90.2	88.4	86.3	83.9	81.4	78.5	75.5	72.4	68.9	65.5	62.0	58.3	54.6
	80°	93.7	94.0	93.7	93.1	92.1	90.7	89.0	87.1	84.9	82.5	79.9	77.0	74.0	70.9	67.4	64.2	60.8	57.1	53.6
	85°	93.7	93.7	93.3	92.4	91.2	89.7	87.8	85.8	83.4	81.0	78.3	75.4	72.4	69.3	65.9	62.8	59.3	55.7	52.4
West	90°	93.7	93.5	92.8	91.7	90.3	88.6	86.6	84.4	82.0	79.4	76.6	73.7	70.7	67.5	64.3	61.1	57.8	54.3	51.2
	95°	93.7	93.3	92.4	91.1	89.4	87.5	84.8	83.0	80.4	77.7	74.9	71.9	68.9	65.8	62.5	59.4	56.1	52.9	49.8
	100°	93.7	93.0	91.9	90.4	88.5	86.4	84.0	81.5	78.8	76.0	73.0	70.0	67.0	63.8	60.7	57.5	54.3	51.2	48.1
	105°	93.7	92.8	91.5	89.7	87.6	84.8	82.7	80.0	77.2	74.2	71.2	68.1	65.0	61.8	58.7	55.6	52.4	49.5	46.4
	110°	93.7	92.6	91.0	89.0	86.7	84.1	81.4	78.5	75.5	72.3	69.2	66.1	62.9	59.7	56.6	53.5	50.5	47.6	44.7
	115°	93.7	92.4	90.6	88.4	85.8	83.1	80.1	77.0	73.8	70.5	67.2	64.0	60.8	57.6	54.4	51.4	48.5	45.6	42.8
	120°	93.7	92.2	90.1	87.7	85.0	82.0	78.8	75.5	72.1	68.6	65.2	61.9	58.6	55.3	52.2	49.3	46.4	43.6	41.0
	125°	93.7	92.0	89.7	87.1	84.1	81.0	77.5	74.0	70.4	66.8	63.2	59.7	56.3	53.1	50.0	47.1	44.3	41.6	39.1
	130°	93.7	91.8	89.4	86.5	83.4	80.0	76.3	72.5	68.7	65.0	61.2	57.6	54.1	50.9	47.8	44.9	42.1	39.6	37.2
North west	135°	93.7	91.6	89.0	86.0	82.7	79.0	75.2	71.2	67.1	63.2	59.3	55.5	52.0	48.7	45.6	42.7	40.0	37.6	35.4
	140°	93.7	91.4	88.7	85.5	82.0	78.1	74.1	69.9	65.7	61.5	57.4	53.5	49.9	46.5	43.4	40.6	38.0	35.7	33.6
	145°	93.7	91.3	88.4	85.0	81.4	77.3	73.1	68.7	64.3	59.9	55.6	51.6	47.8	44.4	41.3	38.5	36.1	33.9	32.0
	150°	93.7	91.2	88.1	84.6	80.8	76.6	72.2	67.6	63.0	58.4	54.0	49.9	46.0	42.5	39.4	36.6	34.3	32.3	30.5
	155°	93.7	91.1	87.9	84.3	80.3	76.0	71.4	66.7	61.9	57.2	52.7	48.5	44.6	40.9	37.6	34.9	32.6	30.7	29.0
	160°	93.7	91.0	87.7	84.0	79.9	75.4	70.7	65.8	60.9	56.2	51.7	47.5	43.4	39.7	36.3	33.3	31.1	29.2	27.7
	165°	93.7	90.9	87.5	83.7	79.6	75.0	70.1	65.1	60.1	55.4	50.9	46.7	42.6	38.8	35.3	32.3	29.8	28.0	26.6
Nord	170°	93.7	90.8	87.4	83.6	79.3	74.6	69.7	64.6	59.6	54.9	50.4	46.1	42.0	38.2	34.6	31.5	28.9	27.0	25.8
	175°	93.7	90.8	87.4	83.4	79.1	74.4	69.4	64.2	59.3	54.6	50.0	45.7	41.6	37.8	34.3	31.1	28.4	26.5	25.2
	180°	93.7	90.8	87.3	83.4	79.1	74.3	69.3	64.1	59.1	54.4	49.9	45.6	41.5	37.7	34.2	31.0	28.3	26.3	25.1

B

Checklist for Planning, Installing, and Operating a Photovoltaic Plant

1. **Suitability of the roof**
   - *Roof orientation:* What yield can be expected?
     - Comparison with existing plants (e.g. with *www.sunnyportal.com*).
     - Yield prognosis as done in Exercise 10.2.
   - *Shading:* Are there serious shading problems now or in the future?
   - *Age of the roof:* Will it need repairs in the next 20 years?
   - *Roof statics:* Can the roof bear the planned photovoltaic installation?
2. **Approvals**
   - *Construction approval:* Is this necessary (e.g. due to protection of historic buildings)?
   - *In case of a rented roof:* Close a roof utilization agreement with the owner.
   - *Connection to the grid:* Make an application to the public utility.
3. **Obtain and check quotes from solar installation companies**
   - *Module manufacturer:* Should be an established, known manufacturer.
   - *Performance tolerance of the module:* Acceptable is a maximum tolerance of ±3%.
   - *Warranty conditions:* Where is the place of jurisdiction?
   - Inverter:
     - Product guarantee should be more than 5 years.
     - European Efficiency should be more than 96% (see Section 7.4).
     - Design factor should be a maximum of 1 (see Section 7.5).
   - Increasing self-consumption (see Section 8.3):
     - Energy management system to control domestic loads should be included.
     - Does an electric heating rod or a heat pump make sense?
     - Is a battery storage system also offered?
     - Do you plan to buy an electric vehicle in future? If yes: Build a large PV plant!
   - Cable dimensioning:
     - Line losses should be a maximum of 1% (see Section 6.3.2).
4. **Financing and insurance**
   - Investment appraisal:
     - Determine object return according to Section 10.2.2.
   - Close a credit agreement with a bank.
   - Insurances:
     - Take out an operator liability insurance.
     - Take out an elementary damage or all-risks insurance.

Photovoltaics – Fundamentals, Technology, and Practice, Second Edition. Konrad Mertens.

5. **During installation**
   - Presorting the modules (if desired, see Section 6.2.3).
   - Careful working of installers:
     - Is the rear-side foil of the module damaged?
     - Is the module being stood upon (microcracks)?
     - Is the roof damaged?
     - Are the string cables clearly marked (numbering)?
6. **After installation**
   - Creating a startup protocol.
   - Acceptance of the system documents, including at least:
     - Roof sketch with modules drawn in and string arrangement plan.
     - Circuit diagram of the whole installation.
     - Datasheets of modules and inverter.
     - Information on the mounting system.
     - Information of length of warranty for modules and inverters.
   - Installation check (possibly):
     - Peak power measurement (see Section 9.3).
     - Bright thermography measurement (see Section 9.4).
7. **During operation**
   - Function check (every 2 weeks):
     - Check whether inverter shows feed-in operation (status MPP).
   - Yield control (every 4–8 weeks):
     - Reading the meter.
     - Comparison with online databases.
   - Soiling check (annually):
     - Check whether solid dirt layers have formed on the module borders.
   - Mechanical check (annual or after a big storm):
     - Check whether the installation rattles with movement.

C

Physical Constants/Material Parameters

Important Physical Constants

Boltzmann constant	$k = 1.3807 \times 10^{-23}$ J $K^{-1} = 8.6175 \times 10^{-5}$ eV K^{-1}
Elementary charge	$q = 1.6022 \times 10^{-19}$ A s
Gravity	$g = 9.81$ m s^{-2}
Planck's constant	$h = 6.6261 \times 10^{-34}$ W s^2
Solar constants	$E_S = 1367$ W m^{-2}
Stefan–Boltzmann constant	$\sigma = 5.6705 \times 10^{-8}$ W $(m^2\ K^4)^{-1}$
Speed of light in vacuum	$c_0 = 2.9979 \times 10^8$ m s^{-1}

Material Parameters of Silicon

Bandgap	$\Delta W_G = 1.12$ eV (at T = 300 K)
	$\Delta W_G = 1.17$ eV (at T = 0 K)
Mobility of electrons	$\mu_N = 1400$ cm^2 $(V\ s)^{-1}$
Mobility of holes	$\mu_P = 450$ cm^2 $(V\ s)^{-1}$
Refractive index at 600 nm	$n = 3.9$
Diffusion constant of electrons	$D_N = 35$ cm^2 s^{-1}
Diffusion constant of holes	$D_P = 12$ cm^2 s^{-1}
Effective density of state	$N_0 = 3 \times 10^{19}$ cm^{-3}
Melting point	$\vartheta_{Melt} = 1414$ °C

Photovoltaics – Fundamentals, Technology, and Practice, Second Edition. Konrad Mertens.

References

1 Kaltschmitt, M. et al. (2007). *Renewable Energy: Technology, Economics and Environment*. Springer.

2 International Energy Agency, International Energy Annual (2006). http://www.eia.gov/iea/ (last accessed 16 December 2017).

3 United Nations, Department of Economic and Social Affairs, Population Division (2015). World Population Prospects: The 2015 Revision, Key Findings and Advance Tables. Working Paper No. ESA/P/WP.241.

4 International Energy Agency, Key World Energy Statistics (2016). http://www.eia.gov/iea/ (last accessed 16 December 2017).

5 Bundesministerium für Wirtschaft und Technologie (2010). Energie in Deutschland, aktualisierte Ausgabe.

6 Monnin, E., Steig, E.J., Siegenthaler, U. et al. (2004). EPICA dome C ice core high resolution holocene and transition CO_2 data. ftp.ncdc.noaa.gov/pub/data/paleo/icecore/antarctica/epica_domec/edc-co2.txt (accessed 29 August 2013).

7 Neftel, A., Friedli, H., Moor, E. et al. (1994). Historical carbon dioxide record from the Siple Station ice core. http://cdiac.ornl.gov/trends/co2/siple.html (accessed 1 April 2017).

8 Mertens, K. et al. (2015). LowCost-Outdoor-EL: Kostengünstige umfassende Vorort-Qualitätsanalyse von Solarmodulen. 30th Symposium Photovoltaische Solarenergie, Staffelstein (März 2015).

9 Lübbert, D. et al. (2006). Uran als Kernbrennstoff – Vorräte und Reichweite, Infobrief WF VIII G – 069/06, Wissenschaftlicher Dienst des Bundestages.

10 Quaschning, V. (2017). website in German www.volker-quaschning.de/datserv (accessed 1 April 2017).

11 Adams, W. and Day, R. (1877). The action of light on selenium. *Proceedings of the Royal Society of London Series A* **25**: 113.

12 Green, M. (2002). Photovoltaic principles. *Physica E* **14**: 11–17.

13 Becquerel, A.E. (1839). Mémoire sur les effets électriques produits sous l'influence des rayons solaires. *Comptes Rendus de l'Academie des Sciences* **9**: 561.

14 Wagner, A. (2015). *Photovoltaik Engineering – Handbuch, Entwicklung und Anwendung*. Springer.

15 Chapin, D., Fuller, C.S., and Pearson, G.L. (1954). A new silicon p–n junction photocell for converting solar radiation into electrical power. *Journal of Applied Physics* **25** (5): 676–677.

Photovoltaics – Fundamentals, Technology, and Practice, Second Edition. Konrad Mertens.

16 Fatemi, N. (2005). Performance of high-efficiency advanced triple junction solar panels for the LILT mission DAWN. 31st Photovoltaic Specialists Conference and Exhibition, Lake Buena Vista, FL (3–7 January 2008). Jet Propulsion Laboratory, California Institute of Technology: Pasadena, CA, United States.

17 Grochowski, J. (1997). Minderertragsanalysen und Optimierungspotentiale an netzgekoppelten Photovoltaikanlagen des 1000-Dächer-Programms, Themen 96/97, Forschungsverbund Sonnenenergie, Köln.

18 Mertens, K. et al. (2016). LowCost-Outdoor-EL: Significant Improvements of the method. Proceedings of 32nd European Photovoltaic Solar Energy Conference, Munich.

19 NASA Langley Research Center (2013). Surface meteorology and solar energy, a renewable energy resource web site, release 6.0. http://eosweb.larc.nasa.gov/cgi-bin/sse/register.cgi (last accessed 16 December 2017).

20 Palz, W., Greif, J., and Scharmer, J. ed. (1998). *European Solar Radiation Atlas*. Springer.

21 Deutscher Wetterdienst (2012). Strahlungskarten der Mittelwerte für Deutschland: Jahr – flächendeckende mittlere Jahressumme (1981–2010). www.dwd.de/Solarenergie 10.2.2012 (accessed 1 September 2013).

22 Deutsche Gesellschaft für Sonnenenergie (2013). *Planning and Installing Photovoltaic Systems*, 3e. Berlin: DGS.

23 RWE (2010). *RWE Bau-Handbuch*, 14ee. Frankfurt: EW Medien und Kongresse GmbH.

24 Goetzberger, A., Knobloch, J., and Vob, B. (1998). *Crystalline Silicon Solar Cells*. Wiley.

25 Müller, R. (1995). *Grundlagen der Halbleiter-Elektronik*. Springer.

26 Kittel, C. (2013). *Einführung in die Festkörperphysik*. Oldenbourg.

27 McCandless, B. and Sites, J. (2003). Cadmium telluride solar cells. In: *Handbook of Photovoltaics Science and Engineering*. Wiley.

28 Lewerenz, H.-I. et al. (1995). *Photovoltaik – Grundlagen und Anwendungen*. Springer.

29 Roxlo, C. (1983). Comment on the optical absorption edge in a-Si:H. *Solid State Communications* **47** (12): 985–987.

30 Hecht, E. (2017). *Optics*. Pearson.

31 PVEducation (2017). Website on the general properties of silicon. http://pveducation.org/pvcdrom/materials/general-properties-of-silicon (accessed 1 April 2017).

32 von Roos, O. (1978). A simple theory of back surface fields (BSF) solar cells. *Journal of Applied Physics* **49** (6): 3503–3511.

33 Roth, T., Hohl-Ebinger, J., Warta, W. et al. (2008). Improving the accuracy of suns-VOC measurements using spectral mismatch correction. 33rd Photovoltaic Specialists Conference, San Diego, CA, USA.

34 Schott Solar, DNR 500293-04(2005). *Data Sheet Multicrystalline Solarcell*. Schott Solar GmbH.

35 Green, M. (1995). *Silicon Solar Cells – Advanced Principles & Practice*. Sidney: University of New South Wales.

36 Green, M. (1982). *Silicon Solar Cells – Operating Principles, Technology and System Applications*. Prentice-Hall Inc.

37 Bosch Solar (2010). Data sheet of Bosch Solar Cell M-3BB, Position May.
38 Hovinen, A. (1994). Fitting of the solar cell I/V-curve to the two diode model. *Physica Scripta* **T54**: 175–176.
39 Reference Solar Spectral Irradiance 2004 ASTM G-173, Online. http://rredc.nrel.gov/solar/spectra/am1.5/ASTMG173/ASTMG173.html (accessed 1 April 2017).
40 Häberlin, H. (2012). *Photovoltaics System Design and Practice.* Wiley.
41 Swanson, R. (2005). Approaching the 29% limit efficiency of silicon solar cell. Photovoltaic Specialists Conference, Conference Record of the 31st IEEE, pp. 889–894.
42 Ristow, A. (2001). Screen-printed back surface reflector for light trapping in crystalline silicon solar cells. Proceedings of the 17th European Photovoltaic Solar Energy Conference, Munich, pp. 1335–1338.
43 Ransome, S. (2004). Quantifying PV losses from equivalent circuit models, cells, modules and arrays. Preprint of poster presented at 19th PVSEC, Paris.
44 Sinton, R., Kwark, Y., Swirhun, S., and Swanson, R.M. (1985). Silicon point contact concentrator solar cells. *IEEE Electron Device Letters* **EDL-6** (8): 405–407.
45 Green, M. (2009). The path to 25% silicon solar cell efficiency: history of silicon cell evolution. *Progress in Photovoltaics – Research and Applications* **17** (3): 183–189.
46 Suntech (2009). Mass production of the innovative plutoTM solar cell technology, White Paper, eu.suntech-power.com/de/technologie/pluto-zelle.html (accessed 19 December 2009).
47 Wang, Z., Han, P., Lu, H. et al. (2012). Advanced PERC and PERL production cells with 20.3% record efficiency for standard commercial p-type silicon wafers. *Progress in Photovoltaics: Research and Applications* **20**: 260–268.
48 Dullweber, Th. et al. (2014). Fine line printed 5 busbar PERC solar cells with conversion efficiencies beyond 21%. 29th European Photovoltaic Solar Energy Conference and Exhibition, pp. 621–626.
49 Alsema, E. de Wildy-Scholten, M.J., and Fthenakis, V.M. (2006). Environmental impacts of PV electricity generation – a critical comparison of energy supply options. 21st European Photovoltaic Solar Energy Conference and Exhibition, Dresden, Germany.
50 Sollmann, D. (2009). Sechs Neuner sind das Ziel. *Photon* **4/2009**: 42–45.
51 Basore, P. (1994). Defining terms for crystalline silicon solar cells. *Progress in Photovoltaics: Research and Applications* **2** (2): 177–179.
52 Podewils, C. (2009). Diamantdraht zum Sägen. *Photon* **4/2009**: 77.
53 Beneking, A. (2012). Neuer Kontakt. *Photon* **2/2012**: 58–61.
54 Chu, S.K. and Siemer, J. (2012). Zinn statt Silber für die Rückseite. *Photon* **9/2012**: 58.
55 Rutschmann, I. (2008). Der Ruf nach Qualität. *Photon* **03/2008**: 52–56.
56 Glunz, S., Benick, J., Biro, D. et al. (2010). n-type silicon – enabling efficiencies > 20% in industrial production. 35th PVSC, Honolulu, HI.
57 Zeman, M. (2006). Advanced amorphous silicon solar cell technologies. In: *Thin Film Solar Cells – Fabrication, Characterization and Applications* (ed. J. Poortmans and V. Arkhipov). Wiley.
58 Repmann, T. (2003). Stapelsolarzellen aus amorphem und mikrokristallinem Silizium. Dissertation. Berichte des Forschungszentrums Jülich.

59 Gao, M. (2002). Optical band gap and electrical properties of a-Si:H, Workshop University of Lanzhou, China, 04.02.2002.

60 Liang, I. et al. (2006). Hole-mobility limit of amorphous silicon solar cells. *Applied Physics Letters* **88**: 063512.

61 Street, R. (2000). *Technology and Applications of Amorphous Silicon*, Springer Series in Materials Science, vol. **37**. Springer.

62 Staebler, D. and Wronski, C. (1977). Reversible conductivity changes in discharge produced amorphous Si. *Applied Physics Letters* **31**: 292.

63 Carlson, D. and Rajan, K. (1998). Evidence for proton motion in the recovery of light-induced degradation in amorphous silicon solar cells. *Journal of Applied Physiology (Bethesda, Md: 1985)* **83**: 1726–1729.

64 Yang, I., Banerjee, A., and Guha, S. (1997). Triple-junction amorphous silicon alloy solar cell with 14.6% initial and 13.0% stable conversion efficiencies. *Applied Physics Letters* **70** (22): 2975–2977.

65 Burgelman, M. (2006). Cadmium telluride thin film solar cells – characterization, fabrication and modeling. In: *Thin Film Solar Cells – Fabrication, Characterization and Applications* (ed. J. Poortmans and V. Arkhipov). Wiley.

66 Gupta, A., Parikh, V., and Compaan, A.D. (2006). High efficiency ultra-thin sputtered CdTe solar cells. *Solar Energy Materials & Solar Cells* **90**: 2263–2271.

67 Mitchell, K., Eberspacher, C., Ermer, J., and Pier, D. (1988). Single and tandem junction $CuInSe_2$ cell and module technology. 20th IEEE Photovoltaic Specialists Conference, Las Vegas, Conference Record, Vol. 2, pp. 1384–1389.

68 Poortmans, J. and Arkhipov, V. ed. (2006). *Thin Film Solar Cells – Fabrication, Characterization and Applications*, 239. Wiley Figure 6.2.

69 Green, M. (2016). Solar cell efficiency tables (version 49). *Progress in Photovoltaics: Research and Applications* **25**: 3–13.

70 Nanosolar Inc. (2007). High performance thin-film photovoltaics using low-cost process technology. 17th Annual Photovoltaic Science and Engineering Conference (PVSEC) Fukuoka, Japan.

71 Green, M. (2001). Crystalline silicon solar cells. In: *Clean Electricity from Photovoltaics* (ed. M.D. Archer and R. Hill). Imperial College Press.

72 Mishima, T., Taguchi, M., Sakata, H., and Maruyama, E. (2010). Development status of high-efficiency HIT solar cells. *Solar Energy Materials & Solar Cells* doi: 10.1016/j.solmat.2010.04.030.

73 Flamand, G. and Poortmans, J. (2006). Towards highly efficient 4-terminal mechanical photovoltaic stacks. *III–Vs Review* **19** (7): 24–27.

74 Karam, N. et al. (2001). Recent developments in high-efficiency $Ga_{0.5}In_{0.5}P$/GaAs/Ge dual- and triple-junction solar cells – steps to next-generation PV cells. *Solar Energy Materials & Solar Cells* **66**: 453–466.

75 Grätzel, M. (2006). Nanocrystalline injection solar cells. In: *Thin Film Solar Cells – Fabrication, Characterization and Applications* (ed. J. Poortmans and V. Arkhipov). Wiley.

76 Qiao, Q. (2015). *Organic Solar Cells: Materials, Devices, Interfaces, and Modeling*. CRC Press.

77 Park, N.-G., Grätzel, M., and Miyasaka, T. (2016). *Organic-Inorganic Halide Perovskite Photovoltaics*. Springer International Publishing.

78 Umweltbundesamt (2010). Nationale Trendtabellen für die deutsche Berichterstattung atmosphärischer Emissionen (Schwermetalle) 1990–2008, Stand: 15.02.2010, Dessau.

79 Schwarzburger, H. (2010). Kein Persilschein für Silizium. *Photovoltaic* **02/2010**: 58–64.

80 Kreutzmann, A. (2008). Vorteil wacker. *Photon* **5/2008**: 30–35.

81 BINE Informations Service. Recycling von photovoltaikmodulen. Projektinfo.www.bine.info (last accessed 16 December 2017).

82 Müller, A. (2009). Das zweite Leben. *Sonne Wind und Wärme* **08/2009**: 182–187.

83 Gellings, R. (2006). Vom regen in die traufe. *Photon* **6/2006**: 52–62.

84 Iken, J. (2010). Abhängigkeiten und alternativen. *Sonne Wind & Wärme* 42–49.

85 Mann, S.A., de Wild-Scholten, M.J., Fthenakis, V.M. et al. (2013). The energy payback time of advanced crystalline silicon PV modules in 2020: a prospective study. *Progress in Photovoltaics: Research and Applications* doi: 10.1002/pip.2363.

86 Bunk, O. (2002). Positive umweltbilanz – anlagen amortisieren sich nach wenigen monaten. *Windblatt – Das Enercon Magazin*, pp. 12–13.

87 Fraunhofer ISE (2017). Photovoltaics Report, updated: 17 November 2016. www.ise.fraunhofer.de/content/dam/ise/de/documents/publications/studies/Photovoltaics-Report.pdf (accessed 1 April 2017).

88 Hagmann, G. (2015). *Leistungselektronik*, 5. issue, AULA-Verlag.

89 Femia, N. (2005). Optimization of perturb and observe maximum power point tracking method. *IEEE Transactions on Power Electronics* **20** (4): 963–973.

90 Laschinski, I. (2008). Systemspannung, TCO-Korrosion und Generatorerdung bei Dünnschichtmodulen. *IHKS Fach Journal* 129–131.

91 Sahan, B. (2010). *Wechselrichtersysteme mit Stromzwischenkreis zur Netzanbindung von Photovoltaikgeneratoren*. Kassel University Press GmbH.

92 Burger, B. (2005). Auslegung und Dimensionierung von Wechselrichtern für netzgekoppelte PV-Anlagen. 20th Symposium Photovoltaische Solarenergie, Staffelstein.

93 Verband der Elektrizitätswirtschaft – VDW – e.V. (2011). Erzeugungsanlagen am Niederspannungsnetz – Technische Mindestanforderungen für Anschluss und Parallelbetrieb von Erzeugungsanlagen am Niederspannungsnetz, August 2011. VWEW Energieverlag GmbH

94 Blitzschutz – Teil 3 (2009). Schutz von baulichen Anlagen und Personen – Sheet 5: Blitz- und Überspannungsschutz für PV-Stromversorgungssysteme, VDE 0185305-3 Sheet 5, October 2009.

95 Schlumberger, A. (2006). Der Tanker bewegt sich – ein Stück. *Photon* **10/2006**: 106–109.

96 Ladener, H. (1999). *Solare Stromversorgung*, 3e. Ökobuch Verlag.

97 Ketterer, B., Karl, U., Möst, D., and Ulrich, S. (2009). Lithium-Ionen Batterien: Stand der Technik und Anwendungspotenzial in Hybrid-, Plug-In Hybrid- und Elektrofahrzeugen, *Wissenschaftliche Berichte*, FZKA 7503, Forschungszentrum Karlsruhe.

98 Wohlfahrt-Mehrens, M. (2007). Materialien für zukünftige Lithium-Ionen Batterien – Entwicklungen und Perspektiven. IMF Seminar, Forschungszentrum Karlsruhe.

99 Sterner, M. and Stadler, I. (2014). *Energiespeicher – Bedarf, Technologien, Integration*, 1. Auflagee. Springer Vieweg.
100 Korthauer, R. (2013). *Handbuch Lithium-Ionen-Batterien*. Springer Vieweg.
101 Mears, D. (2005). Overview of NAS Battery for Load Management, CEC Energy Storage Workshop.
102 Quaschning, V. (2005). *Understanding Renewable Energy Systems*, 1ee. London: Earthscan.
103 Gerhardt, N. et al. (2014). *Geschäftsmodell Energiewende – Eine Antwort auf das "Die-Kosten-der-Energiewende" – Argument, Fraunhofer-Institut für Windenergie und Energiesystemtechnik*. Kassel: IWES.
104 Rothert, M., Bukvíc-Schäfer, A.S., Kreutzer, N. et al. (2012). Ein Jahr Felderfahrung: PV-Anlagen mit Speicherlösung zur Eigenverbrauchserhöhung. 27th Symposium Photovoltaische Solarenergie, Staffelstein.
105 Wiese, R. (2010). Empowering a rural revolution. *Renewable Energy World Magazine*, 15 April 2010.
106 Strauß, P. et al. (2009). *Netzferne Stromversorgung und weltweite Elektrifizierung, Themen 2009*, 94–101. Forschungs-Verbund Erneuerbare Energien.
107 SMA AG (2010). *Produktkatalog Sunny Family 2010/2011 – the future of solar technology*. Kassel: SMA AG.
108 Photovoltaische Einrichtungen – Part 9 (2007). Leistungsanforderungen an Sonnensimulatoren, German edition of the IEC 60904-9:2007, VDE 0126-4-9.
109 Verfahren zur Umrechnung von gemessenen Strom-Spannungs-Kennlinien von photovoltaischen Bauelementen aus kristallinem Silizium auf andere Temperaturen und Einstrahlungen, German edition of the EN 60891 (1994).
110 Photovoltaische Einrichtungen (2010). Verfahren zur Umrechnung von gemessenen Strom-Spannungs-Kennlinien auf andere Temperaturen und Bestrahlungsstärken, Deutsche Fassung der EN 60891:2010, VDE 0126-6.
111 Mertens, K. (2008). Feldstudie zur tatsächlichen Leistung von Photovoltaikanlagen mittels Peakleistungsmessgerät. 23rd Symposium Photovoltaische Solarenergie, Staffelstein.
112 Omega Newport Electronics GmbH (2011). Emissionsfaktoren – Technische Hintergrundinformationen. www.omega.de (accessed 1 April 2017).
113 Köntges, M. et al. (2008). Elektrolumineszenzmessung an PV-Modulen. *ep Photovoltaik aktuell* **7–8/2008**: 36–40.
114 Mertens, K., Stegemann, Th., and Stöppel, T. (2012). LowCost EL: Erstellung von Elektrolumineszenzbildern mit einer modifizierten Standard-Spiegelreflexkamera. 23rd Symposium Photovoltaische Solarenergie, Staffelstein.
115 Mertens, K., Stegemann, Th., and Stöppel, T. (2012). LowCost-EL – Preisgünstige Erstellung von Elektrolumineszenzbildern mit einer Spiegelreflexkamera, Der Bausachverständige. Annual 8th (4 August 2012), pp. 38–39.
116 Köntges, M., Kunze, I., Kajari-Schröder, S. et al. (2010). Quantifying the risk of power loss in PV modules due to micro cracks. 25th European Photovoltaic Solar Energy Conference, Valencia, Spain (6–10 September 2010), pp. 3745–3752.
117 Lausch, D. et al. (2014). Explanation of potential-induced degradation of the shunting type by Na decoration of stacking faults in Si solar cells. *Solar Energy Materials and Solar Cells* **120**, Part A: 383–389.
118 Antony, E.A. (2009) *Photovoltaik für Profis*, Solarpraxis AG.

119 Kreutzmann, A. (2006). Dünnschicht als Preisbrecher. *Photon* **12/2006**: 100–108.

120 Mertens, K. (2005). Kindgerechte Visualisierung von Photovoltaikerträgen. 20th Symposium Photovoltaische Solarenergie, Staffelstein.

121 Quaschning, V. (2000). *Systemtechnik einer klimaverträglichen Elektrizitätsversorgung in Deutschland für das 21,* Jahrhundert. Fortschritt-Berichte, VDI Series 6, No. 437,. VDI Verlag.

122 Bauernverband, D. (2014). Situationsbericht 2014/15 – Trends und Fakten zur Landwirtschaft. www.bauernverband.de/situationsbericht (accessed 1 April 2017).

123 Quaschning, V. (2011). Bewertung von Methoden zur Bestimmung des PV-Anteils sowie von Ausbauszenarien und Einflüssen auf die Elektrizitätswirtschaft. 26th Symposium Photovoltaische Solarenergie, Staffelstein.

124 Fachagentur nachwachsende Rohstoffe e.V (2010). *Biogas – Basisdaten Deutschland*. Info-flyer.

125 Hoffmann, V. (2008). Damals war's – Ein Rückblick auf die Entwicklung der Photovoltaik in Deutschland, Sonnenenergie, November–December 2008, pp. 38–39.

126 pvXchange (2017). Price index of pv modules. http://www.pvxchange.com/priceindex (accessed 1 April 2017).

127 Weber, E. (2009). Entwicklung des PV-Marktes aus Sicht der Forschung. 24th Symposium Photovoltaische Solarenergie, Staffelstein.

128 www.ag-energiebilanzen.de (accessed 1 April 2017).

129 Burger, B., Fraunhofer ISE (2014). Fraunhofer ISE Electricity production from solar and wind in Germany in 2014. www.energy-charts.de (accessed 1 April 2017).

130 Bundesministerium für Umwelt (2012). BMU-Schlussbericht, Langfristszenarien und Strategien für den Ausbau der erneuerbaren Energien in Deutschland bei Berücksichtigung der Entwicklung in Europa und global, 29 März 2012.

131 Sachverständigenrat für Umweltfragen (2011). *Wege zur 100% erneuerbaren Stromversorgung*. Sondergutachten.

132 Neupert, U., Euting, T., Kretschmer, T. et al. (2009). *Energiespeicher – Technische Grundlagen und energiewirtschaftliches Potenzial*. Fraunhofer IRB Verlag.

133 Hartmann, N. et al. (2012). *Stromspeicherpotenziale für Deutschland*. Universität Stuttgart.

134 Huggins, R.A. (2010). *Energy Storage*. New York: Springer.

135 Umweltbundesamt (2010). Energieziel 2050 – 100% Strom aus erneuerbaren Quellen.

Further Information on Photovoltaics

Information on the Internet

https://home.solarlog-web.de/1.html	Yield database for data loggers of the Solare Datensysteme Company
re.jrc.ec.europa.eu/pvgis/apps4/pvest.php	Interactive solar radiation cards with yield estimates (old version)
re.jrc.ec.europa.eu/pvgis.html	Interactive solar radiation cards with yield estimates (new version)
www.dgs.de	Site of the Deutsche Gesellschaft für Solarenergie e.V.
www.photon.info	Module and inverter database (reached via *Publishing/Databases*)
www.photovoltaikforum.com	Forum on photovoltaic with module, inverter and company database
www.pveducation.org/pvcdrom	Interesting and instructive information and applets on photovoltaics
www.pv-ertraege.de	Yield database of the Aachen Association for the Promotion of Solar Power (SFV)
www.satel-light.com	Individual creation of an radiation map of the whole of Europe
www.sfv.de	Information of the Aachen Association for the Promotion of Solar Power (SFV)
www.solarserver.de	Comprehensive information on the subject of solar energy
www.sonnenertrag.eu	Yield database, company-independent
www.sunnyportal.de	Yield database for inverters of the SMA AG Company
www.volker-quaschning.de	Comprehensive information on the subject of renewable energies
www.textbook-pv.org/links.html	Additional interesting links can be found on this site

Photovoltaics – Fundamentals, Technology, and Practice, Second Edition. Konrad Mertens.

Journals

Global Solar Technology	*www.globalsolartechnology.com*
Photon	*www.photon.info*
Photovoltaics International	*www.solarmediastore.com/journals.html*
Photovoltaic Production	*www.photovoltaic-production.com*
Photovoltaik	*www.photovoltaik.eu*
pv magazine	*www.pv-magazine.com*
PV-Tech	*www.pv-tech.org*
Solar Energy	*www.journals.elsevier.com/solar-energy*
Solar Industry	*www.solarindustrymag.com*
Solar Pro	*www.solarprofessional.com*
Solar Today	*www.ases.org/solar-today-magazine*
Solarthemen	*www.solarthemen.de*
Sonnenenergie	*www.sonnenenergie.de*
Sun & Wind Energy	*www.sunwindenergy.com*

Recommended Books

Häberlin, H. (2012). *Photovoltaics: System Design and Practice.* Wiley Detailed description of the system technology of photovoltaic plants (inverters, lightning protection, etc.).

Haselhuhn, R. and Hartmann, U. (2013). *Planning and Installing Photovoltaic Systems.* DGS Deutsche Gesellschaft für Sonnenenergie. Comprehensive work for the practical person with descriptive illustrations.

Quaschning, V. (2016). *Understanding Renewable Energy Systems.* London: Earthscan. Good and comprehensive introduction to the subject of renewable energies.

Index

Printed and bound by CPI Group (UK) Ltd, Croydon, CR0 4YY

27/10/2024

14580198-0001